DNA FINGERPRINTING
in
PLANTS AND FUNGI

Kurt Weising • Hilde Nybom
Kirsten Wolff • Wieland Meyer

CRC Press
Boca Raton   Ann Arbor   London   Tokyo

Library of Congress Cataloging-in-Publication Data

DNA fingerprinting in plants and fungi / Kurt Weising ... [et al.].
   p.   cm.
   Includes bibliographical references (p.) and index.
   ISBN 0-8493-8920-8
   1. DNA fingerprinting of plants. 2. DNA fingerprinting of fungi. I. Weising, Kurt.
QK981.45.D54 1994
581.87'3282—dc20   94-38070
                                                      CIP

PREFACE

It is now almost a decade since Alec Jeffreys and his colleagues discovered minisatellite repeat families in the human genome. This class of repetitive DNA provided the molecular basis to discriminate between human individuals at an unprecedented level. The new strategy of "DNA fingerprinting" soon gained broad application, targeting problems as diverse as forensic casework, monitoring cell line identity, and the analysis of genetic diversity in a population.

Plant and fungal scientists felt highly attracted by the new technique, and DNA fingerprinting was established in plants and fungi at the end of the 1980s. Our own involvement in the field led to numerous requests from researchers, plant breeders, taxonomists and population biologists for protocols, computer programs, references, and general know-how. The idea to combine all this information in a book on "DNA fingerprinting in plants and fungi" was therefore first conceived in 1992.

Initially, we intended to treat the hybridization-based methods only, using nonspecific minisatellites and simple sequence repeats as probes. In the meantime, however, DNA fingerprinting strategies based on PCR evolved, and spread "RAPDly" through many laboratories, including ours. The decision to include this new technology in the scope of the book proved to be a laborious task: the number of references on PCR fingerprinting with arbitrary primers grew as exponentially as the number of DNA target fragments in a PCR. Nevertheless, we managed to cover the literature quite thoroughly until the end of 1993.

This book is primarily intended to provide a benchtop manual of all steps of hybridization- and PCR-based DNA fingerprinting, including the evaluation of results. The description of the technology is complemented by a significant amount of background information. Additional topics are an extensive survey of applications of DNA fingerprinting in plants and fungi, and a comparison of different types of molecular markers. Marker technology has become an increasingly popular and dynamic field of research, and new and promising strategies are developed continuously. Very recent approaches are introduced and discussed in the perspectives section.

We hope this book helps to introduce molecular marker technology in general, and DNA fingerprinting in particular, to laboratories involved in plant and fungal research, breeding, and diagnostics.

Kurt Weising, Ph.D.
Hilde Nybom, Ph.D.
Kirsten Wolff, Ph.D.
Wieland Meyer, Ph.D.

ABOUT THE AUTHORS

Kurt Weising, Ph.D., is a postdoctoral research fellow in the Plant Molecular Biology Group of the Department of Botany at the Johann Wolfgang Goethe University in Frankfurt. Dr. Weising received his M.S. and Ph.D. degrees in 1981 and 1987, respectively, from the Institute of Biology at the University of Frankfurt. He has done postdoctoral research at the School of Biological Sciences, University of Auckland, New Zealand, and at the Department of Botany, University of Frankfurt.

Dr. Weising is a member of the "Gesellschaft für Züchtungsforschung," Göttingen, Germany and the International Society For Plant Molecular Biology, Cologne. He has been the recipient of research grants from the Fazit Foundation, Frankfurt, the Fritz Thyssen Foundation, Cologne, and the German Research Council, Bonn.

Dr. Weising has published more than 30 papers, mainly focusing on plant chromatin structure, plant genome analysis, and DNA fingerprinting. His current major research interests relate to plant microsatellite research, including the functional significance of microsatellites in the plant genome as well as their application as molecular markers.

Hilde Nybom, Ph.D., is employed as a research officer at the Department for Horticultural Plant Breeding-Balsgård at the Swedish University for Agricultural Sciences.

Dr. Nybom obtained her training at the University of Lund, Sweden, receiving an M.S. degree in 1977 and the Ph.D. degree in 1987 at the Department of Systematic Botany. She obtained a postdoctoral grant from the Swedish Natural Science Research Council enabling her to spend a year at the Biology Department at Washington University, St. Louis, USA. She was appointed docent in plant breeding in 1992.

Dr. Nybom is a member of the Lund Botanical Society and the International Society for Horticultural Science, where she serves in the Biotechnology committee. She has received several research grants from the Swedish Natural Science Research Council and from the Swedish Agricultural Science Research Council. She has given invited guest lectures in various countries around the world, and has taken part in international conferences on subjects relating to ecology, molecular biology, and plant breeding.

Dr. Nybom is the author of some 40 papers, mostly on ecology, genetics, and systematics. Her current major research interests relate to the evolution of reproductive systems, and their effects on population structure. For this work as well as applied plant breeding research, molecular methods have proved very useful.

Kirsten Wolff, Ph.D., is a scientific researcher at the Institute of Evolutionary and Ecological Sciences of the University of Leiden, The Netherlands.

Dr. Wolff obtained her training at the University of Groningen, The Netherlands, receiving her M.S. in 1982 from the Department of Genetics and Botany. She received her Ph.D. degree (fellowship form NWO) in population genetics in 1988 from the Department of Genetics at the University of Groningen. After having worked at Washington University in St. Louis on a fellowship from NWO, and a short postdoctoral period at the University of Groningen, she was appointed as scientific researcher at the University of Leiden, stationed at TNO Nutrition and Food Research in Zeist, The Netherlands.

Dr. Wolff is a member of the European Society for Evolutionary Biology, the Genetical Society, and the Society for the Study of Evolution.

Dr. Wolff is the author of more than 20 papers. Her current research interests are the interplay of ecophysiological adaptation, plasticity, and the mating system on population differentiation and speciation processes.

Wieland Meyer, Ph.D., is a research assistant in the Department of Microbiology at Duke University Medical Center in Durham, North Carolina, USA.

Dr. Meyer received his diploma in biology at the Humboldt University of Berlin, Germany, in 1986. Since then, he has done yeast research at the Research Center of Biotechnology in Berlin. He was a scientific co-assistant at the IFZ Research and Developing Society, Ltd., Berlin, where he applied DNA fingerprinting with mini- or microsatellite specific probes to questions of fungal identification, taxonomy, and systematics. In 1992, he obtained his Ph.D. degree from the Department of Biology, Institute of Genetics, Humboldt University of Berlin. He became a postdoctoral fellow in the Mycology Laboratory at Duke University Medical Center in 1992. Here, he utilized several fingerprinting techniques for epidemiological studies of pathogenic fungi.

He is a member of the International Society of Human and Animal Mycology (ISHAM).

Dr. Meyer has published more than 20 papers and is author or co-author of several book chapters on DNA-hybridization as well as PCR-based-fingerprinting of filamentous fungi and medically important yeasts. His current research interests are in molecular taxonomy of the yeast genera *Candida* and *Cryptococcus* as well as investigating the phylogenetic relationships within the filamentous fungal genera *Trichoderma* and *Mortierella* using rDNA gene sequence analysis.

ACKNOWLEDGMENTS

We acknowledge the help of numerous people during the process of this work. In particular, we want to thank Jacques Beckmann (Bet-Dagan, Israel), Bengt-Olof Bjurman (Kristianstad, Sweden), Birgitta Bremer (Uppsala, Sweden), John Brookfield (Nottingham, England), Norman Ellstrand (Riverside, California, USA), Walter Gams (Baarn, The Netherlands), Susan Gardiner (Palmerston North, New Zealand), Lena Gustafsson (Uppsala, Sweden), Günter Kahl (Frankfurt, Germany), T. G. Mitchell (Durham, North Carolina, USA), Sigi Kost (Frankfurt, Germany), Arthur Leewis (Haren, The Netherlands), Anke Marcinkowski (Frankfurt, Germany), Honor Prentice (Lund, Sweden), Barbara Schaal (St. Louis, Missouri, USA), and Linus Svensson (Lund, Sweden).

TABLE OF CONTENTS

Chapter 1

INTRODUCTION

The analysis of genetic diversity and relatedness between or within different species, populations, and individuals is a central task for many disciplines of biological science. During the past decade, classical strategies of evaluating genetic variability such as comparative anatomy, morphology, embryology, and physiology have been increasingly complemented by molecular techniques. These include the analysis of chemical constituents (e.g., plant secondary compounds) and, most importantly, the characterization of macromolecules. The development of so-called "molecular markers," which are based on polymorphisms found in proteins or DNA, has greatly facilitated research in a variety of disciplines such as taxonomy, phylogeny, ecology, genetics, and plant breeding.

For quite a long period of time, allozymes have been the molecular markers of choice. In recent years, however, attention has increasingly focused on the DNA molecule as a source of informative polymorphisms. Because each individual's DNA sequence is unique, this sequence information can be exploited for any study of genetic diversity and relatedness between organisms. A wide variety of techniques to visualize DNA sequence polymorphisms have been developed in the past few years, and molecular markers have been derived from these techniques (see Chapter 2). In this book, we shall focus on the strategy of *DNA fingerprinting* and its applications to various fields of plant and fungal research.

How is DNA fingerprinting defined? The term was originally introduced by Jeffreys et al.[315] to describe a method for the simultaneous detection of many highly variable DNA loci by hybridization of specific multilocus "probes" to electrophoretically separated restriction fragments (see Chapter 2 for details). In recent years, several modifications of the basic technique have appeared, and related strategies have been developed. Most importantly, DNA polymorphisms became detectable by the polymerase chain reaction (PCR). Some of the new methods are still called DNA fingerprinting, but "DNA profiling," "DNA typing," or more specific terms have also been introduced.

In this book, we will refer to any multilocus approach of visualizing DNA polymorphisms as DNA fingerprinting no matter whether it is based on hybridization or PCR. According to this definition, DNA fingerprints are mainly obtained by either of two strategies.

1. "Classical" *hybridization-based fingerprinting* involves cutting of genomic DNA with a restriction enzyme; electrophoretic separation of the resulting DNA fragments according to size; and detection of polymorphic multilocus banding patterns by hybridization with a labeled, complementary DNA sequence, a so-called "probe".

1

2. ***PCR-based fingerprinting*** involves the *in vitro* amplification of particular DNA sequences with the help of specifically or arbitrarily chosen oligonucleotides ("primers") and a thermostable DNA polymerase; the electrophoretic separation of amplified fragments, and the detection of polymorphic banding patterns by such methods as staining.

For both strategies, a wide variety of technical modifications, primers, and probes have been identified which generate such multilocus patterns. Since this book is primarily dedicated to newcomers in the field, we strongly focus (1) on DNA fingerprint methodology, which is comparatively easy to perform, and (2) on the use of probes and primers, which are accessible without cloning steps. To that end, the majority of protocols described here require neither extraordinary laboratory equipment nor special skills in molecular biology. However, more sophisticated techniques, as well as very recent approaches, are discussed and referred to.

DNA fingerprinting techniques exhibit a great (and as yet not fully explored) potential as a tool for a wide range of areas in plant and fungal research, including genotype identification, population genetics, taxonomy, plant breeding, and diagnostics and epidemiology of plant, animal, and human pathogenic fungi. Accordingly, this book may prove useful for a wide spectrum of scientists, including population geneticists, plant and fungal systematists, ecologists, plant breeders, phytopathologists, medically oriented mycologists and dermatologists, as well as students interested in any of these fields. We imagine the following three-step schedule of events to be quite realistic for a newcomer in the field: (1) the "would-be" fingerprinter negotiates a cooperative project with a molecular biology group; (2) with the help of this book, a first set of fingerprint experiments is carried out and published; and (3) armed with his/her first publication, the new fingerprinter obtains a grant to set up his/her own laboratory for further experiments.

Since not only the applications and methodology, but also the molecular background of DNA fingerprinting are treated, the book may also be of value for molecular biologists and students who are interested in genome analysis in general and in the properties of short tandem-repetitive DNA in particular.

Chapter 2

GENETIC VARIATION AT THE DNA LEVEL

In the first section of this chapter, we shortly summarize current strategies to visualize DNA polymorphisms. In the second and third sections, the molecular background of hybridization- and PCR-based DNA fingerprinting is treated in more detail.

I. A SHORT SURVEY OF MOLECULAR MARKERS

A. PROPERTIES OF MOLECULAR MARKERS
The following properties would be generally desirable for a molecular marker:

1. Highly polymorphic behavior
2. Codominant inheritance (which allows us to discriminate homo- and heterozygotic states in diploid organisms)
3. Frequent occurrence in the genome
4. Even distribution throughout the genome
5. Selectively neutral behavior (i.e., no pleiotropic effects)
6. Easy access (e.g., by purchasing or fast procedures)
7. Easy and fast assay (e.g., by procedures amenable to automation)
8. High reproducibility
9. Easy exchange of data between laboratories

No molecular markers are available yet that fulfill all of these criteria. However, according to the kind of study to be undertaken, one can already choose between a variety of marker systems, each of which combines at least some of the above-mentioned properties.

B. PROTEIN MARKERS
For the generation of molecular markers based on protein polymorphisms, the most frequently used technique is the electrophoretic separation of proteins, followed by specific staining of a distinct protein subclass. While some studies use seed protein patterns, the majority of protein markers are represented by allozymes.

Allozyme electrophoresis has been successfully applied to many organisms from bacteria to numerous animal and plant species since the 1960s (see, e.g., May[437]). The studies have encompassed various fields (e.g., physiology, biochemistry, genetics, breeding) and purposes (population structure, mating system, hybridization, polyploidy, systematics).[270,402,485] Allozyme analysis is relatively straightforward and easy to carry out. A tissue extract is prepared and electrophoresed on a starch or polyacrylamide gel. The proteins of this extract

3

are thereby separated according to net charge and size. After electrophoresis, the gel is stained for a particular enzyme by adding a substrate and a dye under the appropriate reaction conditions, resulting in a band at the position to where the enzyme has migrated. Depending on the number of loci, their state of homo- or heterozygosity, and the enzyme molecule configuration, from one to several bands are visualized. The positions of these bands can be polymorphic and thus informative.

Sometimes isozyme and allozyme analysis are incorrectly treated as interchangeable terms. *Isozymes* are enzymes that convert the same chemical substrate, but are not necessarily products of the same gene. Isozymes may be active at different life stages or in different cell compartments. *Allozymes* are isozymes which are produced by orthologous genes, but which differ in composition by one or more amino acids due to allelic differences.

C. DNA SEQUENCING

Polymorphisms at the DNA level can be studied by several means. The most direct strategy is the determination of the nucleotide sequence of a defined region and the alignment of this sequence to an orthologous region in the genome of another, more or less related organism (see review in Hillis et al.[287]). DNA sequencing provides a highly reproducible and informative analysis of data and can be adapted to different levels of discriminatory potential by choosing appropriate regions of the genome. Currently, DNA sequencing is mainly applied for evaluating medium- and long-distance relatedness in phylogeny, but sometimes it is also used for population studies.[291]

In recent years, the sequencing approach has been greatly facilitated by the advent of PCR,[605] which made it possible to isolate homologous DNA sequences from any organism of interest with unprecedented speed. Primers are designed on the basis of sequence information for conserved parts of the DNA, and the desired target sequences are amplified. The PCR product can then be sequenced either directly or after cloning.[291] In plants, primers are often used to amplify the *rbcL* gene from the chloroplast genome[115] and the 18S ribosomal RNA gene.[260,434] Lately, internal transcribed spacers in the ribosomal genes have proved useful for studies on plant evolution.[31] Universal primers for ribosomal RNA genes in fungi are also available.[776]

Unfortunately, DNA sequencing is comparatively difficult as well as expensive. Traditionally, radioisotopes have been (and still are) used for DNA fragment detection. About 300 nucleotides from one DNA strand are usually analyzed by one set of reactions, resulting in a substantial amount of valuable information, provided that the chosen sequence is sufficiently variable. Lately, sequencing machines have become available, which make use of fluorescence-labeled primers (e.g., Smith et al.;[638] see also Chapter 7, Section I). While the technical equipment is more expensive, DNA sequencing using fluorescence technology is cheaper and easier to perform than the traditional method. Moreover, sequence data are directly transferred into a computer. In the future, the fluorescence-based technique may allow routine sequencing of large numbers of samples and long stretches of DNA.

D. RESTRICTION FRAGMENT LENGTH POLYMORPHISMS (RFLPs)

An alternative means for evaluating DNA sequence variation is the analysis of restriction fragment length polymorphisms (RFLPs). Restriction enzymes are endonucleases which occur in a variety of prokaryotes. Their natural function is to destroy foreign DNA molecules by recognizing and cutting specific DNA sequence motifs, typically consisting of four or six bases. Each enzyme has a particular recognition sequence, and the bacteria usually protect their own DNA from being cut by methylating this sequence.

A large number of different restriction enzymes is commercially available. Digestion of a particular DNA molecule with such an enzyme results in a reproducible set of fragments of well-defined lengths. Point mutations within the recognition sequence as well as insertions or deletions will result in an altered pattern of restriction fragments and may thus bring about a screenable polymorphism between different genotypes (Figure 1).

The methodology is quite similar to hybridization-based fingerprinting, which actually represents a special case of RFLP analysis. Genomic DNA is extracted, digested with restriction enzymes, and separated by electrophoresis on a gel. This gel is Southern blotted onto a membrane, and specific fragments are made visible by hybridization with a labeled probe. Two main differences exist between the RFLP technique and hybridization-based fingerprinting: (1) DNA fingerprinting (in the sense used in this book, see Chapter 1) makes use of multilocus probes, creating complex banding patterns, whereas RFLP probes are usually locus specific, resulting in an easy-to-screen codominant marker behavior; and (2) DNA fingerprinting is mostly performed with non-species-specific probes that recognize ubiquitously occurring sequences such as minisatellites, whereas RFLP probes are species-specific.

1. RFLPs in Nuclear DNA

RFLP analysis of nuclear DNA typically uses species-specific probes which are obtained from a cDNA or genomic library (see below) of the investigated species, or a close relative. To establish a library requires considerable effort, and RFLP studies therefore are performed rarely for wild species. However, for many economically important crop species such as cereals and tomato, and for model organisms such as *Arabidopsis*, hundreds of RFLPs have been identified and used as molecular markers to establish genetic maps. Such maps are important tools for marker-assisted selection in breeding programs and map-based cloning of genes.[503,680,681] Additional application areas for RFLPs are cultivar identification[212,565] and phylogenetic studies.[147,160,650]

Restriction enzymes with a six-basepair (bp) recognition sequence are usually used in combination with agarose gel electrophoresis. For a fine scale study of variation, four-bp cutters and polyacrylamide gels have also been advocated.[213,372] There has been some debate as to whether copied DNA (cDNA) or genomic libraries result in a higher number of RFLPs.[151,281,797] cDNA libraries are constructed from transcribed parts of DNA. For this purpose, mRNA is isolated and *in vitro* transcribed into DNA by reverse tran-

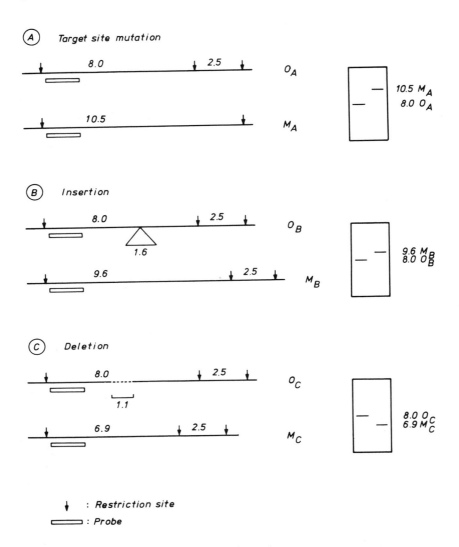

FIGURE 1. Molecular basis of RFLPs. An RFLP can originate from the mutation of a restriction enzyme target site (A) as well as from the insertion (B) or deletion (C) of a piece of DNA between two target sites. After separation of restriction-digested genomic DNA by agarose gel electrophoresis, Southern blotting, and hybridization with a sequence-specific labeled probe, RFLPs are visualized by autoradiography or nonradioactive detection techniques. Hybridization signals derived from the original alleles (O) and the mutated alleles (M) are indicated in the right panel of each figure. Numbers refer to the length of a particular restriction fragment in kilobases.

scriptase. These cDNA molecules are then cloned into a vector and used as probes in subsequent RFLP analyses. For the construction of a genomic library, total DNA is isolated, digested with a restriction enzyme, and cloned into a vector. The choice of the restriction enzyme and the size of the cloned fragments determine the quality of the library. In plants, the methylation-sensitive restriction enzyme *Pst*I is often used to establish a library for RFLP studies, since size selection for small (i.e., unmethylated) *Pst*I fragments usually results in probes with low copy DNA sequences, which are preferred for RFLP analyses.[797]

In addition to the above-mentioned anonymous probes, ribosomal DNA (rDNA) (coding region of 18S, 5.8S, and 25S rRNA) is often used as a source for RFLPs in plants since (1) the same probes can often be applied to different species because of their conserved coding sequences and (2) polymorphisms in rDNA are easy to detect due to the high abundance of these sequences; there are hundreds of tandemly repeated rDNA copies located on one or more chromosomes. Variability is mainly confined to the noncoding rDNA regions of the intergenic spacer, where polymorphisms in the number of short tandem repeats occur. Coding regions, on the other hand, are generally conserved. Numerous publications report on the successful use of rDNA polymorphisms in population biology and systematics.[260,354,392,520,603] Results from rDNA analysis, however, should be treated with caution as concerted evolution and gene conversion may lead to erroneous conclusions. Moreover, different rDNA sequences may occur within the same individual, which further complicates the analysis.

2. RFLPs in Cytoplasmic DNA: cpDNA and mtDNA

Chloroplast DNA (cpDNA) in plants and mitochondrial DNA (mtDNA) in animals and plants are both uniparentally transmitted in contrast to the biparentally transmitted nuclear DNA.[534] In most cases, cpDNA and mtDNA are transmitted maternally; one well-known exception is the paternal transmission of cpDNA in gymnosperms. Both types of DNA are present in several to hundreds of copies per cell. Because of their uniparental inheritance, both types of DNA act as single heritable units, which must be kept in mind when analyzing data from cpDNA or mtDNA studies.

There are two main approaches for studying RFLPs in cytoplasmic DNA. The first is to extract mtDNA and/or cpDNA separately from the nuclear DNA, which can be achieved by ultracentrifugation in density gradients or, alternatively, by differential extraction procedures.[468,688] Cytoplasmic DNA is then digested with restriction enzymes and electrophoresed on agarose or polyacrylamide gels, and RFLPs are directly detected by ethidium bromide or silver staining. The second strategy is to isolate and digest the total DNA of the organism, followed by electrophoresis and Southern blotting of the restriction fragments. The cytoplasmic DNA is then visualized by hybridization with a labeled probe (total cp or mtDNA or a specific cp or mtDNA sequence).

cpDNA RFLPs have been studied extensively in plants.[535,648] Gene order and size of the chloroplast genome, approximately 150 kb, are relatively constant across species. The DNA molecule consists of an inverted repeat as well as one large and one small single copy region. There is usually no need for isolating species-specific probes since most sequences are conserved. Several complete libraries of the chloroplast genome from different species have been made widely available among scientists and can be successfully applied also to unrelated species.[648] RFLP and sequence analyses of the chloroplast genome have proved to be very valuable for taxonomic and phylogenetic studies at an interspecific or intergeneric level.[115] However, only rarely has intraspecific variation been encountered at levels that are high enough for population studies.[674,795]

With mtDNA, a different situation is encountered in plants as compared to animals. In animals, the mtDNA molecule is small, and its gene order is very conservative, whereas the rate of sequence divergence is higher than in nuclear DNA. Its high sequence variability, small genome size, and relatively easy isolation from animal tissues make mtDNA a valuable source for RFLPs in population studies, especially for the analysis of maternal lineages and population history. In plants, on the other hand, mtDNA is much larger, and sequences are not particularly divergent, whereas the general architecture of the molecule is highly variable. As a consequence of intragenomic recombination, different forms and sizes of mtDNA can be found within an individual, which may render RFLP analysis impossible. Therefore, mtDNA RFLPs have been used only rarely in plants, except for specific purposes, e.g., to analyze cytoplasmic male sterility[146,368,751] or to follow seed migration routes in gymnosperms.[156]

E. DNA FINGERPRINTING BASED ON HYBRIDIZATION

The technique of classical DNA fingerprinting is methodologically derived from RFLP analysis and is mainly distinguished from the latter technique by the kind of hybridization probe applied to reveal polymorphisms. To obtain a typical DNA fingerprint, probes are used to create complex banding patterns by recognizing multiple DNA loci simultaneously (see Section II.B). Each of these loci is characterized by more or less regular arrays of tandemly repeated DNA motifs that occur in different numbers at different loci.

Two categories of such multilocus probes are mainly used. The first category comprises cloned DNA fragments or oligonucleotides which are complementary to so-called "minisatellites,"[315] i.e., tandem repeats of a basic motif of about 10 to 60 bp. The second category is exemplified by oligonucleotide probes which are complementary to so-called "simple sequences"[686] or "microsatellites,"[408] i.e., tandem repeats of very short motifs, mostly 1 to 5 bp.

With both kinds of probes, a high degree of polymorphism between related genotypes is usually observed, which has been exploited for numerous studies in diverse areas of genome analysis. The history of classical DNA fingerprinting, as well as the molecular nature of the polymorphic DNA sequences recognized by fingerprint probes, are described in more detail in Section II of this chapter. The underlying methodology is described in Chapter 4, Section III, and the applications for plants and fungi are described in Chapter 5.

F. MOLECULAR MARKERS BASED ON DNA AMPLIFICATION

In 1985, a new technique was introduced which revolutionized the methodological repertoire of molecular biology: the polymerase chain reaction (PCR).[604,605] This technique allows us to amplify any DNA sequence of interest to high copy numbers, thereby circumventing the need of molecular cloning. In order to amplify a particular DNA sequence, two single-stranded oligonucleotide primers are designed, which are complementary to motifs on the template DNA. The primer sequences are chosen so as to allow base-specific binding to the template in reverse orientation. Addition of a thermostable DNA polymerase in a suitable buffer system and cyclic programming of primer annealing, polymerization, and denaturation steps result in the exponential amplification of the sequence between the primer sites (see Section III.A for a more detailed description).

Shortly after its introduction, many variants of the basic PCR strategy were developed (reviewed by Innis et al.[308]), and it was obvious that the technique would also be feasible for the detection of DNA polymorphisms. Initial attempts to reveal DNA polymorphisms made use of *specific primers* complementary to known sequences. These experiments demonstrated that primers, which are complementary to *flanking regions* of minisatellite[319] and simple sequence (=microsatellite) loci,[408,745] yield highly polymorphic amplification products. This type of polymorphism (treated in more detail in Chapter 7) has turned out to be particularly useful for studies on population genetics and genome mapping, especially in humans.

Another strategy made use of *semispecific primers*, which are complementary to repetitive DNA elements. For human genome analysis, an abundant class of randomly interspersed DNA called "Alu repeats" was preferentially used for this purpose, and "Alu-PCR" revealed considerable levels of polymorphism (see, e.g., Ledbetter et al.[394]). Later on, the principle of Alu-PCR (i.e., the use of primers complementary to interspersed repeats) was adopted for other species as well (see, e.g., Kaukinen and Varvio[344]). As an alternative to interspersed repeats, primers complementary to other repetitive sequence elements were also successfully used for the generation of polymorphisms. Such sequences include intron/exon splice junctions,[752] tRNA genes,[767] 5S RNA genes,[366] zinc finger protein genes,[707] as well as mini- and microsatellites.[279,405,459,460] The latter technique is described in more detail in Section III.B.

We would like to briefly mention four interesting applications of PCR with specific and/or semispecific primers: (1) Cullings[133] developed primers complementary to the 5′ end of the 28S ribosomal RNA gene. In combination with a universal plant DNA primer, DNA was amplified from higher plants only, allowing the identification of different plant species and also the detection of plant associates in, for instance, plant/fungal systems. (2) In a similar approach designed for identification of mycorrhizae and rusts, Gardes and Bruns[209] developed two fungal-specific primers for the internal transcribed spacer region in the nuclear ribosomal repeat. (3) Taberlet et al.[676] developed three primer pairs matching tRNA sequences of the cpDNA genome. These primer pairs, complementary to conserved sequences, amplified intervening noncoding (and therefore variable) regions and may prove useful for evolutionary studies of algae and higher plants. (4) Lessa[397] developed primers complementary to specific exons, resulting in the amplification of the intervening introns. These amplification products were checked with denaturing gradient gel electrophoresis (DGGE, see Chapter 7) for polymorphisms and, if polymorphic, sequenced. Genetic variability was estimated from these sequences.

In 1990, several laboratories introduced a strategy that made use of one or two short, GC-rich **primers of arbitrary sequence** to generate PCR amplification products from genomic DNA. This technique, which does not require any sequence information, was called random amplified polymorphic DNA (RAPD) analysis,[782] arbitrarily primed polymerase chain reaction (AP-PCR),[766] or DNA amplification fingerprinting (DAF).[86,87] Typically, fingerprint-like patterns of variable complexity were produced, which distinguished genotypes. Polymorphisms detected by this method were called RAPDs[782] or amplified fragment length polymorphisms (AFLPs).[86,87] As is the case with mini- and microsatellites (see Section II.A) the polymorphic nature of the amplified DNA fragments is paralleled by a polymorphic nomenclature. A common term, multiple arbitrary amplicon profiling (MAAP) has been suggested to describe the common characteristics of all techniques.[88]

The use of PCR for the amplification of polymorphic DNA fragments with the help of arbitrary primers is a main focus of this book (see Section III.C. for the theory; Chapter 4, Section IV for the methodology; and Chapter 5 for the applications). We will mainly use the term RAPD, since this MAAP variant described by Williams et al.[782] has spread more "RAPDLY" than the others, and is used by most laboratories.

G. MISCELLANEOUS MARKERS

In recent years, several new and sophisticated methods have been developed to screen for polymorphisms at the DNA level. For instance, some techniques make use of the conformation dependence of DNA mobility on the primary sequence in DGGE,[705] others allow the internal mapping of minisatellite repeat variants.[320,321] Since some of these techniques may have the potential of a more widespread application in the future, they will be summarized briefly in Chapter 7.

II. MINISATELLITES AND SIMPLE SEQUENCES: THE MOLECULAR BASIS OF CLASSICAL DNA FINGERPRINTING

A. THE NOMENCLATURE OF TANDEM-REPEATED DNA ELEMENTS

The technique of classical, i.e., hybridization-based, DNA fingerprinting highlights certain classes of repetitive, highly polymorphic DNA sequences in eukaryotic genomes, the nomenclature of which is sometimes a bit confusing (see below). **Repetitive DNA** is an integral component of eukaryotic genomes and may comprise up to more than 90% of total DNA in certain plant genomes. According to the way it is organized, repetitive DNA may be classified as either **interspersed** or **tandemly repeated**. In interspersed repeats, the repeated DNA motifs occur at multiple sites throughout the genome. Tandem repeats, on the other hand, consist of arrays of two to several thousand basic motifs which are arranged in a head-to-tail fashion. Though this kind of organization is also exhibited by some genes (e.g., the transcription units for histone and ribosomal RNA), most tandem repeats probably consist of noncoding DNA.

Tandem repeats may be classified according to the length and copy number of the basic repeated element, as well as to their genomic localization

1. **Satellite** DNA was originally described more than 25 years ago. It was named after its separability from bulk DNA by buoyant density gradient centrifugation. Satellites consist of very high repetitions (usually between 1000 and more than 100,000 copies) of a basic motif and form very long, often heterochromatic stretches of DNA. The length of a repeat unit may vary between 2 and several thousand bp, but repeat units of 100 to 300 bp are most commonly observed. Satellites usually occur at a few genomic loci only.

2. The term **minisatellites** was invented in 1985 to describe another family of tandemly organized repeats.[315] This class of DNA consists of shorter motifs (usually 10 to 60 bp) and shows a lower degree of repetition at a given locus. Minisatellites may form "families" with related sequences and occur at many loci in the genome (probably many thousands).

3. Tandem repeats made up from very short (between 1 and about 10 bp) motifs have been called **simple sequences** by Tautz and Renz.[686] Later on, such sequences have also been coined **microsatellites**[408] in continuation of the above nomenclature, "simple repetitive sequences," or "simple tandem repeats."[175] Microsatellites are generally characterized by short motifs, a comparatively low degree of repetition, and dispersed distribution over many genomic loci (the highest number of loci observed with the shortest motifs).

4. To complete the armada of differently sized tandem repeats, Nakamura et al.[488] created the term **midisatellite** for a DNA category combining typical properties of satellites (i.e., a long array of repeats at a single genomic locus) and minisatellites (i.e., variable number of a tandemly repeated 40-bp sequence).

A common property of mini- and microsatellite tandem arrays is that identical or related motifs occur at multiple genomic sites, i.e., these sequences share the properties of both tandemly repeated as well as interspersed DNA. Moreover, different mini- and microsatellites often occur intermingled with each other in a particular stretch (see, e.g., Weber[744] Armour et. al.;[16,18] Figure 2). Together with the accumulation of point mutations within repeat units, the

A. MINISATELLITE REPEAT UNIT CONSENSUS SEQUENCES

ACAGGGGTGTGGGG	human	Bell et al. (1982)
GGAGGTGGGCAGGAXG	human	Jeffreys et al. (1985a)
GAGGGTGGXGGXTCT	M13 phage	Vassart et al. (1987)
AGAGGGCGGCGG	rice	Daly et al. (1991)

B. CLONED SIMPLE REPETITIVE SEQUENCES

..TATA(GATA)$_{10}$(GACA)$_{10}$TTTG..	*Elaphe radiata*	Epplen et al. (1982)
..(AT)$_{17}$(GT)$_{13}$(AT)$_{15}$(GTAT)$_8$AT(GT)$_{16}$..	mouse	Love et al. (1990)
..(CT)$_{30}$(CA)$_{30}$(GA)$_4$..	mouse	Love et al. (1990)
..TCAGT(AT)$_7$(AC)$_8$ATTGA..	*Piper reticulatum*	Condit and Hubbell (1991)
..TTGAA(TA)$_7$CATA(CA)$_{10}$TAT..	*Virola sebifera*	Condit and Hubbell (1991)
..CGGCGACGA(CGG)$_8$TGGCGG..	*Oryza sativa*	Zhao and Kochert (1993)
..(AT)$_3$GT(AT)$_4$(GT)$_8$..	*Brassica napus*	Lagercrantz et al. (1993)

FIGURE 2.   Examples of cloned minisatellite (A) and simple repetitive sequence stretches (B).

intermingling of different types of repeats may result in DNA sequences which are **cryptically simple**,[687] i.e., their repeat structure is more or less concealed. Some examples of naturally occurring minisatellites and simple sequences are shown in Figure 2.

Since tandem repeats in general, and mini- and microsatellite-like sequences in particular, are characterized by highly variable copy numbers of identical or closely related basic motifs, this class of DNA polymorphism was designated "VNTR" (variable number of tandem repeats),[489] which was a further contribution to the polymorphic nomenclature of these polymorphic sequences.

For reasons of simplicity, we shall mainly refer to the terms minisatellites and simple sequences for the categories of DNA sequences treated in this book. A more comprehensive review of the definition and nomenclature of tandemly repetitive DNA sequences has recently appeared.[685]

B. DNA FINGERPRINTING WITH TANDEM-REPETITIVE PROBES

1. Human Minisatellites

Highly polymorphic loci based on tandem repeats were detected in the human genome in the early 1980s.[39,810] However, it was not until 1985 that genetic fingerprinting of humans was achieved by the detection of **multiple hypervariable DNA loci simultaneously**.[315,316] Alec Jeffreys and his colleagues identified a family of tandem-repeated DNA sequences (which they called minisatellites) that was characterized by a common, GC-rich core sequence of 10 to 15 bp (exemplified by a tandem repeat of four 33-bp motifs within an intron of the human myoglobin gene). Hybridization of restriction-digested human DNA with a probe derived from this core sequence under nonstringent conditions detected several extremely variable minisatellites simultaneously and thus created an individual-specific DNA fingerprint. It was clearly demonstrated that a variable number of tandem repeats at a given locus provided the molecular basis of the observed length polymorphisms. The basic strategy of DNA fingerprinting with multiallelic, multilocus hybridization probes complementary to variable numbers of tandem repeats is outlined in Figures 3 and 4.

This first generation of human minisatellite probes (e.g., 33.6 and 33.15, identified by Jeffreys and coworkers) also was shown later to cross-hybridize to a variety of mammalian,[314] avian,[83,773] fungal,[64] and plant genomes,[137,589] thereby revealing individual-specific patterns in most animal species.

2. Other Minisatellites

Since Jeffreys' pioneering work, many minisatellite loci have been cloned and sequenced not only from humans (e.g., Nakamura et al.,[489] Wong et al.[803,804]), but also from other mammals (e.g., Georges et al.,[215] Kashi et al.[343]), birds (e.g., Gyllensten et al.,[250] Hanotte et al.[264,265]), insects (e.g., Haymer et al.[276]), fish (e.g., Bentzen and Wright[42]), and plants (e.g., Broun and Tanksley,[69] Dallas et al.[138]). In 1987, it was found by chance that hybridization probes derived from an internal repeat sequence of the protein III gene of bacteriophage M13 reveals hypervariable loci in humans and animals.[726] Later on, the M13 probe was shown to hybridize to minisatellite-like sequences in the genome of a wide variety of organisms (see, e.g., Ryskov et al.[600]), including plants,[588] fungi,[453,454] protozoa,[590] and even bacteria.[302] The M13 repeat has now become one of the most frequently used minisatellite probes for DNA fingerprinting in plants and fungi (see Chapter 5). The localization of the minisatellite repeats in the genome of the M13 phage is illustrated in Figure 5.

The M13 phage is not the only exotic source for minisatellite-detecting probes. For example, repeat sequences from three viruses have been successfully used for DNA fingerprinting of sheep,[128] and even the gene for a glycine-rich eggshell protein of the parasitic trematode *Schistosoma mansoni* was shown to recognize minisatellite-like sequences in the human genome.[547,548] Furthermore, new minisatellite-detecting probes seem to be comparatively easy to design. In an impressive series of experiments, Vergnaud and coworkers have demonstrated that probing the human genome with synthetic tandem repeats (STRs) of **arbitrarily** chosen oligonucleotides (basic unit 14 to 20 bp) very often results in polymorphic fingerprints, provided that nonstringent hybridization conditions are applied.[727–729] In an extension of this study, Mariat and Vergnaud[429] showed the majority of these probes to be informative for selected mammalian, bird, and fish genomes also.

3. Simple Sequence Probes

The existence of tandem repeats consisting of very short (1 to 6 bp) sequence motifs was recognized in the early 1970s [e.g., $(TAGG)_n$ repeats in satellite DNA of a hermit crab[632]]. Since then, a large number of studies have been undertaken on the occurrence and distribution of this kind of DNA in human, animal, fungal, plant, and bacterial genomes (see, e.g., Beckmann and Weber,[38] Greaves and Patient,[240] Hamada et al.,[258] Lagercrantz et al.,[384] Miklos

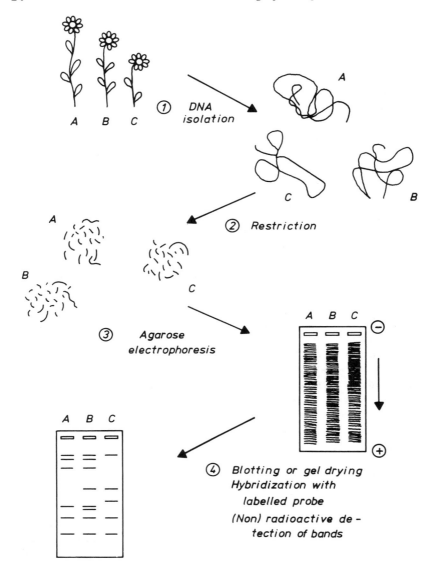

FIGURE 3. Strategy of hybridization-based DNA fingerprinting. Genomic DNA is isolated from the organism of interest (1) and digested with a suitable restriction enzyme (2). The restriction fragments are separated by gel electrophoresis (3). The gel is then denatured, neutralized, and either dried or blotted onto a membrane (4). Multiple polymorphic restriction fragments are detected by hybridization with a labeled multilocus probe (e.g., a cloned minisatellite) and visualized by autoradiography or nonradioactive techniques (4).

et al.,[463] Stallings et al.,[659] Tautz and Renz,[686] Tautz et al.,[687] and Weising et al.[757,759]). The general outcome of these studies was that these so-called simple sequences[686] or microsatellites[408] are ubiquitous components of most, if not all,

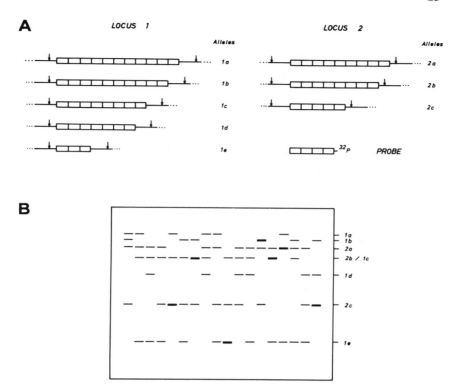

FIGURE 4. Hybridization of restriction-digested genomic DNA with multiallelic, multilocus, "variable number of tandem repeat" probes results in DNA fingerprints. (A) Architecture of genomic loci recognized by "fingerprint" probes. Each block represents a simple motif of about 2 to 50 nucleotides. These blocks are tandemerized in variable copy numbers representing five (1a–e) and three (2a–c) RFLP alleles in locus 1 and 2, respectively, in the present example. Vertical arrows indicate conserved restriction sites outside the tandem repeats. A ^{32}P-labeled tetramer of the motif serves as a hybridization probe. (B) In this fingerprint analysis, genomic DNA of 18 individuals of a population are cut at the restriction sites indicated by arrows in (A). Fragments are electrophoresed, blotted onto a membrane, and hybridized with the tetrameric probe. With one exception, all individuals show different banding patterns.

eukaryotic genomes, while they are rare or absent in prokaryotes (see, e.g., Gross and Garrard[243]).

The pioneering work of Jörg Epplen and colleagues showed that synthetic hybridization probes complementary to simple sequence motifs [such as (GATA)$_4$] could be successfully applied to DNA fingerprinting.[8,609] The authors used in-gel hybridization with labeled oligonucleotides, a method exhibiting several advantages over the conventional blotting techniques, such as speed, "100% transfer efficiency", and circumvention of prehybridization steps. Originally applied to human DNA, oligonucleotide probes complementary to simple repetitive sequences are now increasingly used as multilocus probes, revealing hypervariable target regions in animals, plants, and fungi (reviewed by Epplen et al.,[179] see Chapter 5).

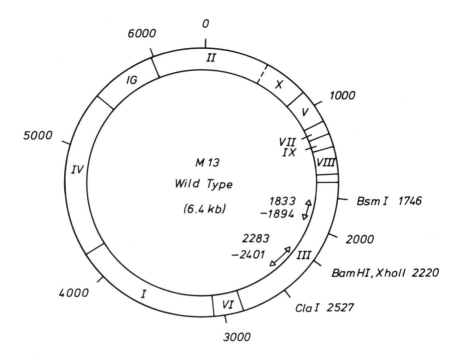

FIGURE 5. The genome of the wild-type M13 phage. Arabic numbers indicate map positions in base-pairs. Over 90% of the viral genome codes for the genes I to X. Double-headed arrows show the locations of two minisatellite tandem repeat blocks within the protein III gene. IG: intergenic region.

While the probes used in this "oligonucleotide fingerprinting" approach probably recognize microsatellite-like target sequences, the nature of the observed polymorphisms is not yet clear. Since sequenced microsatellites are generally not much longer than about 100 bp, and the size of the fingerprint fragments detected by simple sequence oligonucleotide probes usually ranges from 1 to more than 10 kb, the observed polymorphisms are probably not only based on variable numbers of microsatellite repeat units. Cloning experiments have shown that different types of simple sequences are often intermingled with each other and with other types of (tandem as well as interspersed) repeats (see, e.g., Armour et al.,[16] Broun and Tanksley,[69] Kaukinen and Varvio,[344] Zischler et al.[840]). It remains to be determined whether the concerted action of a mixture of different classes of repeats or an as yet unknown mechanism is responsible for the generation of polymorphisms in oligonucleotide finger-printing.

4. Factors Affecting the Level of DNA Fingerprint Polymorphism

A multitude of studies has shown that the pattern complexity as well as the level of polymorphism revealed by DNA fingerprinting with both minisatellite and simple repetitive oligonucleotide probes depends on (1) the investigated species (sometimes even the population), (2) the repeated sequence motif used for hybridization, and (3) to some extent also on the restriction enzyme (see, e.g., Zeh et al.[830]). It is still not clear which features (if any) are essential for a fingerprint probe to reveal high or low levels of polymorphism. Consequently, the optimal combination of species, probe, and enzyme has to be determined empirically for each purpose (Figure 6). This is particularly true for different simple sequence-complementary probes, where species-, variety-, or individual-specific patterns may be observed within the same genus (see Sharma et al.,[623] Weising et al.,[761] and Figure 7).

Interestingly, individual-specific patterns have been more frequently reported for animal than for plant species. This may be partially explained by the occurrence of different reproductive strategies in plants as compared to animals. Whereas sexual reproduction and outcrossing are the rule in most animal populations, selfing and vegetative propagation are widespread phenomena in plants (see also Chapter 5, Sections II and III). Moreover, many plant species investigated by DNA fingerprinting to date are crops that have been domesticated and "streamlined" for thousands of years, (e.g., by inbreeding), thereby reducing genetic variability. Elevated levels of inbreeding are reflected by low levels of variability in multilocus fingerprints not only in plant, but also in animal species.[266,376,570]

Taken together, the level of polymorphism exhibited by a particular probe within a particular species is obviously dependent upon at least three factors: (1) the reproductive strategies exhibited by (or forced upon) the species under investigation (i.e., selfing, outcrossing, vegetative propagation, inbreeding in isolated or captive populations, domestic animals and crop plants), (2) the molecular properties (e.g., mutation rate) of the target sequence and its surroundings, and (3) (to a smaller extent) the choice of restriction enzyme.

C. CHROMOSOMAL LOCALIZATION OF MINISATELLITES AND SIMPLE SEQUENCES

The genomic distribution of GC-rich minisatellites detected by the human 33.6 and 33.15 probes has been studied extensively, not only in the human, but also in several mammalian and bird genomes. Segregation analyses showed that the majority of bands visualized in fingerprints are assorted independently to the progeny.[317,773] However, linkage between different minisatellite loci was also observed.[17,214,215,765] In some cases, minisatellite "haplotypes" were detected, which consist of up to ten tightly (i.e., 100%) linked bands, e.g., in dogs,[314] swans,[450] and several species of gallinaceous birds.[266]

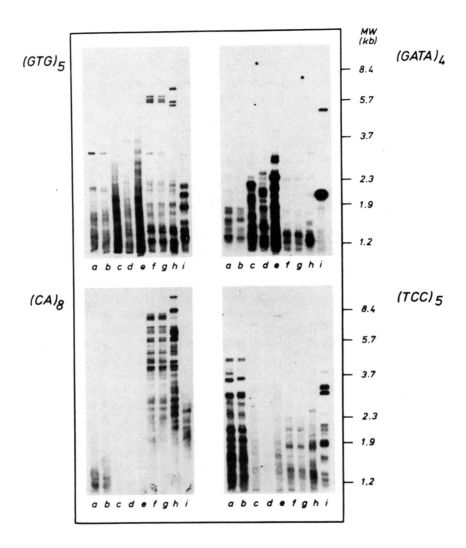

FIGURE 6. Screening for optimal probe/species combinations by oligonucleotide fingerprinting using genomic DNA from individual plants of two date palm cultivars (lanes a and b, *Phoenix dactylifera*); three species of sugar cane (lane c, *Saccharum spontaneum*; lane d, *S. robustum*; lane e, *S. officinarum*); two genotypes of banana (lanes f and g, *Musa* sp. AAA genome; lane h, ABB genome); and one papaya cultivar (lane i, *Carica papaya*). DNA samples were digested with *Hin*fI, separated on an agarose gel, and in-gel hybridized consecutively with each of the ^{32}P-labeled oligonucleotide probes (GTG)₅, (GATA)₄, (CA)₈, and (TCC)₅. Size markers (kb) are indicated.

Sequence analyses have shown that minisatellite arrays are often intermingled with other kinds of repetitive DNA, especially with interspersed repeats (see above). In humans, *in situ* hybridization revealed a prevalent localization of minisatellites close to telomeres.[597,728] Such a distribution was, however, not found in other species (e.g., mouse,[339] cattle,[215] and tomato[69]).

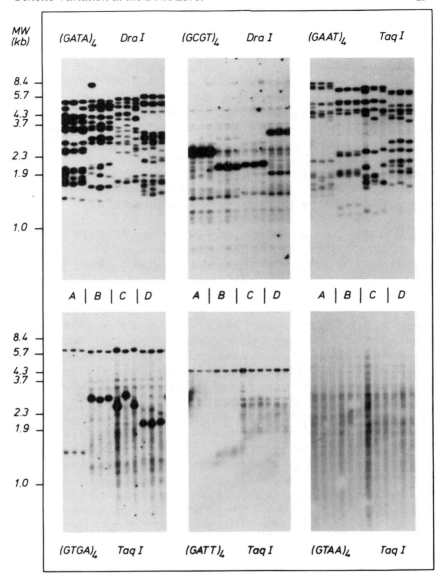

FIGURE 7. Oligonucleotide fingerprinting of chickpea (*Cicer arietinum*) using different hybridization probes. The genomic DNA of three individual plants of each of the four chickpea accessions (A, B, C, D) was digested with *Taq*I or *Dra*I, separated on an agarose gel, and in-gel hybridized with ^{32}P-labeled probes. Size markers (kb) are indicated. The patterns range from highly variable [e.g., (GAAT)$_4$] to completely monomorphic [e.g., (GATT)$_4$]. Accession-specific fingerprints are observed with (GCGT)$_4$.

Genetic mapping experiments were also carried out with PCR-amplified microsatellites (see also Chapter 7). These studies indicated that, e.g., (CA)$_n$ repeats are evenly dispersed throughout the human,[762] mouse,[154] and rice genome,[809] albeit some local clustering occurs. In concordance with these results, *in situ* hybridization with simple sequence-specific oligonucleotide probes showed that microsatellite motifs are quite homogeneously distributed throughout the human genome [e.g., $(CAC)_n$ repeats[837]].

To summarize our present state of knowledge, the currently available data suggest a clustered rather than dispersed distribution of minisatellites throughout the genome, with a preference for regions that are generally rich in repetitive DNA. Simple sequence repeats (=microsatellites), on the other hand, appear to be more evenly dispersed throughout eukaryotic genomes. However, it should be noted that data have been collected for only a few repeat motifs and a few organisms (i.e., little data are available yet for plants and fungi). Therefore, it may be too early to draw general conclusions about the chromosomal distribution of minisatellites vs. simple sequences.

D. MECHANISMS RESPONSIBLE FOR MINISATELLITE AND SIMPLE SEQUENCE POLYMORPHISM

Minisatellites and simple sequences are often characterized by high meiotic mutation rates (up to 5%; see, e.g., Jeffreys et al.[318]), which mainly concern the number of repeats. However, internal heterogeneity of the repeated units[320] as well as somatic mutations are also observed.[15,349] Interestingly, mutation rates appear to be positively correlated with the total size of the array, not only in minisatellites,[18,239] but also in simple sequence stretches.[99,744] Both types of repeats might remain invariant for long periods of time if they contain a few repeated motifs only. However, as soon as the tandem copy number exceeds a certain threshold, the chance for further mutation is greatly enhanced (see Richards and Sutherland[575] for a discussion of these so-called "dynamic mutations"). In concordance with these observations, high molecular weight bands within a multilocus fingerprint are often more variable than bands occurring in the low molecular weight range.

The molecular basis of both minisatellite and simple sequence variability is still a matter of discussion. Possible mechanisms include replication slippage, transposition, recombinational events and/or unequal exchange between sister chromatids at mitosis/meiosis or between homologous chromosomes at meiosis, and gene conversion (reviewed by Jarman and Wells,[312] Jeffreys et al.,[323] Richards and Sutherland,[575] Wolff et al.[801]). The **slippage** hypothesis implicates slipped-strand mispairing of the newly replicated strand during the replication process.[398,399] *In vitro* experiments have shown that replication slippage may actually result in considerable amplification of a given simple sequence repeat.[613]

Several lines of evidence have lent support to the **recombination** hypothesis: (1) a variety of minisatellite core sequences share homology to the bacterial recombination signal chi,[315,316] (2) minisatellite-like sequences have been found at sites of meiotic crossing over,[107] (3) both mini- and microsatellites behave as recombinational hot spots in transfected mammalian cells.[733,734]

In a detailed study on a particularly variable human minisatellite, Wolff et al.[799–801] observed no exchange of flanking markers in a newly created allele, thus ruling out unequal exchange between homologous chromosomes as a mutational mechanism in this case. In another human minisatellite locus, MS32 (reviewed by Jeffreys et al.[323]), a strong bias of mutation events toward the 5′ end of the array was observed, suggesting the existence of a mutational hot spot. Some mutant alleles were shown to contain segments from both parental alleles, providing evidence for interallelic exchange. Flanking markers, on the other hand, were not exchanged. Therefore, it was suggested that the major mutational process was gene conversion involving the nonreciprocal transfer of repeat units from a donor allele into the 5′ end of a recipient allele.

Taken together, recombinational processes as well as replication slippage may certainly contribute to the creation of minisatellite and simple sequence variability. However, other (as yet unidentified) mechanisms may also be involved, especially in case of the "explosive" amplification of trinucleotide-based microsatellites associated with some human genetic diseases,[99,575] and in case of the polymorphisms that underly "oligonucleotide fingerprints" (see Section II.B.3). Structural analysis of mutated vs. parental alleles provides the most obvious way to gain more information about the mutational mechanisms. In this respect, the availability of transgenic systems will be particularly informative, since successive deletion of flanking DNA will allow us to locate mutational hot spot regions more precisely. Studies on "transminisatellitic mice" are already on their way.[323]

E. FUNCTIONAL SIGNIFICANCE OF MINISATELLITES AND SIMPLE SEQUENCES

Though having been investigated since their initial discovery, the functional impact of simple sequences and minisatellites for eukaryotic genomes is still poorly understood. Whereas both types of sequences comply with the concept of "selfish" DNA[157,528] in being able to amplify and propagate in the absence of counterselective pressure, possible structural and functional roles neverthe-less have been implicated and discussed. Because the functional significance of this category of DNA is not the central aspect of this book, putative functions will be only briefly summarized. For more detailed treatments of this topic, the interested reader may refer to several recent reviews[178,180,312,731] as well as to the references cited below.

1. Telomeres

Telomeric repeats, which are found at the ends of eukaryotic chromosomes, represent a "special version" of simple sequences. A basic motif of 4 to 10 bp, which shows a marked base asymmetry (one strand is GA-rich and the other strand is CT-rich), is repeated several hundred to thousand times and forms a single-stranded 5′ overhang at the very ends of each chromosome. Telomeres are formed (and prolonged) by the action of a specific DNA polymerase called telomerase. Using an internal RNA molecule as the template, this enzyme adds new telomeric repeat units to the end of existing telomeres. Telomeric sequences represent a rare example of simple sequences having clearly defined functions; they protect the chromosomal ends from degradation and fusion processes and compensate for DNA loss due to incomplete replication of chromosomal ends. Moreover, a functional impact of telomeric repeats on nuclear architecture and cell senescence is discussed. The structure and functions of telomeres have been reviewed by Blackburn.[53,54]

2. Centromeres

The majority of chromosomal centromeric regions probably consists of repeated sequences. A simple repeated element, $(GGAAT)_n$, was recently identified in yeast and human centromeric regions.[234] This repeat exhibits an unusual DNA conformation and has a high affinity for specific nuclear proteins. The finding of highly reiterated repeats reminiscent of degenerate telomere sequences also in plant centromeric regions (*Arabidopsis thaliana*[574]) suggests that simple repeats may be a general structural component of functional centromeres.

3. Transcription

With some exceptions (e.g., multigene families, ribosomal and tRNA genes, transposable elements), repetitive DNA is generally thought to be transcriptionally silent. While this is probably also true for most (but not all, see, e.g., Swallow et al.[672]) minisatellites, short stretches of simple sequences (usually tri- or hexanucleotide repeats) may well occur in transcribed and even translated regions of DNA. For example, in a variety of genes, repeated triplets code for particular amino acids which make up most of the respective protein (e.g., GGT repeats in glycine-rich, and CCA repeats in proline-rich plant proteins[120]).

Transcription of simple sequence motifs has received much attention during the past 2 years, since a sudden amplification of transcribed or even translated trinucleotide repeats to high copy numbers ("dynamic mutation") was shown to be closely associated with a variety of human genetic diseases (reviewed by Caskey et al.[99] and Richards and Sutherland[575]). Except for providing large numbers of identical amino acids within specific proteins, the functional impact of simple sequence transcription is as yet unclear. Several putative roles, such as sex determination[305] and the control of biological rhythms,[774] have been discussed.

4. Transcriptional Regulation

It has been shown by several investigators that minisatellites or simple sequences located in DNA control regions are able to enhance or reduce the transcription rate of neighboring genes.[224,259,419,490,653] For example, poly(CA) was found to enhance the expression of a transgene in transfected mammalian cells,[259] and two $(GA)_n$ motifs were identified as important stimulating elements of a *Drosophila* heat shock gene promoter.[224,419] A negative effect on transcription was observed by Naylor and Clark.[490] In their study, the presence of a $(CA)_n$ repeat in the promoter region abolished transcription of a reporter gene in transfected mammalian cells.

5. Recombination

As already mentioned above (Section II.D), minisatellites as well as simple sequences [especially $(CA)_n$ repeats] have been attributed a functional role as recombinational hot spots (e.g., Wahls et al.[733,734]), a view which is nevertheless controversially discussed (see above).[312,323,799–801]

6. Replication

Simple sequence repeats have been detected at putative replication origins of the slime mold *Physarum polycephalum*.[527] Opstelten et al. put forward a general hypothesis in which slippage of simple sequences may produce locally unpaired regions of DNA that are recognized by replication initiation factors.

7. General Considerations

Through which mechanisms can minisatellites and simple sequences influence elementary cellular processes such as recombination or transcription of neighboring genes? Two main mechanisms have to be considered: (1) some simple sequences (e.g., of the $[CA]_n$ or $[GA]_n$ type) may exert their biological effects (e.g., on chromatin structure) by adopting unusual DNA conformations (see Vogt[731] for a discussion) and (2) some repeats may be recognized by regulatory proteins. Sequence-specific binding of nuclear proteins to minisatellites[116,117,735,815] and simple repeats[180,221,820] was frequently observed, either to single- or double-stranded DNA. In a detailed study, Richards et al.[576] demonstrated the specific binding of human nuclear proteins to four different di- and ten trinucleotide repeats. Hence, protein binding to simple repeats appears to be a general phenomenon. In case of recombination hot spots, single-strand-specific DNA-binding proteins might initiate or otherwise assist in recombinational processes (see Collick et al.[117] and Yamazaki et al.[815] for a discussion). In case of gene regulatory regions, minisatellite- or simple sequence-binding proteins might trigger (or abolish) the generation of the transcription initiation complex and/or an active chromatin structure (see Lu et al.[419] for a discussion). The latter hypothesis is supported by the demonstration of sequence-specific interactions of defined, gene regulatory proteins with minisatellites[699,700] as well as with simple sequences (e.g., the "GAGA" factor[221,419]).

Taken together, we believe that the majority of minisatellites and simple sequences do indeed behave selfishly by having evolved strategies which ensure their genomic survival. Nevertheless, there is considerable evidence that certain repeats located at particular genomic positions have acquired one or more of the specific functions discussed above. Future research will shed more light on the mechanism(s) responsible for mediating transcriptional regulation or the initiation of recombination by simple tandem repeats.

III. DETECTION OF DNA POLYMORPHISM BY PCR-BASED FINGERPRINTING

A. PRINCIPLE OF THE PCR

The PCR is a versatile technique that was invented in the mid-1980s.[604] Since the introduction of thermostable DNA polymerases in 1988,[605] the use of PCR in research and clinical laboratories has increased tremendously. The method is based on the enzymatic *in vitro* amplification of DNA. Starting from a very low amount of template DNA (mostly in the nanogram range), millions of copies of one or more particular target DNA fragments are produced which can be electrophoresed and visualized by staining or autoradiography. PCR is characterized by its high speed, selectivity, and sensitivity. Its application in diverse variants and for many different purposes has opened up a multitude of new possibilities in molecular biology. The present section describes the general principle of the technique. In the following sections, we will introduce PCR with tandem repeat primers and arbitrary primers as special variants of the PCR technique which are useful for DNA fingerprinting.

In a typical PCR, three temperature-controlled steps can be discerned, which are repeated in a series of 25 to 50 cycles. A reaction mix consists of

1. A buffer, usually containing Tris-HCl, KCl, and $MgCl_2$
2. A thermostable DNA polymerase which adds nucleotides to the 3' end of a primer annealed to single-stranded DNA
3. Four deoxynucleotides (dNTPs: dATP, dCTP, dGTP, dTTP)
4. Two oligonucleotide primers
5. Template DNA

The reaction mix is usually overlaid with mineral oil to prevent evaporation and condensation of water onto the walls and lid of the tube.

The selectivity of the reaction is determined by the choice of the primer(s). Primers are single-stranded pieces of DNA (oligonucleotides) with sequence complementarity to template sequences flanking the targeted region. To allow for exponential amplification, the primers must anneal in opposite directions, so that their 3' ends face the target (see Figure 8). Amplification is most efficient when the two primer binding sites are not further apart than about

---

FIGURE 8. Principle of the PCR. A target DNA sequence is exponentially amplified with the help of flanking primers and a thermostable DNA polymerase. The reaction involves repeated cycles, each consisting of a denaturation, a primer annealing, and an elongation step. Primers are represented by black boxes. 5' ends are indicated by open circles, and 3' ends are indicated by closed circles. In the initial stages of the reaction, both longer and shorter reaction products are generated. Only the short fragment flanked by both primers (indicated by two small asterisks in cycle 3) is amplified exponentially, and therefore, it predominates the final product almost exclusively. See text for a detailed description of the reaction.

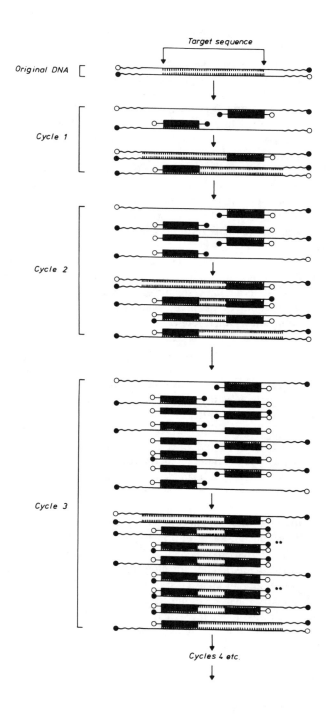

4 kb. However, amplification products of up to 10 kb can be obtained under optimal conditions.

The principle of the cycling reaction is outlined in Figure 8. In the first step, the template DNA is made single stranded by rising the temperature to 94°C (**denaturing step**). In the second step, lowering of the temperature to about 25 to 65°C (depending on primer sequence and experimental strategy) results in primer annealing to their target sequences on the template DNA (**annealing step**). The primers will preferably hybridize to binding sites that are identical or highly homologous to their nucleotide sequence, although some mismatches (especially at the 5′ end) are usually allowed. For the third step, a temperature is chosen where the activity of the thermostable polymerase is optimal, i.e., usually 72°C (**elongation step**). The polymerase now extends the 3′ ends of the DNA-primer hybrids toward the other primer binding site. Since this happens at both primer-annealing sites on both DNA strands, the target fragment is completely replicated (cycle 1).

In the next cycle, the two resulting double-stranded DNAs are again denatured, and both the original strand as well as the product strand now act as a template. Repeating these three-step cycles 25 to 50 times results in the exponential amplification of the target (=amplicon) between the 5′ ends of the two primer binding sites. Other, longer products are also generated (see Figure 8). However, since these fragments are only linearly amplified, their relative amount in the final product is negligible.

In the initial stage of PCR development, the Klenow fragment of the DNA polymerase I of *Escherichia coli* was used for DNA amplification. However, since this enzyme denatures irreversibly at the melting temperature of 94°C, new enzyme had to be added after each cycle. In 1988, the DNA polymerase of a thermophilic bacterium, *Thermus aquaticus*, was introduced by Saiki et al.[605] This enzyme, called *Taq* polymerase, withstands the high temperature to a certain extent. The use of higher temperatures has two important advantages. First, annealing conditions can be made more specific. Second, the thermostability of the enzyme allows us to perform the PCR in an automated thermocycler.

Several parameters influence the specificity of PCR, i.e., the temperature profile of the thermocycler, the annealing temperature, the activity and amount of the polymerase, concentrations of primers, template DNA and Mg^{2+}, and the presence of additional chemicals such as dimethylsulfoxide (see Chapter 4, Section IV.A.2 for details). The annealing temperature is usually chosen as high as possible to prevent unspecific amplification. The first thermocyclers used for PCR actually consisted of three waterbaths for heating and cooling the reactions (see, e.g., Lassner et al.[390]), and there are still people who prefer this technique to more modern machines. Whereas the second generation used water for cooling, now most thermocyclers cool and heat using a Peltier element.

One main reason for the versatility of the PCR technique is that any kind of primers can be chosen, depending on the purpose of the study. On the one extreme, a particular DNA sequence can be amplified by a pair of **specific**

primers, the sequences of which are designed based on sequence information. Such a strategy is useful for gene isolation or for the analysis of transferred genes in transgenic organisms. Specific primers based on unique sequences that flank simple tandem repeats (=microsatellites) are also increasingly used for mapping purposes and population genetics (see Chapter 7). On the other extreme, anonymous DNA sequences can be amplified by **arbitrary primers**. Although the choice of the nucleotide sequence of such a primer (pair) is completely arbitrary, defined pieces of DNA are usually amplified under appropriate experimental conditions. This strategy, described in detail in Section III.C, is most useful for such applications as genome mapping and DNA fingerprinting.

Between these extremes, there are several possibilities to design **semispecific primers**. For example, some strategies make use of primers complementary to known repetitive DNA sequences. These sequences include interspersed repeats, e.g., in Alu-PCR[394] and SINE-PCR,[393] intron-exon splice junctions,[752] 5S RNA genes,[366] tRNA genes,[676,767] and zinc finger protein genes.[707] The use of short tandemly repeated sequences as primers (Section III.B) also falls into this category.

Since specific, semispecific, and arbitrary primers can also be used in any combination, their potential for different experimental purposes is endless. In this book, we focus on techniques which do not require molecular cloning and/ or extended sequence information. Therefore, we will not go into detail for any of the specific and semispecific approaches, except for the use of simple tandem repeats or minisatellites as primers (see below) and the amplification of microsatellite sequences by specific flanking primers (Chapter 7). There are several comprehensive handbooks and review articles available which describe (semi)specific PCR applications in greater detail (e.g., Arnheim and Erlich;[19] Erlich,[182] Erlich et al.,[183] Innis et al.,[308] and McPherson et al.[447]).

B. PCR WITH MINISATELLITE AND SIMPLE SEQUENCE PRIMERS

In this section, we describe the use of oligonucleotides, consisting of simple sequence repeats, or the core sequence of minisatellite repeats as the single primer in the PCR. These oligos are traditionally used as probes for conventional DNA fingerprinting based on hybridization (Section II). Their application as PCR primers was first described by Lieckfeldt et al.[406] and Meyer et al.[459,460] The core sequence of the wild-type phage M13 (GAGGGTGGXGGXTCT, see Figure 5), as well as the simple sequences $(CA)_8$, $(CT)_8$, $(CAC)_5$, $(GTG)_5$, $(GACA)_4$, and $(GATA)_4$, were used as PCR primers to distinguish, e.g., between different isolates of *Cryptococcus neoformans*, a human pathogenic fungus (see Chapter 5, Section V, for details[460]).

Though quite new, this technique has yielded distinctive, variable, multifragment profiles from all fungal and plant species tested so far (see, e.g., Matsuyama et al.[436] and Neuhaus et al.[500]). A similar approach was taken by

Heath et al.[279] who used various VNTR core sequences as primers (including M13) to study fish, birds, and humans. However, though differences between species were found, intraspecific variation was not observed. Most recently, Zietkiewicz et al.[835] have described a new variant of the technique using 5'- or 3'-anchored simple sequence repeat oligonucleotides as primers. In these oligonucleotides, a $(TG)_n$ or $(CA)_n$ repeat is followed by two to four arbitrarily chosen nucleotides which trigger site-specific annealing. By electrophoresing radiolabeled PCR products from a variety of eukaryotic taxa, the authors observed interspecific as well as intraspecific polymorphisms.

Just as with RAPDs, PCR with minisatellite and simple sequence primers provides a convenient method of genetic identification and differentiation in plants and fungi at various levels ranging from species to individuals. While the molecular basis of the underlying polymorphisms has not been addressed yet, the use of these oligonucleotides as PCR primers instead of hybridization probes offers one important advantage. It combines the high degree of DNA polymorphism detected by conventional multilocus probes in DNA finger-printing experiments with the technical simplicity and speed of the PCR method, facilitating large-scale experiments.

C. PCR WITH ARBITRARY PRIMERS

In this section, we will describe the use of arbitrary primers for DNA fingerprinting. These primers are available in kits or separately from commercial companies (like Operon Technologies, Inc. and Pharmacia LKB) or can be synthesized on a DNA synthesizer.

The amplification of anonymous polymorphic DNA fragments is based on standard PCR methodology, with the distinction that usually only one primer, with an arbitrary nucleotide sequence, is used. However, the simultaneous use of two different primers is also possible and often yields additional information.[87,768,782,784] Several people have come up with different names for the use of arbitrary primers in PCR, each having a different protocol (reviewed by Caetano-Anollés et al.[90]). Three main streams can be distinguished: (1) RAPD, (2) DAF, and (3) AP-PCR. Other terms and acronyms have been suggested. The term MAAP seems to describe the characteristics of all three previous terms adequately.[88,89] Some investigators have combined different elements of the main streams as mentioned above (see, e.g., Collins and Symons[119]). Most of the methodology presented in this book will deal with RAPDs, although many remarks will also apply to the other two techniques. For reasons of simplicity and because the RAPD technique is most widespread, we always use the term RAPD when the fragments generated by one of these techniques are treated in a general sense.

All three techniques have in common that anonymous DNA sequences are amplified using arbitrarily chosen primers. While nothing is known about the identity and the sequence context of a particular amplification product, its presence or absence in different organisms can serve as a highly informative character for the evaluation of genetic diversity and relatedness. In some cases, fingerprint-like patterns are produced. A typical setup of a RAPD experiment is outlined in Figure 9.

The RAPD procedure was invented by Williams et al.[782] It is technically the simplest variation of the arbitrarily primed PCR methods. Primers with ten nucleotides and a GC content of at least 50% are generally used. The amplification products are separated on an agarose gel and detected by staining with ethidium bromide. Primers with lower GC content usually do not yield amplification products. Because a G–C bond consists of three hydrogen bridges and the A–T bond of only two, a primer-DNA hybrid with less than 50% GC will probably not withstand the temperature at which polymerization takes place (72°C). Therefore, the DNA-primer hybrid will have melted before the polymerase has started polymerization. Since the nucleotide order within the 10-mer primers is arbitrary, no prior knowledge of DNA sequences is needed, and the primers can be universally used for eukaryotes and prokaryotes. Figure 10 shows an example of an ethidium bromide stained gel with RAPD fragments.

DAF was introduced by Caetano-Anollés et al.[86] The authors used short primers, often only five to eight nucleotides long, with either low or high stringency annealing steps and a two-temperature instead of the standard three-temperature cycling program. Resulting fragments were separated on polyacrylamide gels and visualized by silver staining. Reaction conditions could be tailored to obtain a desired level of pattern complexity.

AP-PCR was introduced by Welsh and McClelland.[766] Oligonucleotides of 20 or more nucleotides, originally designed for other purposes, were used as primers (e.g., the M13 universal sequencing primer; the pBS reverse sequencing primer; and the Kpn-R, KA, KB, KM, KR, KX, and KZ primers). Two cycles with low stringency (allowing for mismatches) were followed by 30 to 40 cycles with high stringency. $\alpha[^{32}p]$ dCTP was included in the last 20 to 30 cycles. Radiolabeled products were separated by polyacrylamide gel electrophoresis and made visible by autoradiography. The AP-PCR variant of the arbitrary PCR method is used the least, compared to RAPD and DAF. It is also the most complicated method and uses radioisotopes. However, it can be simplified by separating the fragments on agarose gels and using ethidium bromide staining for visualization.

As for ordinary PCR, very little DNA is needed (10 to 400 ng), and therefore, methods using arbitrary primers can also be applied in cases where there are only small amounts of DNA available. For example, Benito et al.[41] showed that a small piece of cereal endosperm is sufficient for 60 RAPD reactions; the seed is still viable after this analysis. Brown et al.[74] demonstrated that RAPDs can even be performed with single tobacco protoplasts.

To obtain an amplification product with only one primer, there must be two identical (or highly similar) target sequences in close vicinity to each other: one site on one strand and the other site on the other strand, in the opposite direction. The distance between both sites should not exceed a few kilobases, since smaller fragments are more efficiently amplified than larger ones. The number of fragments which can be theoretically expected from one primer,

RAPD technology

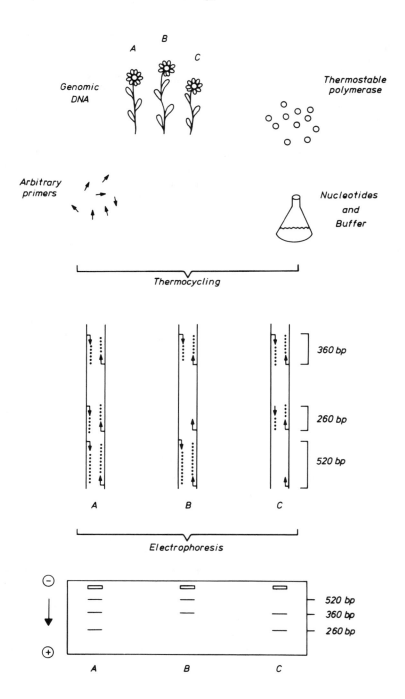

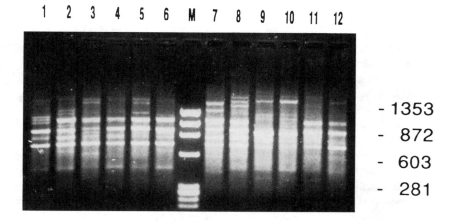

FIGURE 10. RAPD fragment patterns obtained from chrysanthemum cultivars and representatives of related *Dendranthema* species. Amplification products were separated on an agarose gel and stained with ethidium bromide. The primer is 5′-CGG TCA CTG T-3′. Lanes 1 to 6 are *D. grandiflora* cultivars CH6, CH7, CH11, CH14, CH55, and CH57; lanes 7 to 12 are *D. japonicum, D. boreale, D. yoshinaganthum, D. pacificum, D. shiwogiku,* and *D. grandiflora* cultivar CH 53. Size markers (bp) are indicated.

annealing with 100% homology, can be calculated from primer length and the complexity of the target genome, supposing that the nucleotides are present in equal proportions. Williams et al.[784] gave the equation $b = (2000 \times 4^{-2n}) \times C$, where b is the expected number of fragments per primer, n is the primer length in nucleotides, and C is the complexity of the organism (i.e., genome size in basepairs per haploid genome). For example, in a plant species such as maize (genome size of 6×10^6 kb), 10.9 fragments per 10-nucleotide primer are expected that have a 100% homology between primer and template, whereas for yeast (1.6×10^4 kb), only 0.029 fragments are expected.

The results of many investigations, however, suggest that the number of fragments per primer is largely independent of the genome complexity of the investigated organism. Thus, plants with large genomes such as onion[779] and conifers[96] did not exhibit more complex RAPD fragment patterns than plants with comparatively small genomes, such as tomato[361] and *Arabidopsis*.[572,573]

FIGURE 9. Strategy of PCR with arbitrary primers. Genomic DNA from the organism of interest (A, B, C), one (or two) primer of arbitrary sequence, a thermostable DNA polymerase, and a suitable buffer containing Mg^{2+} ions are combined into a reaction tube and subjected to PCR (usually 45 cycles). The arbitrary primers anneal to anonymous target sequences of the template DNA. If two primers anneal in an opposite direction and at a suitable distance from each other, the fragment between the two primers is amplified (illustrated by dotted lines). Amplification products are separated by gel electrophoresis and visualized by, e.g., ethidium bromide staining. A variety of mechanisms (e.g., primer target site mutation) may result in presence/absence polymorphisms of amplified fragments. See text for a detailed description.

Similarly, the ploidy level of a plant does not appear to influence the number of fragments per primer.[794] This obvious independence of RAPD fragment number from genome size may be explained by mismatch and primer competition. While most RAPD primers are likely to exhibit 100% homology to their binding sites in maize, primer binding sites will usually have 90% or less homology in yeast.

To investigate this phenomenon, Williams et al.[784] performed some interesting RAPD experiments in which DNA samples of two organisms were pooled in different ratios. First, DNAs of two individuals from the same species were analyzed. Both individuals exhibited a characteristic banding pattern when examined separately. If both template DNAs were pooled in different ratios prior to amplification, the bands derived from each individual were amplified in proportion to the amount of their genomic DNAs added. However, some bands were only poorly amplified and detected only if the specific DNA was added in excess. In a second experiment, soybean DNA (having a high complexity genome) was pooled with DNA of a cyanobacterium (having a low complexity genome). Interestingly, all amplified RAPD fragments originated from the soybean genome, even if the cyanobacterial DNA was added in excess. The conclusion from these experiments was that the amplification reaction is determined in part by competition for priming sites in the genome. Primers will preferably bind to target sites with a higher degree of homology. These are more likely to be available in a more complex genome.

Caetano-Anollés et al.[89,90] performed some elegant experiments to study the mechanisms and processes of amplification occurring during the DAF reaction. The authors showed that the first eight nucleotides (from the 3' end) are crucial for the generation of a particular band. If the eight 3'-most nucleotides were identical, primers with eight, nine, or ten nucleotides resulted in identical or highly similar banding patterns on silver-stained polyacrylamide gels. Patterns generated by related primers of five to eight nucleotides were different in complexity and length distribution. Interestingly, decreasing primer length also decreased the number of products, while the mean size of the amplified bands was increased.

A working model was proposed to explain these findings,[89,90] and it illustrates the competitive nature of PCR with arbitrary primers. According to this model, DNA amplification is modulated at two levels. First, primer target sites are selected in a **template screening phase**. The selectivity in this stage is determined by primer sequences and influenced by reaction conditions. Bonafide as well as mismatch annealing may occur, resulting in a complex family of primary amplification products. In subsequent rounds of amplification, the newly formed molecules may interact in diverse ways. Since an amplified single-stranded DNA molecule has palindromic ends, it can self-anneal to form a hairpin loop. The model suggests that competition occurs between single-stranded template DNA, primers, and the terminal palindromic sequences to form either double-stranded DNA, a primer-target DNA complex, or an intramolecular hairpin loop in the single-stranded DNA, respectively. The model suggests further that the different types of molecules tend to establish an equilibrium, and only a subset of template target sites is amplified to high copy numbers.

The hairpin symmetry may not be confined to the primer sequence itself, but may also extend into internal regions of the fragment, thereby further stabilizing the hairpin loop. For efficient amplification of a given fragment to occur, hairpin-loop formation must be outcompeted by the primer/template duplex. It appeared that very short primers, five or six nucleotides long, form less stable hybrid molecules with single-stranded DNA than longer primers do. A higher frequency of hairpin-loop formation would thus explain the lower complexity of banding patterns obtained with shorter primers. On the other hand, it was found that large hairpin loops (formed by long fragments) are less stable than shorter loops. Consequently, large hairpin loops are probably less effective as competitors, which would explain why the size distribution of amplification products from very short primers is biased toward longer fragments.

In plant and fungal cells, DNA is not only contained in the nucleus, but also in the cytoplasm. Plants possess a chloroplast (cpDNA) and a mitochondrial genome (mtDNA), fungi only have an mtDNA. These cytoplasmic genomes are much smaller than the nuclear genome. Plant cpDNA has a size of about 150 kb, and the size of plant mtDNA ranges from 200 to 2400 kb, whereas fungi have a mitochondrial genome of 18 to 121 kb.[620] Using the formula given previously,[784] it seems obvious that chances are low for RAPD fragments to result from cytoplasmic DNA. To test for this, Thormann and Osborn[693] hybridized labeled cpDNA and mtDNA, respectively, to blots of RAPD fragments amplified from *Brassica* species. They found that 5% of the fragments hybridized to mtDNA, while less than 0.5% were homologous to cpDNA. Kazan et al.[348] found that one out of ten fragments amplified from *Stylosanthes* DNA originated from the chloroplast genome. These findings, together with theoretical considerations given above, suggest that the majority (95%) of RAPD markers originates from nuclear DNA. In an interesting set of experiments, Lorenz et al.[413] compared fragment patterns derived from separately isolated *Beta vulgaris* DNA of the nuclear genome, chloroplast genome, mitochondrial genome, and total genome, respectively. Reproducible DNA fingerprints could be obtained from both organellar DNAs using various primers, with the M13 core sequence being the most informative. Most important, it was shown that some fragments derived from mtDNA, and to a lesser extent from cpDNA, were represented also in the fragment pattern derived from total DNA, though they were lacking in the nuclear DNA. These results imply that a possible contribution of cytoplasmic amplification products should be considered when, e.g., aberrant inheritance of RAPD fragments is observed.

Polymorphisms detected by the RAPD technology can theoretically result from several types of events: (1) insertion of a large piece of DNA between the two annealing sites may render the original fragment too large to be amplified, resulting in its loss; (2) the deletion of a DNA fragment carrying one of the two primer annealing sites also results in the loss of a fragment (see Figure 9); (3) a nucleotide substitution may affect the annealing of one of the two primers at a given site because of changes in homology, which can lead to a presence/absence of polymorphism or to a change in fragment size; or (4) insertion or deletion of a small piece of DNA can lead to a change in size of the amplified fragment. In practice, however, size changes are rarely observed. Instead, a particular RAPD fragment is usually present (allele A) or absent (allele a). This allele distribution is typical for a dominant marker. A fragment is seen in the homozygous (AA) as well as in the heterozygous (Aa) situation, and only the absence of the fragment reveals the underlying genotype (aa). In the case of codominant markers, on the other hand, the heterozygous state (AA') can be discerned from both homozygous states (AA and A'A'). Williams et al.[782] and Fritsch and Rieseberg[201] found at least 95% of RAPD fragments to behave as dominant markers, while the remaining less than 5% behaved codominantly, i.e., as two alleles with a different size. Echt et al.[173] found no codominant RAPD fragments using 19 different primers.

For many applications, the dominant behavior of RAPD fragments is a disadvantage. For example in population genetics, allele frequencies cannot be estimated since the homozygote (AA) cannot be discerned from the heterozygote (Aa). The same is true for mapping studies using segregating F_2 families. Since the AA cannot be discriminated from the Aa individuals, valuable information is lost.

The dominant character of RAPD fragments is not problematic in haploid situations, which are frequently encountered in mosses and fungi. A haploid situation can also be exploited in gymnosperms. Here, the haploid megagametophyte is strictly maternal, derived from a single meiospore, and large enough to allow for RAPD analysis.[309,702] The maternal genotype for each marker can be deduced from the genotypes of the haploid megagametophyte. A plant heterozygous for a RAPD marker (Aa) will show a segregation ratio in the megagametophytes of 1:1 for presence (A) and absence (a) of this marker, whereas a marker in a homozygous state will occur in all megagametophytes.

It was generally observed that the genomic copy number of an amplified DNA sequence cannot be deduced from intensity differences of RAPD fragments.[173] Therefore, it is usually also impossible to discriminate between homozygosity and heterozygosity of a dominant allele. Devos and Gale[153] analyzed wheat alien addition lines in which one to four copies of specific chromosomes were present. They found no intensity differences of RAPD fragments between monosomic and tetrasomic addition lines. On the other hand, some authors reported that it is possible to analyze the relative presence of RAPD amplification products in a DNA mixture. For example, it was shown that in a mix of DNA of offspring from dragonflies the relative presence of specific paternal markers could be determined, and paternal success could be analyzed.[251,252] The authors noted, however, that a marker will only be observed if it is amplified from at least 20% of the DNA mix. Knowledge of the diagnostic markers as well as prior calibration experiments therefore are necessary, e.g., by making "synthetic offspring."

Chapter 3

LABORATORY EQUIPMENT

Most of the equipment required to carry out the methods described in this book is already available in many biochemical, mycological, and plant biology laboratories. However, some items have relatively specific uses and are not generally found in a nonmolecular biology laboratory.

The purpose of this chapter is to provide the researcher with information on necessary equipment. First, we must consider to what extent hybridization-and/or PCR-based fingerprinting will be performed in the laboratory and what resources are already available. If the objective is to perform an occasional DNA fingerprinting experiment, borrowing may be more realistic than purchasing high-quality equipment for each step of the process. On the other hand, if fingerprinting is to be the main focus of the laboratory, the list provided below will give an idea of the equipment required.

In order to help the reader to set priorities when purchasing equipment to set up a new laboratory, the importance of each item is classified as follows, with application range as well as available alternatives given within parentheses:

xxx   Essential
xx   Highly recommended
x   Helpful or optional
PCR   Needed for PCR experiments only
(x)   Required for tissue culture and/or cloning experiments

Incubators

xxx   Water baths (for growing fungal and plant cells, labeling probes, hybridization, incubation, and restriction analysis)
xxx   Dry incubator/rotary shaker (for growing fungal and plant cells, washing Southern blots, etc.)
x   Thermostat-controlled metal heating blocks which hold test tubes and/or microcentrifuge tubes (for small-scale DNA extraction and restriction analysis)
PCR   Thermo cycler (for automated PCR reactions)
xx   Hybridization oven with rotating tubes (for Southern blot and gel hybridizations; alternatively, but not preferably, hybridization may be carried out in sealed bags in a thermo-controlled rotary shaker or water bath)

Plant, plant cell, and mycelia growth equipment

(x)   Sterile bench, laminar flow hood, and/or biological safety cabinet (for micropropagation, transfer of fungi, etc.)

(x)   If a constant-temperature room available, a rotary shaker with clamps
        for 100-, 250-, and 500-ml and 1- or 2-l Erlenmeyer flasks
(x)   If a constant-temperature room not available, a closed-environment
        rotary shaker equipped as above
(xx)   Incubator for culture tubes or petri dishes (for mycelia growth)
x   Phase contrast microscope

Sterilization

xxx   Autoclave, leakproof autoclaving bags, and sterilization tape (to ster-
        ilize solutions, glassware, or other items)
xxx   Filters and filtration units (0.22 μm pore size; to sterilize
        nonautoclaveable solutions, e.g., vitamins, glucose, etc.)
xx   Oven (e.g., to sterilize glassware)
x   Ultraviolet (UV) light source
(x)   Bunsen burner or equivalent (to sterilize inoculation loops)

Water purification

xxx   Deionization/water purification cartridges and/or glass distillation
        apparatus
xxx   Storage containers

Centrifuges

x   Ultracentrifuge (20,000 to 80,000 rpm) with fixed-angle, vert-
        ical, and/or swinging-bucket rotors (e.g., for CsCl-DNA purifi-
        cation)
xx   Low-speed centrifuge (10,000 to 20,000 rpm) floor model or tabletop-
        refrigerated centrifuge for larger tubes (for large-scale DNA extrac-
        tion)
xxx   One or more microfuges (optionally refrigerated) that hold standard
        1.5- and/or 0.5-ml microfuge tubes (for small-scale DNA extraction,
        PCR, etc.)
x   Refractometer (to adjust solution concentrations, e.g., for CsCl gradi-
        ents)

Refrigeration and material storage

xxx   Refrigerator 4°C (to store, e.g., solutions and fungal stock cultures or
        DNA)
xxx   Freezer –20°C (to store, e.g., solutions, especially small aliquots of
        PCR solutions, DNA, and DNA-modifying enzymes)
xx   Freezer –70 to –80°C (to store frozen plant or fungal tissue samples
        and for autoradiography)

x   A cold room is very useful for performing reactions and centrifugations at 4°C

x   Ice machine

xx   Isothermic storage containers for liquid nitrogen and dry ice

x   Lyophilizer (to freeze-dry fungal or plant material, e.g., for long-term storage)

Safety and measurement of radioactivity

xxx   Laboratory coats and gloves

xxx   Fume hood

xxx   Plexiglass radiation shielding and containers (essential when working with strong β-emitters such as ^{32}P)

xxx   Proper storage and waste containment systems

xxx   Contamination monitor (Geiger counter)

x   Liquid scintillation counter (for exact calculation of radioactivity incorporated in a probe)

Supplies (glass and plastic ware)

(xxx) Culture tubes and caps

(xxx) Petri dishes

(xxx) 100-, 250-, and 500-ml or 1-, 2-, 3-, or 4-l Erlenmeyer flasks (for mycelia and cell growth)

xxx   Glass reagent bottles, various sizes

xxx   Common laboratory glassware (flasks, beakers, graduated cylinders, funnels, etc.)

xxx   Glass or disposable plastic pipettes (0.2, 1.0, 5.0, 10.0, and 25.0 ml)

xx   Pasteur pipettes

xxx   Adjustable micropipettes and disposable pipette tips (0.5 to 10, 10 to 100, 0.5 to 100, and 200 to 1000 μl)

It is advisable to designate one set of pipettes for each full-time researcher and one set for PCR work exclusively. Pipettes or pipette tips can be plugged with cotton on one side to prevent aerosol contamination, which is especially important for PCR work as well as for safety reasons when working with human pathogens.

x   Mechanical (Peleus™ balls) and electrical or battery-driven pipetting devices for 1-, 5-, 10-, and 25-ml pipettes

x   Polyallomer or ultraclear ultracentrifuge tubes (for ultracentrifugation only)

xx   Polypropylene bottles (1 l and 500 and 250 ml) and adapters

xx   Polypropylene snap-cap and/or sterile screw-cap tubes (15 and 50 ml) (for organic extractions)

xx   Glass centrifuge tubes (e.g., Corex™ tubes) (15 and 30 ml) and adapters

xxx   Microfuge tubes (0.5, 1.5, and 2.0 ml)

x   Microtiter plates

DNA quantitation

xx   Spectrophotometer (UV and visible light)

x   Portable UV source (254 nm)

Electrophoresis equipment

xxx   Power supply: medium voltage (up to 500 V) for agarose gels

x   Power supply: high voltage (up to 3000 V) for polyacrylamide gels

xxx   One or more submarine horizontal gel apparatus (at least 15 × 20 cm; preferably 25 × 30 cm) for agarose gels

xx   Submarine minigel apparatus for agarose gels

x   Vertical gel apparatus for polyacrylamide gels

xxx   Combs, spacers, glass plates, clamps and/or gel sealing tape

xx   UV-transparent trays (for photographing stained gels)

xx   Gel dryer and vacuum pump (for gel drying only)

x   Vacuum transfer unit or electrotransfer unit (can be used as a quicker alternative to the classical Southern blot technique)

x   Peristaltic pumps for recirculation of gel buffers (generally not needed for water-cooled gel apparatus)

Darkroom equipment

xxx   UV transilluminator (302 nm or 366 nm wavelength; to visualize DNA fragments)

xxx   UV-protective shielding for eyes and skin

xxx   35 mm camera and/or Polaroid™ MP4 camera or equivalent

xxx   Red filter (for UV photography of ethidium bromide-stained gels)

xxx   35 mm black and white film, or Polaroid high-speed film [type 665 (negative/positive) or type 667 (positives only)]

xx   35 mm film development equipment

xxx   X-ray film [e.g. or Biomax XAR, (Kodak™), SB-5 (Fuji™), or ß-max (Amersham™)]

xxx   X-ray cassettes, intensifying screens

xxx   X-ray automatic developer or hand developer/fixer baths and plastic trays (an automatic developer is only recommended if large quantities are processed)

x   X-ray film hangers

General laboratory equipment

xx	Microwave oven (e.g., for melting agar and agarose)
xxx	pH meter and pH indicator papers
xxx	Thermometers
xxx	Magnetic heater/stirrer
x	Vortex mixer
xxx	Balances (analytical and top loading)
x	Laboratory timers and stopwatches
xx	Mortars and pestles (for grinding cell material)
xx	Homogenizer (e.g., Ultra-Turrax™)
x	Desiccator (for storing dried tissue and/or hydrophilous chemicals or reagents)
x	Dishwashing machine
xx	Vacuum concentrator (e.g., SpeedVac™; e.g., for concentrating DNA-samples before loading on an agarose gel)
xx	UV-crosslinker (for fixation of DNA to nylon membranes, alternatively this can also be done by oven baking)
(xx)	Inoculation loops and needles
x	Heat sealer and plastic bags (alternatively used instead of a hybridization oven)
x	Dialysis clips and tubing (used for cleaning DNA from higher salt concentration, e.g., after CsCl-centrifugation)
xx	Disposable columns (to separate a labeled probe from the unincorporated nucleotides; alternatively, Pasteur pipettes or disposable pipette tips plugged with glass wool can also be used)
xx	Plastic wrap (e.g., Saran Wrap™)
x	Aluminum foil (to cover flasks before autoclaving)
xx	Plastic dishes or glass trays (for washing filters and gels)
x	Syringes and needles
xxx	Nylon hybridization membranes
xx	Whatman™ 3MM filter paper or equivalent (for Southern transfer)
xx	Alcohol/waterproof lab markers
xx	Lightbox (for viewing autoradiograms)
xx	Calculator
xx	Personal computer, software packages, and printer
x	Scanning densitometer, video camera, or image analyzer (for scanning gels and autoradiographs)
x	Phosphorimager including screens, cassettes, and image eraser (an alternative to autoradiography)

Chapter 4

METHODOLOGY

I. SAFETY PRECAUTIONS

Working with DNA often involves the use of hazardous chemicals and equipment. The following recommendations are intended to protect the experimenter's health and should be read in advance of starting laboratory work.

1. Chloroform-isoamyl alcohol, phenol, and solutions containing high concentrations (0.2% and more) of β-mercaptoethanol are often used in DNA isolation protocols. These compounds are toxic upon inhaling and should be handled in a fume hood.
2. Acrylamide and bisacrylamide are neurotoxic. Solutions containing these compounds should be handled with gloves and with great care. Contamination of working areas (e.g., during casting of polyacrylamide gels) should be avoided. Acrylamide solutions should be polymerized by the addition of N,N,N',N'-tetramethylenediamine (TEMED) and ammonium persulfate before disposal (see Section IV.C).
3. Ethidium bromide, generally used for staining of DNA after gel electrophoresis or cesium chloride (CsCl) centrifugation, is a powerful mutagen and carcinogen. Solutions and gels containing ethidium bromide should always be handled with gloves, and contamination of working areas must be avoided. Ethidium bromide-containing waste should be collected and disposed of separately in a legal and environmentally safe way. Sambrook et al.[606] describe several ways to handle waste containing ethidium bromide.
4. Experiments involving the use of radioactive isotopes (e.g., the powerful β-emitting isotope ^{32}P) should be carried out in a separate laboratory and behind appropriate shielding (e.g., 1 cm of plexiglass). Gloves should be worn throughout, and a hand monitor must be routinely used to check for radioactive contamination of the experimenter, the material, and the working area. Radioactive waste has to be collected and disposed of in a legal and environmentally safe way.
5. As ultraviolet (UV) light is highly mutagenic and destructive, eyes and skin should be protected by wearing glasses and protective cloth, e.g., when photographing gels on a transilluminator. UV light should be switched off immediately after taking the photograph. Longer or frequent use of the UV transilluminator makes air circulation a necessity to prevent buildup of toxic ozone from the UV light. It is much less hazardous for your eyes to discuss results sitting over a Polaroid™ photo than over a UV transilluminator.

6. Good standard laboratory practice should be observed when working with bacteria and fungi. It is advisable always to work with sterile material and solutions and to wear gloves. Genetic manipulation experiments must be carried out according to the regulations in each country.
7. Any manipulation of fungi should be carried out in a biological safety hood to avoid contaminating the laboratory with light spores and because some fungi may be detrimental for humans.

II. DNA ISOLATION AND PURIFICATION

A. ISOLATION OF PLANT DNA

A variety of problems may be encountered during the isolation and purification of high molecular weight DNA from plant species. These include (1) partial or total DNA degradation due to the presence of endogenous nucleases; (2) co-isolation of highly viscous polysaccharides, which render the handling of samples difficult; and (3) co-isolation of polyphenols and other secondary plant compounds, which cause damage to DNA and/or inhibit restriction enzymes and *Taq* polymerases. In a considerable number of plant species, DNA preparations tend to be brown colored due to the oxidation of polyphenols to quinonic compounds. These, in turn, are powerful oxidizing agents that damage DNA and proteins.[412] As a consequence, yields of high molecular weight DNA from plants are often poor.

Because the biochemical composition of plant tissues and species varies considerably, it is virtually impossible to supply a single isolation protocol which is optimally suited for each plant species. Even closely related species may require quite different isolation procedures. Optimization strategies most importantly concern the ingredients (and also the pH) of the extraction buffer. However, a variety of essential compounds which are protecting DNA from degradation are contained in the majority of extraction buffers. For example, EDTA (ethylenediamine tetraacetate) is generally included in DNA isolation buffers and storage solutions, since this compound chelates bivalent cations and thereby inhibits metal-dependent DNases. Reducing agents such as β-mercaptoethanol are also usually included to inhibit oxidization processes, which either directly or indirectly cause damage to DNA.

When choosing a particular protocol, one should also keep in mind that different kinds of experiments demand different levels of DNA purity. For example, DNA fingerprint analyses based on restriction and hybridization experiments require relatively pure high molecular weight DNA, while polymerase chain reactions (PCR) may also work on rather crude and/or somewhat degraded DNA samples.[45,77,176,362,420] Accordingly, a large variety of different DNA isolation protocols have been developed. In this chapter, we provide four benchtop protocols, which are routinely used in our laboratories for a variety of plant species. Annotated references for alternative, more specialized, or PCR-oriented methods are compiled in Tables 1A through 1F. New plant DNA isolation techniques (mostly dealing with problematic species), as well as new

TABLE 1A
DNA Isolation Protocols Based on CTAB Buffers

Ref.	Species	Remarks
486	*Vigna* (mung bean)	DNA-CTAB complexes are precipitated by lowering the NaCl concentration; DNA is purified by CsCl centrifugation
603	*Hordeum vulgare*	Authors introduced precipitation of the CTAB-DNA complex by isopropanol instead of lowering the salt concentration
584, 585	Wide range of taxa	Minipreparation based on Murray and Thompson;[486] suitable for fresh, herbarium, and even mummified tissues
163	*Solanum, Glycine*	Several field preservation methods for plant tissues are compared; desiccation is recommended
558	Wide range of taxa	Several field preservation methods for plant tissues are compared; desiccation is recommended
390	*Lycopersicon*	Minipreparation; tissue homogenization is performed by means of a viral sap extractor
164, 165	Wide range of taxa	Protocol is specifically designed for use with fresh tissue
743	*Cuphea*	Modification of the basic protocol (e.g., phenol extraction of CTAB-DNA complexes)
757	Wide range of taxa	Modification of the basic CTAB protocol of Doyle and Doyle;[164] DNA is then purified via ammonium actetate treatment, CsCl centrifugation, gel filtration on Sepharose, or ion exchange chromatography
211	*Musa acuminata, Ipomoea batatas,* and other taxa	Authors report on the influence of different concentrations of β-mercaptoethanol on DNA yield
661	Red algae; various species	Method specifically designed for red algae; final purification step includes ion exchange chromatography on Qiagen tips
81	Wide range of taxa	DNA is purified via low melting agarose gel electrophoresis
342	*Nelumbo* (lotus)	High concentrations of β-mercaptoethanol are used (5%)
295	*Saccharum*	Nuclei are isolated first, then lysed with CTAB buffer
587	*Podophyllum, Polyalthia,* and *Taraxacum*	Field-collected tissue is stored in a saturatedNaCl-CTAB solution; final DNA purification step involves preparative low-melting-point agarose gel electrophoresis

TABLE 1A (continued)
DNA Isolation Protocols Based on CTAB Buffers

Ref.	Species	Remarks
468	Wide range of taxa	Modification of the basic CTAB protocol of Saghai-Maroof et al.;[603] numerous variations are reviewed
7	*Linum usitatissimum*	Modifications include the use of high concentrations of β-mercaptoethanol (5%), polysaccharide removal by ethanol precipitation of DNA from 2*M* NaCl, and DNA purification by Chelex or gel electrophoresis
692	*Prunus persica*	DNA is isolated from leaves slowly dried at room temperature
737a	Gossypium	DNA is isolated from a small portion of single seeds; usful for PCR analysis

TABLE 1B
DNA Isolation Protocols Based on Potassium Acetate/SDS Precipitation of Proteins and Polysaccharides

Ref.	Species	Remarks
148	Wide range of taxa	Original description of this strategy of DNA isolation; polysaccharides and proteins are removed by SDS/potassium acetate precipitation; a CTAB variant is also described
677	*Oryza sativa*, *Lycopersicon*, and other taxa	Plant tissue is dried in a food dehydrator; isolation protocol is based on Dellaporta et al.[148]
145	*Abelmoschus* (okra)	Specifically designed for plant tissues which are rich in **viscous polysaccharides**; isolation from dark-grown tissue; isolation protocol is based on Dellaporta et al.[148]
724	*Ipomoea*	Modification of the procedure of Dellaporta et al.[148]
331	Malvaceae, Bombacaceae, Moraceae	Specifically designed for plant tissues rich in **polyphenols**: PVP included in the buffer system of Dellaporta et al.[148]
468	Wide range of taxa	Modification of the protocol of Dellaporta et al.[148], PVP-360 included in the isolation buffer
522a	*Lycopersicon*, *Solanum, Vicia*	Modification of the protocol of Dellaprota et al.[148]

TABLE 1C
Crude DNA Minipreparation Protocols (Specifically Designed for PCR Analysis)

Ref.	Species	Remarks
176	*Brassica napus*	DNA is ethanol precipitated from crude leaf extracts, redissolved, and directly used for PCR
388	*Triticum, Trifolium, Nicotiana tabacum*	Plant leaf material is squashed onto a nylon membrane, washed, eluted, and directly used for PCR
45	*N. tabacum*	Leaf and root pieces are directly used for PCR
150	Wide range of taxa	Protoplasts are isolated from small leaf disks, lysed, the DNA precipitated, redissolved, and used for PCR
77	*B. napus, Helianthus annuus*	DNA samples are isolated within microtiter plates, embedded in agarose, and used for PCR
420	*N. tabacum, Glycine max, Zea mays*	Protocol includes a combination of glass bead homogenization, shock freezing, and boiling
523	*Oryza sativa, Z. mays*	Freezing/boiling procedure
823	*O. sativa*	Crude extracts of seedlings are directly used for PCR
468	Wide range of taxa	Modification of the procedure of Edwards et al.[176]
296	*Porphyra perforata* (red algae)	Softening of cell walls by LiCl treatment; crude extracts are precipitated by ethanol
297	*B. napus*	Nondestructive protocol using cotyledon fragments from microspore-derived embryos; based on the method of Dellaporta et al.[148]
362	*Lycopersicon*	Tissue is boiled in alkaline buffer
41	*Hordeum vulgare, Secale cereale*	Crude minipreparation based on the method of Dellaporta et al.[148]
114	*O. sativa, Triticum aestivum* (seeds)	Half seeds (nonground) are treated with a buffer containing proteinase K or a chelate resin; the supernatant is directly used for PCR
112	Wide range of taxa	High salt (2 M NaCl) extraction buffer
74	*N. tabacum, T. aestivum*	RAPDs from single lysed protoplasts and microcolonies; freezing/thawing procedure
737b	Wide range of taxa	Tissue is ground in NaOH and used directly for PCR

TABLE 1D
DNA Isolation Protocols Involving Isolation of Nuclei

Ref.	Species	Remarks
787	*Nicotiana tabacum*	Nuclei are stabilized by polyamines and purified via Percoll step gradient centrifugation
738	*Pisum sativum*	Tissue is treated with ether; nuclei are stabilized by hexylene glycol and purified via Percoll step gradient centrifugation; DNA is purified via CsCl centrifugation
127	*Nicotiana, Glycine*	Nuclei are purified via Percoll step gradient centrifugation
126	*Theobroma cacao*	Protocol is specifically designed for tissues rich in **polyphenols**; PVP, BSA, and DIECA are included in the extraction medium
295	*Saccharum*	Isolated nuclei are lysed with CTAB/SDS
118	*Vitis vinifera*	Tissue is homogenized in reaction tubes using a motor-driven metal homogenizer
150	Wide range of taxa	Nuclei minipreparation via protoplasts, specifically designed for PCR analysis
438	Kelp (Laminariales)	Protocol is specifically designed for brown algae; final DNA purification via gel filtration on Sepharose spun columns
543	*Gossypium*	Protocol is specifically designed for tissues rich in **polyphenols** and **polysaccharides**; PVP, ascorbic acid, and DIECA are included in nuclear isolation buffer; nuclei are lysed with CTAB

TABLE 1E
Isolation Protocols for Very High Molecular Weight DNA Suitable for Pulsed-Field Gel Electrophoresis (>100 kb)

Ref.	Species	Remarks
249	*Arabidopsis*	DNA is isolated via protoplasts, either embedded in agarose plugs or in solution
717	*Lycopersicon*	DNA is isolated via protoplasts embedded in agarose plugs
111	*Triticum, Secale, Hordeum*	DNA is isolated via protoplasts embedded in agarose plugs
246	*Triticum, Secale, Nicotiana*	Liquid nitrogen-powdered tissue is embedded into agarose plugs, and DNA isolation is performed within the agarose

TABLE 1E (continued)
Isolation Protocols for Very High Molecular Weight DNA Suitable for Pulsed-Field Gel Electrophoresis (>100 kb)

Ref.	Species	Remarks
32	*Arabidopsis, Nicotiana*	Liquid isolation procedure gives higher yields than isolation via protoplasts
245	Poaceae, Fabaceae	Extension of the protocol of Guidet et al.;[246] can be applied for crushed seeds and flours

TABLE 1F
Miscellaneous DNA Isolation Protocols

Ref.	Species	Remarks
749	*Vicia faba*	Small scale 5-h procedure involving CsCl centrifugation in a tabletop centrifuge; applicable for plants, algae, yeast, mammals, insects, and bacteria
275	Wide range of taxa	Extensive phenol, PVP, and PEG treatment of extract; DEAE sephacel column chromatography
451	*Nicotiana, Zea mays, Helianthus annuus*	Protocol specifically designed for protoplasts and tissue-cultured cells, also suitable for other tissues; **polysaccharides** are removed by precipitation with 0.1 vol ethanol
371	*Abies alba, Picea abies*	Minipreparation from 5 mg of dormant bud tissue
93	*Gossypium hirsutum*	A guanidine-hydrochloride buffer is used; DNA is purified via ion exchange chromatography
299	*Betula*	Specifically designed for plant tissues rich in **polyphenols**; high molarity urea-phosphate buffer; inclusion of DIECA and PVP
428	*Fragaria*	Specifically designed for (DNA and RNA) isolation from tissues rich in **polysaccharides** and **polyphenols**; differential solubility of these compounds as compared to DNA or RNA in 2-butoxy-ethanol is exploited
247	*P. abies* and a wide range of other taxa	Specifically designed for tissues rich in **terpenoids** and **polyphenols**; acidic extraction medium; PVP; cysteine; DNA purification on RPC-5 columns
327, 328	Wide range of taxa	Cell walls are solubilized by inclusion of potassium or sodium salts of ethyl xanthogenate in the extraction buffer; small amounts of fresh tissue can be processed without homogenization

TABLE 1F (continued)
Miscellaneous DNA Isolation Protocols

Ref.	Species	Remarks
832	Wide range of taxa, also including fungi	Benzyl chloride is used in the extraction medium since it reacts with -OH residues in polysaccharides
833	*A. alba* (needles)	Miniprep version of the protocol of Guillemaut and Maréchal-Drouard;[247] tissue is homogenized together with tungsten carbide beads in a shaking mill within Eppendorf tubes
607	Rhodophyta (red algae)	PCR-oriented miniprep for fresh and dried algal materials; gel purification of crude DNA

modifications of existing procedures, are appearing in almost every issue of the "Plant Molecular Biology Reporter".

The standard protocol used in our laboratories is based on the cetyltrimethylammonium bromide (CTAB) procedure[164,584,603] of total DNA isolation (see Table 1a). Three alternative procedures are provided in addition: (1) a minipreparation based on the CTAB method of Lassner et al.[390]; (2) a variation of the potassium acetate/sodium dodecyl sulfate (SDS) precipitation technique originally described by Dellaporta et al.[148]; and (3) a method involving isolation of nuclei prior to DNA, modified from Willmitzer and Wagner.[787] The basic variant of each of these protocols yields relatively crude DNA preparations, which are nevertheless sufficiently pure for restriction and PCR analyses in many plant species. In other cases, further purification might be necessary to ensure complete restrictability and/or reproducibility of PCR experiments. To that end, we provide a protocol for DNA purification by CsCl centrifugation.

1. Collection and Preservation of Plant Tissue

The quality as well as the yield of a plant DNA preparation are to a considerable extent influenced by the condition of the starting material. Whenever possible, material harvested immediately before DNA isolation should be used. Season of harvesting can also be important, since the amounts of polysaccharides, polyphenols, and other undesirable compounds within plant tissues may vary strongly throughout the course of the year.[541] In temperate areas, harvesting young leaves in springtime is probably beneficial for most species.

Fresh tissue, which has to be stored for a short period of time (up to several hours or even days, depending on species and tissue), should be kept cool but not frozen (e.g., wrapped in paper and kept on ice). Storage for longer periods of time requires either freezing directly in −80°C, shock freezing with liquid nitrogen followed by storage at −80°C (−20°C is not advisable), or freeze drying (i.e., lyophilization). While fresh or frozen material usually gives higher yields and a somewhat more intact DNA, lyophilized tissue is easier to handle, requires no liquid nitrogen for efficient homogenization, and can be stored for years at room temperature under desiccated conditions.

In many ecologically or taxonomically oriented studies, the plant species of interest grow at remote locations, where it is neither possible to isolate DNA directly from fresh material nor to shock freeze or lyophilize the tissue. A variety of strategies have been tested to optimize field collection and preservation of plant material in the absence of laboratory facilities.[163,271,407,558,587,677,692] The general outcome of these studies was that chemical treatments (e.g., with ethanol, methanol, or other different organic solvents) are unsuitable for tissue preservation, the only exception being the storage of leaf tissue in concentrated NaCl-CTAB solutions.[587] On the other hand, rapid drying by means of silica gel or anhydrous calcium sulfate generally turned out to be an appropriate means of preserving field-collected tissue for DNA isolation, not only for higher plants,[81,407] but also for red and brown algae.[438,607] More recently, even slow drying of leaf tissue was shown to result in acceptable DNA preparations from some species.[692]

We have found the following modification of the silica gel technique very useful to preserve leaf material from several species of the paleotropical tree genus *Macaranga* collected under tropical rainforest conditions in East Malaysia. The only equipment needed is a batch of silica gel (containing a moisture indicator dye) in a waterproof bag and a set of 50-ml Falcon™ tubes.

Method:

1. Harvest the youngest leaves you find, and roll them to a cylindric shape (about 3 g of leaf material per tube).
2. Transfer the leaf cylinder to a Falcon™ or other close-fitting polypropylene tube so that it fits to the inner wall.
3. Fill the tube with dry silica gel up to the rim.
4. Close tightly. Change the silica gel every 6 h.

After the second change, the leaf material is usually completely desiccated. It may be stored in the same tube filled to one third with silica gel. Using the CTAB procedure (see Section II.A.2), we successfully isolated DNA from the dried material several months later. It is, however, generally advisable to process the samples as soon as possible upon return to the laboratory. Silica gel can easily be regenerated (i.e., redried) under field conditions, e.g., using a cooker and a pan.

2. Isolation Protocols
a. CTAB Procedure

In our hands, a modified CTAB procedure based on the protocols of Saghai-Maroof et al.,[603] Rogers and Bendich,[584] and Doyle and Doyle[164] is the method of choice for obtaining good-quality total DNA from many plant species and also from fungi.[757,758] CTAB is a cationic detergent which solubilizes membranes and forms a complex with DNA. After cell disruption and incubation with hot CTAB isolation buffer, proteins are extracted by chloroform-isoamyl alcohol, and the CTAB-DNA complex is precipitated with isopropanol. The DNA pellet resulting after centrifugation is washed, dried, and redissolved. Depending on the species, additional purification steps may or may not be necessary in order to remove RNA, polysaccharides, polyphenols, and other contaminating substances. Here, we include RNase treatment and ammonium acetate precipitation, which removes RNA and some polysaccharides.

Solutions:

Liquid nitrogen
Isolation buffer: 2% w/v CTAB, 1.4 M NaCl, 20 mM EDTA, 100 mM Tris-HCl, pH 8.0, 0.2% β-mercaptoethanol (added just before use).
Chloroform-isoamyl alcohol (24:1) (see Safety precautions)
RNase A solution: 10 mg/ml RNase A in 10 mM Tris-HCl, 15 mM NaCl, pH 7.5; boiled for 15 min, cooled to room temperature, and stored at –20°C
100% Isopropanol
Washing solution: 70% ethanol, 10 mM ammonium acetate
TE buffer: 10 mM Tris-HCl, 1 mM EDTA, pH 8.0
7.5 M ammonium acetate

Method:

1. Grind up to 3 g of fresh or 0.5 g of lyophilized or dried plant material to a fine powder using liquid nitrogen and a mortar and pestle. Liquid nitrogen is not essential for lyophilized and dried tissue, but facilitates the grinding procedure considerably. For fibrous material, precutting of the leaves with scissors and/or the addition of some sterilized quartz sand may also help powdering. In case of fresh tissue, make sure that the material will not thaw before being added to the isolation buffer. Otherwise, cellular enzymes may rapidly degrade the DNA. If liquid nitrogen is not available, fresh tissue can also be ground in prewarmed isolation buffer. However, this procedure may cause a reduction of DNA yield and quality.

2. Transfer the powder as fast as possible into 15 ml of prewarmed (60°C) isolation buffer in a capped polypropylene tube. Suspend clumps with the help of a spatula.

3. Incubate for 30 min at 60°C in a water bath. Mix gently every 10 min. The optimal incubation temperature and time may vary between different materials and/or species.

4. Add 1 vol of chloroform-isoamyl alcohol, and cap the tube and extract for 10 min on a rotary shaker or by hand. Mixing should be done gently but thoroughly enough to ensure emulsification of the phases.

5. Centrifuge for 10 min (5000 × g, room temperature). Depending on the desired purity of DNA preparations, the (upper) aqueous phase may then be reextracted one to several times with fresh chloroform-isoamyl alcohol.

6. Transfer the final aqueous phase to a glass centrifuge tube using a large-bore pipette. Add heat-treated RNase A to a final concentration of 100 μg/ml. Mix, and incubate at room temperature for 30 min.

7. Add 0.6 vol of ice-cold isopropanol, cover with parafilm and mix gently but thoroughly by inverting the tube several times. At this stage, the DNA-CTAB complex may precipitate as a whitish network. In this case, spool the precipitate out of the solution using a glass hook (e.g., a bent Pasteur pipette), transfer it to the washing solution (Step 8), and let dry (Step 9). If the sample appears flocculent, cloudy, or water clear after mixing with isopropanol (which is often the case), collect the precipitate by low-speed centrifugation (10 min, 2000 × g, 4°C). If a pellet is visible, continue with Step 8. If not, place the solution at –20°C for 30 min to overnight, and centrifuge again, perhaps at higher speed.

8. Add 20 ml of washing solution, gently agitate the pellet for a few minutes, and collect by centrifugation (10 min, 5000 × g, 4°C). Residual CTAB is removed by this step.

9. Invert the tubes and drain on a paper towel for about 1 h. Take care that pellets do not slip down the glass wall. Pellets should neither contain residual ethanol nor should they be too dry. In both cases, redissolving may be difficult.

10. Add an appropriate volume of TE buffer (e.g., 1 ml, depending on the subsequent purification steps chosen), and let the pellets dissolve at 4°C without agitation. Intactness of DNA and the presence of contaminating polysaccharides are the main determinants of the duration of this step. High molecular weight DNA may take several hours to dissolve.

Ammonium acetate treatment (optional):

11. Add 0.5 vol of 7.5 *M* ammonium acetate solution, mix, and chill on ice for 15 min.

12. Centrifuge for 30 min (10,000 × g, 4°C). Transfer supernatant to a new tube.

13. Add 2 vol of 96% ethanol, mix by inversion, and store for 1 h at –20°C.

14. Centrifuge for 10 min (5000 × g, 4°C), wash pellet in 70% ethanol, and centrifuge again.

15. Drain pellet, and dissolve in an appropriate volume of TE buffer.

The protocol given above can easily be scaled down to milligram amounts of plant tissue (see also Rogers and Bendich[584]). If less than 300 mg of fresh tissue are used, tissue homogenization (using an exactly fitting conical pestle) as well as all incubation, precipitation, and centrifugation steps are best performed within microfuge tubes. The method is also highly suitable for fungal DNA.

Several variations of the CTAB protocol have been published (Table 1A). Many of these concern the elimination of polyphenols, which render the DNA preparations brown colored, and unaccessible for restriction enzymes. The detrimental influence of polyphenols and their oxidation products can be counteracted by different strategies: (1) inclusion of polyphenol absorbants such as bovine serum albumin (BSA) and soluble polyvinylpyrrolidone (e.g., PVP-40) in the isolation buffer, usually at concentrations of 1 to 2%;[81,126,468,584] (2) inclusion of phenoloxidase inhibitors such as diethyldithiocarbamic acid (DIECA) in the isolation buffer;[299,543] and (3) inhibition of polyphenol oxidation by including elevated concentrations of antioxidants (such as sodium bisulfite, cysteine, ascorbic acid, or β-mercaptoethanol) in the isolation buffer. For example, Kanazawa and Tsutsumi[342] found that as much as 5% β-mercaptoethanol (instead of 0.2% in the standard method) was needed to yield clean DNA preparations from *Nelumbo* (waterlily). In this case, neither 2% PVP-40 nor 2% BSA were sufficient to prevent browning. The inclusion of 5% β-mercaptoethanol in the isolation buffer was also recommended by Aldrich and Cullis[7] for flax, while 1% β-mercaptoethanol was found optimal for *Ipomoea* species.[211] Interestingly, the latter authors also found that using more than 0.1% β-mercaptoethanol greatly reduced the recovery of DNA from banana leaves. Therefore, we recommend that the effectiveness of varying the concentration of reducing compounds should be empirically tested using 0.2% β-mercaptoethanol as a convenient starting point.

Other modifications of the basic technique concern the removal of polysaccharides, which might render DNA preparations highly viscous. According to Doyle and Doyle,[165] DNA can be successfully isolated from plants rich in polysaccharides by simply increasing the CTAB concentration in the isolation buffer. A simple way to remove polysaccharides was recently described by Fang et al.[189] and Aldrich and Cullis.[7] The crude DNA pellet obtained after the final washing step (Step 9, see above) is dissolved in TE buffer containing 2 M NaCl and reprecipitated with 2 vol of ethanol. During ethanol precipitation in the presence of high salt, many polysaccharides remain dissolved and are removed from the DNA preparation. Polysaccharides are also efficiently removed by CsCl centrifugation (see below).

Still another strategy to reduce the amount of undesired contaminants (and actually the original strategy of CTAB-based DNA isolation procedures) makes use of the fact that DNA-CTAB complexes are soluble in high salt only. Consequently, instead of using isopropanol, these complexes can also be precipitated by lowering the NaCl concentration to less than 0.7%.[486,584,585] Polyphenols, residual proteins, and many polysaccharides remain in the supernatant. After centrifugation, the pellets are redissolved in buffers containing > 1.0 M NaCl and further purified by, e.g., reprecipitation with ethanol or CsCl centrifugation.

b. DNA Minipreparation

If only small amounts of material are available or many samples have to be processed simultaneously, a minipreparation is the method of choice. In addition to numerous DNA isolation methods which were specifically designed for small amounts of tissue (e.g., for PCR analysis, Table 1C), most "regular" protocols such as those given above may also be scaled down to a miniprep performed in Eppendorf tubes. Here, we give a protocol based on the CTAB method of Lassner et al.[390] The plant material may be ground in liquid nitrogen in advance and either kept frozen or lyophilized. Alternatively, fresh material may be ground directly in the Eppendorf tube with the aid of a small (disposable) grinder (Eppendorf-Netheler-Hinz GmbH or Bel-arts Products) fitting exactly in the tube. This can be done with or without liquid nitrogen, quartz sand, or finely ground glass particles.

Solutions:

Extraction buffer: 140 mM sorbitol, 220 mM Tris-HCl, pH 8.0, 22 mM EDTA, 800 mM NaCl, 0.8% CTAB, 1% sarkosyl, 0.2% β-mercaptoethanol (added just before use)
Chloroform-isoamyl alcohol (24:1) (see Safety precautions)
100% Isopropanol
70% Ethanol
TE buffer: 10 mM Tris-HCl, 1 mM EDTA, pH 8.0

Method:

1. Combine into an Eppendorf tube: 1 ml of extraction buffer, 0.4 ml of chloroform-isoamyl alcohol, and plant tissue (approximately 30 to 50 mg dried tissue or 300 to 500 mg fresh tissue).
2. Incubate with shaking at 55°C for 10 min.
3. Centrifuge in a microfuge (5 min, 12,000 × g).
4. Transfer supernatant to a fresh Eppendorf tube.
5. Precipitate nucleic acids by mixing with 1.2 vol of isopropanol.
6. Centrifuge in a microfuge (10 min, 12,000 × g).
7. Decant supernatant, and wash pellet twice with 1 ml 70% ethanol.
8. Drain pellet, and dissolve in 50 µl TE.

This method will usually yield enough DNA for numerous PCR reactions.

c. Potassium Acetate Procedure

This strategy of plant DNA isolation was introduced by Dellaporta et al.[148] Though being somewhat more time consuming, it represents a reasonable alternative to the CTAB procedures. Its key step relies on the simultaneous precipitation of proteins and polysaccharides by high concentrations of potassium acetate in the presence of SDS. No organic extractions are required, which is a considerable advantage over other methods.

Solutions:

Liquid nitrogen (optional)
Isolation buffer: 100 mM Tris-HCl, pH 8.0, 50 mM EDTA, 500 mM NaCl, 10 mM β-mercaptoethanol (added just before use)
20% SDS
5 M potassium acetate
100% isopropanol
TE buffer: 10 mM Tris-HCl, 1 mM EDTA, pH 8.0
3 M sodium acetate, pH 8.0

Method:

1. Grind up to 3 g of fresh or 0.5 g of lyophilized or dried plant material to a fine powder, using liquid nitrogen and a mortar and pestle; see also Step 1 of the CTAB method.
2. Transfer the powder as fast as possible into 15 ml of prewarmed (65°C) isolation buffer in a capped polypropylene tube. Suspend clumps with a spatula.
3. Add 1 ml of 20% SDS, and mix thoroughly. Incubate for 20 min at 65°C in a water bath. Mix every 10 min.
4. Add 5 ml of 5 M potassium acetate, mix thoroughly, and incubate on ice for 30 min. In this step, proteins and polysaccharides are precipitated together with the insoluble potassium dodecyl sulfate, leaving the nucleic acids in solution.
5. Centrifuge for 20 min (20,000 × g, 4°C). Filter the supernatant (which might still contain some floating, particulate material) through Miracloth® into a new centrifuge tube. Precipitate nucleic acids by the addition of 0.7 vol of ice-cold 100% isopropanol followed by gentle mixing.
6. Incubate for 1 h to overnight at –20°C.
7. Centrifuge for 20 min (20,000 × g, 4°C). Discard supernatant.
8. Invert the tubes, and drain on a paper towel for about 1 h. Take care that pellets do not slip down the glass wall.
9. Redissolve pellets in 700 µl of TE buffer (overnight at room temperature or 1 h at 65°C). Transfer the sample to a microfuge tube, and reprecipitate nucleic acids by adding 0.1 vol of 3 M sodium acetate and 0.7 vol of ice-cold isopropanol followed by mixing.
10. Incubate for 1 h to overnight at –20°C.
11. Pellet nucleic acids in a microfuge (10 min, 12,000 × g). Wash pellet with 70% ethanol, and centrifuge again.
12. Invert the tubes, and drain on a paper towel for about 1 h. Dissolve pellets in an appropriate amount (e.g., 100 µl) of TE buffer.

As is the case with the CTAB procedures, a variety of modifications of the basic "Dellaporta" technique have been published. Many of these deal with problematic species rich in polyphenols and/or polysaccharides (Table 1B), and similar strategies as described above have been followed to remove such compounds.[145,331] If desired, the crude DNA preparation obtained after Step 12 may be further purified by, e.g., CsCl centrifugation.

d. DNA Preparation via Nuclei

Isolation of plant DNA via nuclei may help circumvent several of the problems mentioned in the previous sections: (1) the majority of undesirable constituents such as cytoplasmic polysaccharides, polyphenols, and degrading enzymes are removed by the first centrifugation step;[126] (2) as long as the DNA is enclosed in intact nuclei, it is largely protected from shearing forces as well as from the action of nucleases; and (3) cross-contamination of nuclear DNA with organellar DNA is avoided. This final point should be taken into consideration when the experimental aim is random amplified polymorphic DNA (RAPD) analysis. RAPDs are generated not only from nuclear, but sometimes also from mitochondrial and chloroplast, DNA.[413,693] The nuclear vs. organellar origin of a RAPD fragment determines its marker behavior in, e.g., linkage analysis. On the other hand, the exclusion of mitochondrial and chloroplast DNA is probably not very relevant for hybridization-based DNA fingerprinting, since long arrays of short tandem repeats of DNA have not yet been described in the genomes of these organelles.

A major disadvantage of this approach is that extraction of nuclei from dried or frozen material is usually less efficient than from fresh material. However, if fresh material is available, DNA isolation via nuclei is the method of choice for obtaining high-quality DNA. A considerable variety of plant DNA isolation methods via nuclei have been published, only a small collection of which are compiled in Table 1D. Here, we present a shortened and modified version of a method originally described by Willmitzer and Wagner[787] for the isolation of nuclei from tissue-cultured tobacco cells. In this method, tissues are disrupted in a buffer containing polyamines, which stabilize nuclear structures. Nuclei are then purified by differential centrifugation using Percoll® (Pharmacia LKB) solutions, and DNA is prepared by lysis of the nuclei followed by proteinase K digestion. The protocol described below is for 10 g of leaf material, but can be scaled down to about 3 g and scaled up to more than 50 g.

Solutions:

Isolation buffer: 250 mM sucrose, 10 mM NaCl, 10 mM MES, pH 6.0, 5 mM EDTA, 0.15 mM spermine, 0.5 mM spermidine, 20 mM β-mercaptoethanol, 0.1% BSA, 0.6% NP-40. Note: make this buffer as a 5X concentrated solution (but keeping NP-40 at 0.6%). Sterilize by filtration, and store in a tightly closed bottle in a refrigerator. Dilute with cold distilled water prior to use, and adjust NP-40 concentration to 0.6%. NP-40 is a nonionic detergent. MES (morpholinoethane sulfonic acid) is a buffer reagent.

Floating buffer: Prepare immediately before use. Under constant stirring, add 90 g of cold Percoll® to 12 g of 5X concentrated isolation buffer. Readjust the pH to 6.0 by dropwise addition of 1 M HCl. Do this carefully, since overtitration might result in irreversible precipitation of Percoll®.

Silicone emulsion (e.g., Serva) or 1-octanol (optional)

TE buffer: 10 mM Tris-HCl, 1 mM EDTA, pH 8.0

200 mM EDTA

10% Sarkosyl

Proteinase K (2 mg/ml)

Method (Note: Steps 1 through 8 are performed at 4°C):

1. Cut 10 g of freshly harvested tissue into 2- to 3-cm pieces, and collect these in a sterilized 200-ml Erlenmeyer flask on ice.
2. Add 50 ml of cold isolation buffer (5ml/g of tissue) and a drop of silicone emulsion or 1-octanol (to prevent foaming). Homogenize in an Ultra-Turrax® homogenizer (Janke and Kunkel GmbH and Co.) with increasing speed. The homogenization step should be rather brief to prevent the suspension from warming. For leaf material, 4 × 15 s is usually sufficient. Homogenization can also be performed using mortar and pestle. In this case, silicone emulsion is not needed, but some quartz sand should be added which helps disrupting the tissue.
3. Filter the slurry through four layers of cheesecloth.
4. Centrifuge the filtrate in a swinging-bucket rotor for 10 min (2000 × g, 4°C). Nuclei and some starch are pelleted by this step.
5. Discard the supernatant. Resuspend the pellet in 20 to 40 ml of isolation buffer by pipetting up and down. Centrifuge as above.
6. Decant and discard the supernatant. Resuspend the pellet in 20 to 40 ml of floating buffer, and centrifuge in a swinging-bucket rotor for 10 min (5000 × g, 4°C). In this step, starch and most residual cell wall materials will sediment, while nuclei float on the top of the Percoll® solution.
7. Collect the floating nuclei with the help of a Pasteur pipette or a spatula, and resuspend in 20 to 40 ml of isolation buffer. Centrifuge as in Step 4.
8. Resuspend the pellet in TE buffer. Lyse nuclei by adding 0.1 vol of EDTA, sarkosyl, and proteinase K stock solutions to yield final concentrations of 20 mM, 1%, and 200 µg/ml, respectively. At this stage, the suspension becomes highly viscous.
9. Incubate under gentle agitation for 10 min at 60°C, followed by 2 h at 37°C.
10. Centrifuge for 15 min (5000 × g, room temperature) to remove remaining cell wall debris. The supernatant contains high molecular weight nuclear DNA, which may be either directly precipitated by adding 0.1 vol of 3 M sodium acetate and 2 vol of ethanol or further purified by, e.g., CsCl centrifugation.

3. DNA Purification via CsCl Density Gradient Centrifugation

The DNA preparations obtained by one of the methods described above may still contain considerable amounts of RNA, polysaccharides, proteins which are tightly bound to DNA, and other contaminants. Nevertheless, the level of purity may be sufficient for PCR with specific and arbitrary primers and, in the case of some species, also for hybridization-based DNA fingerprinting. However, the presence of RNA interferes with the spectrophotometric quantitation of DNA, polysaccharides render the sample viscous, and both residual proteins and polysaccharides may inhibit the action of restriction enzymes. Further purification, therefore, will be desirable or even imperative in a number of plant species.

A variety of procedures have been developed to purify DNA from crude preparations, including, e.g., CsCl centrifugation (see below), preparative electrophoresis in low melting point agarose,[81,587,607] gel filtration, and ion exchange chromatography.[202,757] Any of these methods can be combined with any of the isolation procedures given above. The reader may also refer to alternative, less commonly used procedures (see Tables 1A through F).

Though CsCl centrifugation is somewhat expensive and time consuming and a (mini)ultracentrifuge is needed, it is the method of choice when high-quality DNA is desired, e.g., for DNA fingerprint analyses of recalcitrant species. The principle of this technique is the "banding" of DNA in a CsCl gradient which forms during ultracentrifugation. As a first step, the DNA solution is adjusted to a final density of 1.55 g/ml, which may be controlled in a refractometer. Concentration adjustment can be done by either adding solid CsCl (1 g of CsCl for each ml of DNA solution) or, as described below, by mixing with a 1.5× concentrated stock (i.e., 6.6 M CsCl). An intercalating dye (usually ethidium bromide) is then added, which forms a fluorescent complex with DNA and allows visualization of the DNA band after centrifugation. During the run, the DNA-dye complex will form a sharp band that is positioned according to the density of the DNA sample (which is positively correlated to its GC content). Depending on the amount of DNA, this band will be either visible with the naked eye or can be located with the help of UV light. Contaminating proteins are dissociated from DNA by the high CsCl concentration and float on the surface of the gradient, the RNA is pelleted, and polysaccharides are generally also separated from DNA. After centrifugation, the DNA band is removed from the gradient, the dye is extracted, and the DNA is freed from CsCl by ethanol precipitation or dialysis.

The technique may be inadequate when large numbers of samples have to be processed. Problems may also occur when very low yields are expected, since banded DNA of less than 10 µg is hard to locate.

Solutions:

Gradient buffer: 6.6 M CsCl, 50 mM Tris-HCl, pH 8.0, 10 mM EDTA, 150 µg/ml ethidium bromide (see Safety precautions)
Light mineral oil

TE buffer: 10 m*M* Tris-HCl, 1 m*M* EDTA, pH 8.0
TE-saturated 1-butanol: mix 1-butanol with TE buffer until two phases appear
  after standing. Use the upper (butanolic) phase.
96% ethanol
70% ethanol

Method:

1. Transfer crude DNA samples obtained by one of the methods described above to an ultracentrifuge tube, adjust the desired volume by an appropriate amount of TE buffer, and mix with exactly 2 vol of gradient buffer. CsCl solutions are very heavy. To avoid overloading, volumes of DNA and CsCl solutions should be calculated to fill the tubes up to two thirds only. Refer to the operating instructions of your ultracentrifuge.

2. Overlay the solution with light mineral oil in order to fill the ultracentrifuge tube completely. Balance the tubes carefully. For fixed-angle and vertical rotors, seal the tubes in a sealing device, and make sure that they are vacuum proof. Load the rotor according to the operator's instructions.

3. Centrifuge at 16°C for 24 h at 100,000 × g in a vertical or fixed-angle rotor or for 48 h at 100,000 × g in a swinging-bucket rotor. We prefer a swinging-bucket rotor since DNA bands will not be contaminated by the RNA pellet, the use of sealed tubes is not necessary, and DNA is more easily removed from the tube. A two-step centrifugation, 16h at 210,000 xg in a fixed-angle rotor often gives a better resolution for fungal DNA.

4. After centrifugation, remove the tubes from the rotor, and fix them in a support. Place a tray beneath the support in order to avoid bench contamination. DNA should be visible as a red-colored band under daylight conditions. If very low amounts of DNA are present, locate the fluorescent band under long-wave UV light, and mark its position with a pen.

5. If a sealed tube was used, two hypodermic wide-gauge needles are needed for removing the band. Pierce one needle (which allows air to come in) into the top of the tube where the mineral oil is located. Pierce a second needle attached to a syringe through the tube wall, approach the DNA band from beneath, and carefully remove it by slight suction. There are usually hard and soft zones in the ultracentrifuge tube walls. Use the soft zone for piercing the hole. Take great care not to pierce through both walls and into your finger tips. Proceed with Step 7.

6. If a nonsealed standard ultracentrifuge tube was used, remove the mineral oil layer by pipetting, and use a 2-ml syringe attached to a wide-gauge hypodermic needle to remove the DNA band from the top. This technique is less risky than piercing, both in terms of injury and contamination with ethidium bromide.

7. Extract the ethidium bromide from the DNA in one syringe per sample. All steps are performed within the syringe; tubes and pipettes are not

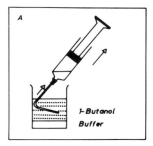 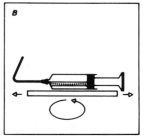 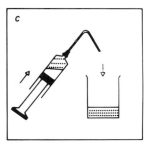

FIGURE 11. Facilitated ethidium bromide extraction from DNA samples using a syringe fitted to a bent hypodermic needle. See text for details.

required (Figure 11). After having removed the DNA from the tube, bend the needle to form a hook, and draw 1 vol of TE-saturated 1-butanol into the syringe, followed by 1 vol of air (Figure 11A). Mix phases by fixing the syringe on a rotary shaker (Figure 11B). After one to several minutes of extraction, place the syringe vertically to separate phases. Remove the upper, organic phase (Figure 11C), and add fresh 1-butanol. Repeat all steps until the aqueous phases are water clear.

8. Remove the needle, and transfer the final aqueous phase into a centrifuge tube. Dilute with 2 vol of TE buffer (to avoid coprecipitation of CsCl), and precipitate the DNA by adding 2 vol of ethanol, followed by gentle but thorough mixing of the phases. As an alternative to precipitation, CsCl may also be removed from the DNA preparation by dialysis.[468]

9. Spin for 30 min (5000 × g, 4°C), wash pellets in 70% ethanol, dry, and redissolve pellets in 100 to 200 µl of TE buffer.

B. ISOLATION OF FUNGAL DNA

Fungal material is either collected from the environment (i.e., from different kinds of substrate, infected host species etc.) or grown and propagated in the laboratory. Fungal species are often associated with other kinds of organisms and sometimes are not recognized easily from their habitus. Therefore, it is generally advisable to initiate *in vitro* cultures from field-collected material prior to DNA isolation in order to ensure the identity of the investigated species. Both filamentous and yeast-like fungi can be grown in liquid media as well as on the surface of (or embedded in) solid agar plates and/or slants. Whereas agar-solidified media are generally preferred for long-term maintenance of cultures, liquid media are used for the fast accumulation of mycelia or cells for DNA isolation.

Note: All microbiological work should be performed in a biological safety hood (see also Safety precautions). All media have to be autoclaved prior to use.

1. Isolation and Cultivation of Fungi from Plant Material, Soil, and Air-Borne Spores

Fungal cultures may be initiated from any kind of substratum. However, depending on the origin of the isolate, the fungal species of interest is often contaminated with other organisms. For instance, soil, plant litter, and various decaying materials harbor a plethora of microorganisms (1 g of soil may contain some hundreds of species!). Therefore, purification of fungal isolates is essential and is usually performed by some kind of dilution (see below). Establishing single spore-derived colonies (with the help of a microscope or a micromanipulator) is the most reliable method. It not only helps reduce the risk of contamination, but it also yields genetically homogeneous material.

Samples should be kept dry or moderately moist, well aerated, and cool during transport and storage. For the isolation of fungal colonies from infected plant material, cuttings are incubated on an appropriate medium solidified by agar. In case of endophytic fungi, plant material may be surface sterilized prior to plating (e.g., by ethanol and/or hypochlorite treatment), which reduces the number of contaminants. Often, mixed cultures consisting of several species develop from the explants. Pure cultures of the species of interest may be obtained by, e.g., cutting off and subculturing segments from marginal regions of a colony or by picking up small amounts of spores with the help of a glass needle or a micromanipulator.

Soil samples are diluted with sterile water, thoroughly agitated, and plated on appropriate media solidified by agar (see below). Individual fungal colonies appear after a certain incubation period. In order to obtain pure cultures of the species of interest, the soil material is repeatedly diluted and replated.

Air-borne spores can be collected by exposing agar plates at the desired sampling site for several minutes or by using specific spore traps.[789] Individual colonies develop after incubation of the spores on an appropriate medium.

2. Isolation and Cultivation of Human and Animal Pathogenic Fungi

Fungal isolates may be obtained from any part of the body, e.g., blood, cerebrospinal fluid, bone marrow, vaginal fluid, stool, throat swabs, urine, sputum, bronchial washes, skin and nail scrapings, and hair. Potential pathogens are most efficiently isolated from affected sites. For example, on diseased skin, the advancing edge of a lesion is usually most suitable for the isolation of the causal agent. Nonsterile specimens such as skin, nail, hair pieces, stool, or vaginal fluid are plated onto agar plates containing antibiotics to inhibit bacterial growth. A standard medium is Sabouraud's agar[601] (2 to 4% glucose, 1% neopeptone, and 2% agar). Cultures are incubated at a temperature of 25 to 30°C. Sterile fluids such as blood and cerebrospinal fluid are diluted into liquid media and incubated for 2 days up to several weeks. After initiation of the cultures, similar purification procedures are applied as described above for fungi growing on/in plant hosts and other substrates. For more detailed information on the isolation and cultivation of human and animal pathogenic fungi, the reader may refer to a standard textbook on medical microbiology or mycology (e.g., McGinnis,[446] Balows et al.,[33] Joklik et al.,[334] and Kwon-Chung and Bennett[382]).

3. Long-Term Maintenance of Fungal Cultures

If possible, fungal strains should be maintained as living cultures for two main reasons: (1) DNA reisolations may become necessary to replace contaminated or degraded DNA samples and/or to confirm previous results and (2) maintenance of cultures provides other researchers with reference strains.

Fungal strains can be maintained in sterile water at room temperature, as agar cultures at 4°C, or in 15% glycerol frozen at −70°C.

a. Storage in Sterile Water

This simple and inexpensive technique works well with, e.g., yeast cultures. Sterile water is the only medium required.

Method:

1. Using a sterile inoculation loop, pick a single colony from the culture plate, and dilute it into 2 ml of sterile water in a screw-cap tube. Work in a biological safety hood.
2. Store at room temperature in a dark place for up to 1 year.

b. Storage in Culture Tubes (Slants)

This technique is recommended for both filamentous and yeast-like fungi. Agar slants are preferred over petri dishes because they allow longer storage periods before a new transfer is necessary.

Media:

MA[51,55] (malt-extract agar): 30 g/l malt extract (Difco Laboratories), 15 g/l agar-agar, pH 7.0

CMA[626] (cornmeal agar): cornmeal extract, 15 g/l agar-agar, pH 6.0. Cornmeal extract is prepared as follows: boil 50 g of ground maize grain wrapped in a cloth in a water bath. After simmering for 1 h, squeeze the extract through the cloth, and adjust to 1 l with sterile water

CZA[135,136,161] (Czapek-Dox agar): Czapek-Dox broth (30 g/l glucose, 5 g/l yeast extract, 3 g/l NaNO$_3$, 1 g/l K$_2$HPO$_4$, 0.5 g/l MgSO$_4$-7H$_2$O, 0.01 g/l FeSO$_4$-7H$_2$O), 15 g/l agar-agar, pH 6.0 to 6.5

YPD-agar[723,763] (yeast-peptone-dextrose agar): YPD medium (20 g/l glucose, 10 g/l yeast extract, 20 g/l peptone), 15 g/l agar-agar

Method:

1. Prepare one of the solid media described above according to the requirements of the fungal species to be grown. Sterilize by autoclaving.
2. Pipette 6-ml aliquots into sterile culture tubes. Work in a biological safety hood.
3. Lean culture tubes against a support at an angle of about 70°.
4. Allow slants to solidify. Unused agar slants may be stored at room temperature or at 4°C for a maximum of 6 months.
5. Using a sterile inoculation loop, transfer a single colony from a culture plate to the slant, and distribute the cells on the agar surface.
6. Incubate for several days at the required temperature (leave caps loose!).
7. When the cultures have grown to the desired size, close the cap as tightly as possible, and store at 4°C (viable for 1 to 2 years).

c. Storage as Frozen Glycerol Stocks

Medium:

Sterile 30% glycerol

Method:

1. Mix 1 ml of the glycerol solution with 1 ml of fungal liquid culture (late-log or early stationary phase; see below and Section II.B.5) in a sterile microfuge screw-cap tube. Work in a biological safety hood.
2. Store at −70°C (viable for about 5 years).
3. Revive by scraping some cells off the frozen surface and streaking these onto agar plates containing an appropriate medium. Keep storage tubes frozen, and return them to −70°C immediately after use!
4. Incubate agar plates for several days at the required temperature.

In liquid culture, the division rate of fungal cells is maximal in the logarithmic phase and drops off toward the stationary phase. To ensure a high growth rate after reviving glycerol stocks or after transfer to a new medium, cells from liquid cultures should be generally transferred in the late-log or early stationary phase.

4. Induction of Sporulation in Fungal Cultures

Inducing sporulation is necessary for two purposes: (1) establishment of single spore-derived cultures and (2) large-scale production of mycelia from filamentous fungi. Since each spore grows into one mycelium, inoculation of a high number of spores into a liquid medium produces a large amount of fungal material in a short period of time. A special nitrogen-deficient medium is often used for high accumulation of spores.

Solutions:

Nitrogen-deficient medium:[345] 10 g/l potassium acetate, 1 g/l yeast extract, 0.5 g/l glucose
Washing solution: 0.1 M NaCl, 0.5% Triton X-100

Method:

1. Pipette 25-ml aliquots of the autoclaved medium into sterile 100-ml Erlenmeyer flasks. Work in a biological safety hood.
2. Lean Erlenmeyer flasks against a support at an angle of about 70°, and allow the agar to solidify.
3. Using a sterile inoculation loop, transfer a single colony from an agar plate to the sporulation flask, and distribute the cells on the agar surface.
4. Incubate for 3 to 4 days at the required temperature (generally 23 to 28°C; leave caps loose).
5. To harvest the spores, wash the agar surface with 1 to 2 ml of sterile washing solution.
6. Spore suspensions can be stored at 4°C and should be used within 2 weeks of harvest. The number of spores may be calculated with the help of a counting chamber (e.g., hemocytometer, Thoma chamber) under microscope.

5. Large-Scale Accumulation of Fungal Material for DNA Isolation

Fungal material should be grown in liquid rather than solidified media prior to DNA isolation for two reasons. First, the use of liquid media allows the accumulation of large amounts of mycelia or cells in a short period of time (see above). Second, polysaccharides (i.e., agar-agar attached to the fungal mycelia) may render DNA isolation difficult (see Section II.A). If culturing on agar plates is necessary for one or the other reason, the agar should be carefully stripped off the mycelia prior to DNA isolation.

Efficient aeration is needed for large-scale growth. To ensure this, the medium should constitute no more than one fifth of the total volume of the (preferably Erlenmeyer) flask, and culturing should be performed on a shaking incubator at 220 rpm. Any media which offer favorable growing conditions for the investigated fungal species can be used. Some commonly used media are described below. However, some species may require very special media. For more detailed information, refer to Constantinescu,[123] Booth,[58] and Papavizas and Davey[538] or to the catalogs of the following culture collections: CBS,[101] IMI,[306] and ATCC.[25,26]

Filamentous fungi and yeasts have fundamentally different growth characteristics and nutrition requirements. While yeast-like fungi are generally characterized by fast growth rates (i.e., short cell cycles) and modest requirements on their growth media, filamentous fungi usually exhibit slow growth rates (i.e., longer cell cycles) and often have highly specific demands on medium composition (e.g., the inclusion of certain trace elements). Procedures for both types of fungi are given below.

In general, the ingredients for liquid media are dissolved in water, aliquoted into Erlenmeyer flasks, and autoclaved. Small amounts may also be filter sterilized, which circumvents the risk of caramelization (e.g., of glucose).

a. Culturing of Filamentous Fungi in Liquid Media

Media:

General fungal medium:[345] 30 g/l glucose, 2 g/l yeast extract, 2 g/l peptone, mineral salts and trace elements mix stock [10 ml/l A, 10 ml/l B, 10 ml/l C, 1 ml/l D (see Table 2)], pH 6.0. Solutions A and B are added before autoclaving the liquid or solid media. Solutions C and D are filter sterilized and added after autoclaving. This rich medium is used for the growth of mycelia of, e.g., species of the genera *Chaetomium, Fusarium, Gibberella, Mucor, Penicillium,* and *Verticillium.*

CYM-complete medium:[567] 20 g/l dextrose, 2 g/l yeast extract, 2 g/l peptone, 1 g/l K_2HPO_4, 0.46 g/l KH_2PO_4, 0.5 g/l $MgSO_4$-$7H_2O$, pH 7.0. This medium is used for the growth of mycelia of, e.g., species of the genera *Absidia, Mucor,* and *Morcerella.*

V8-juice medium:[467,480] 200 ml/l V8-juice (unfiltered multivitamin juice), 3 g/l $CaCO_3$, pH 7.2. This medium is used for the growth of mycelia of, e.g., species of the genus *Leptosphaeria.*

Malt-yeast broth:[274] 3 g/l malt extract, 3 g/l yeast extract, 5 g/l peptone, 10 g/l glucose, pH 7.5. This medium is used for the growth of mycelia of, e.g., species of the genus *Leptosphaeria.*

Czapek-Dox medium:[135,136,161] 30 g/l glucose, 5 g/l yeast extract, 3 g/l $NaNO_3$, 1 g/l K_2HPO_4, 0.5 g/l $MgSO_4$-$7H_2O$, 0.01 g/l $FeSO_4$-7 H_2O, pH 6.0 to 6.5. This medium is used for the growth of mycelia of, e.g., species of the genera *Trichoderma, Aspergillus,* and *Emericella.*

Method:

1. Transfer 100-ml aliquots of the sterile medium into 500-ml sterile Erlenmeyer or baffled bottom flasks. Work in a biological safety hood.
2. Inoculate each flask with a spore suspension (see Section II.B.4; final spore density 10^8/ml). If hyphal inoculum is used, the number of starting hyphae may be increased by preculturing the mycelia in small liquid cultures followed by disruption in a sterile blender.

TABLE 2
Mineral Salts and Trace Elements Mix Stock, According to Kayser[345]

	Solutions	Ingredients
A	Phosphate buffer	1.54 g/l KH_2PO_4
		0.26 g/l K_2HPO_4
B	Mineral salt and	2.00 g/l NH_4NO_3
	trace elements solution	0.40 g/l Na_2SO_4
		0.15 g/l $MgSO_4$-$7H_2O$
		0.02 g/l $MnSO_4$-$4H_2O$
		0.02 g/l $CuSO_4$-$5H_2O$
		0.02 g/l $ZnSO_4$-$7H_2O$
C	Ferric solution	0.15 g/l $Fe(NH_4)_2SO_4$
D	Calcium solution	1.10 g/l $CaCl_2$

3. Incubate for 72 h (caps loose) on a rotary shaker (220 rpm) at the required temperature (most filamentous fungi grow well at 25 to 28°C; however, some species have other requirements).

4. To harvest the mycelia, filter the culture medium under vacuum through a Buchner funnel or centrifuge it for 15 min (2500 × g, room temperature).

5. After harvest, wash mycelia twice with sterile water. Mycelia may now immediately be used for DNA isolation or stored by freezing in liquid nitrogen. Frozen mycelia may be kept at –20°C for a few days or at –70°C for up to 1 year. Alternatively, frozen mycelia may be lyophilized and stored in a dry and dark place for several years at room temperature.

b. Culturing of Yeast-Like Fungi in Liquid Media

Media:

YPD broth:[723,763] 20 g/l glucose, 10 g/l yeast extract, 20 g/l peptone. This rich medium is widely used for yeasts (e.g., *Candida*, *Saccharomyces*, *Cryptococcus*) if no special growth conditions are required.

YNB[777,802] (yeast-nitrogen base broth): 1.7 g/l yeast-nitrogen base without amino acids and ammonium sulfate, 5 g/l $(NH_4)_2SO_4$, 20 g/l glucose. This minimal medium supports growth of several yeast-like fungal species which have no special nutritional requirements.

Method:

1. Transfer 3-ml aliquots of the sterile liquid medium into 10-ml sterile culture tubes. Work in a biological safety hood.

2. Use a sterile loop to inoculate each tube with a single yeast colony from a culture plate.
3. Incubate overnight at 30°C in a shaking incubator at 220 rpm (some yeast species may require other temperatures, e.g., 37°C is optimal for *Cryptococcus neoformans*).
4. Transfer the 3-ml overnight cultures into 50-ml Erlenmeyer flasks containing fresh medium.
5. Incubate for 48 h at 30°C in a shaking incubator at 220 rpm.
6. To harvest the yeast cells, centrifuge the culture medium for 15 min (2500 × g, room temperature).
7. Wash the yeast cells twice with sterile water. The cells may now immediately be used for DNA isolation or stored by freezing in liquid nitrogen. Frozen cells may be kept at –20°C for a few days or at –70°C for up to 1 year.

6. DNA Isolation Protocols

There is no consensus on the optimal stage of fungal cultures for DNA isolation. Some researchers claim that cultures should be derived from the log phase, while others prefer stationary phase cultures. Cell wall disruption is usually easier in the log phase. The efficient disruption of the cell wall is one of the critical steps during DNA isolation. Fungal cell walls contain chitin and are often highly resistant to mechanical forces. Several strategies to destroy the cell walls may be followed. One commonly used technique makes use of sand or small glass beads to grind fresh mycelium in a mortar with a pestle. This is reliable and inexpensive, but many hyphal segments remain intact. A more efficient method (also often used for plant tissue; see Section II.A), which we favor, is to freeze the mycelia or cells in liquid nitrogen prior to grinding with mortar and pestle. It is important to keep the powder frozen by adding liquid nitrogen throughout the procedure. Still another technique involves cell wall-degrading enzymes to generate protoplasts prior to DNA isolation. This method avoids mechanical force and yields highly intact genomic DNA, but is time consuming and expensive.

A large variety of methods to isolate fungal DNA have been described. However, the majority of methods belong to one of the following categories:

1. SDS- or TNS/PAS-based methods.[102,244,395,553,561,654,805,821] This strategy makes use of strong detergents (SDS: sodium dodecyl sulfate, TNS/PAS: triisopropylnaphthalene sulfonic acid/para-aminosalicylic acid) to break the cell and organelle membranes.
2. CTAB-based methods.[486,584,841] In this strategy, the DNA-binding detergent CTAB is included in the isolation buffer. CTAB-based methods are also often used for plants and are described in detail in Section II.A.2. (see also Table 1A).
3. Methods based on protoplast isolation.[142,263] Here, specific mixtures of cell wall-degrading enzymes (e.g., zymolase) are used to prepare proto-

plasts, from which DNA extraction is easier to perform. This technique is very useful for fungal cells with strong cell walls or a large capsule (such as *Cryptococcus neoformans*).

The extraction methods described below are successfully used in our laboratories. Only minimum equipment is required, and DNA is usually sufficiently pure for PCR and restriction experiments. If further purification of the genomic DNA is desired, refer to one of the different strategies outlined in Section II.A. Some of the plant DNA extraction procedures summarized in Tables 1A through F (especially the CTAB methods) will also work for fungi.

a. SDS Procedures

Three different SDS-based procedures are given here, two of which are minipreps. The first protocol is modified after Picknett et al.[553] Following initial cell wall disruption, the fungal cells are lysed by addition of an SDS-containing isolation buffer. Proteins are removed by proteinase K treatment, and RNA is eliminated by RNase.

Solutions:

Liquid nitrogen
Extraction buffer: 25 mM Tris-HCl, 25 mM EDTA, 50 mM NaCl, 1% SDS, pH 8.0
5 M NaCl
Proteinase K: 20 mg/ml
RNase A: 10 mg/ml (to inactivate contaminating DNases, the RNase has to be preincubated for 15 min at 100°C)
Phenol (saturated with 1× TE; see Safety precautions)
Ratio of chloroform to isoamyl alcohol (24:1)
3 M sodium acetate, pH 5.2
96% Ethanol
70% Ethanol
1× TE: 10 mM Tris-HCl, 1 mM EDTA, pH 8.0

Method:

1. Grind 2 g of frozen mycelium in liquid nitrogen using a mortar and pestle. Do not allow the powder to thaw.
2. Suspend the powder in 25 ml of extraction buffer in a centrifuge tube, and mix gently.
3. Incubate for 1 h at 0°C.
4. Add 5 M NaCl to a final concentration of 1 M, mix gently, and incubate for 1 h at 0°C.
5. Centrifuge for 30 min (2500 × g, 4°C) to separate the cell wall and cell membrane fragments from the DNA.

6. Transfer the supernatant to a new tube, and incubate for 10 min at 65°C.
7. Add proteinase K (50 μg/ml final concentration), and incubate for 20 min at 37°C.
8. Add an equal volume of phenol, and shake gently for 2 min.
9. Centrifuge for 15 min (2500 × g, room temperature) to separate the phases.
10. Transfer the aqueous phase to a new tube, add an equal volume of phenol and chloroform (1:1), and shake gently for 2 min.
11. Centrifuge as above.
12. Transfer the aqueous phase to a new tube, add an equal volume of chloroform and isoamyl alcohol (24:1), and shake gently for 2 min.
13. Repeat Step 9, and then transfer the aqueous phase to a new tube.
14. Add RNase A (50 μg/ml final concentration), and incubate for 3 h at 37°C.
15. Add an equal volume of phenol, and shake gently for 2 min.
16. Repeat Steps 9 to 13.
17. Transfer the aqueous phase to a new tube (e.g., 30-ml glass centrifuge tube), and precipitate the DNA by adding 0.03 vol of 3 *M* sodium acetate and 2.5 vol of cold 96% ethanol. Mix well, and incubate for 1 h or overnight at –20°C.
18. Centrifuge for 30 min (14,000 × g, 4°C).
19. Wash the DNA pellet twice with 70% ethanol, and centrifuge for 10 min (14,000 × g, 4°C).
20. Air dry the DNA pellet, dissolve it in ca. 500 μl of 1× TE, and store at +4 or at –20°C.

The second protocol is modified after Lee and Taylor.[395] Following cell wall disruption, cells are lysed, extracted with phenol/chloroform, and the DNA is precipitated with isopropanol. The method can be used for small amounts of mycelia or cells. This makes it an excellent procedure for the preparation of template DNA for PCR experiments.

Solutions:

Liquid nitrogen
Lysis buffer: 50 m*M* Tris-HCl, pH 7.2, 50 m*M* EDTA, 3% SDS, 1% β-mercaptoethanol
Phenol (saturated with 1X TE, see Safety precautions!)
Ratio phenol to chloroform to isoamyl alcohol (25:24:1)
3 *M* sodium acetate, pH 8.0 (pH 5.2 also works)
100% isopropanol
70% ethanol
1× TE: 10 m*M* Tris-HCl, 1 m*M* EDTA, pH 8.0

Method:

1. Grind 0.1 to 0.3 g of mycelia in a microfuge tube in liquid nitrogen using a disposable pestle (a conical grinder, exactly fitting the tube and rotated by hand or an electric potter at 200 rpm for several minutes).
2. Add 400 μl of lysis buffer, and vortex until the mixture becomes homogeneous. If the suspension is too viscous, add more lysis buffer (up to 700 μl).
3. Incubate for 1 h at 65°C.
4. Add an equal volume of phenol, chloroform and isoamyl alcohol (25:24:1), and vortex briefly.
5. Spin in a microfuge for 15 min (or longer, aqueous phase should be clear) at 14,000 × g at room temperature.
6. Transfer the aqueous phase to a new tube, and precipitate the DNA by adding 0.03 vol of 3 M sodium acetate and 0.5 vol of isopropanol. Mix gently but thoroughly.
7. Incubate for at least 30 min at 4°C.
8. Spin in a microfuge for 15 min (14,000 × g, room temperature).
9. Discard the supernatant. Rinse the pellet once with 70% ethanol.
10. Invert the tube, and drain on a paper towel.
11. Air dry the DNA pellet, dissolve it in an appropriate volume of 1× TE (e.g., 100 to 500 μl), and store at +4 or –20°C.

The third protocol (modified after Cenis[102]) avoids the use of hazardous chemicals such as phenol. It is suitable for the extraction of very small amounts of tissue cultured directly in a microfuge tube, which makes it especially useful for the preparation of template DNA for PCR experiments.

Solutions:

Extraction buffer: 200 mM Tris-HCl, pH 8.5, 250 mM NaCl, 25 mM EDTA, 0.5 % SDS
3 M sodium acetate, pH 5.2
100% isopropanol
70% ethanol
1× TE: 10 mM Tris-HCl, 1 mM EDTA, pH 8.0

Method:

1. Grow the fungal culture in 500 μl of liquid media in a 1.5-ml microfuge tube according to one of the methods described above (Section II.B.5).
2. Pellet the mycelia by centrifugation for 5 min at 15,000 × g in a microfuge.

3. Discard the supernatant, wash the pellet with 1× TE, and centrifuge as above.
4. Discard the supernatant, and add 300 µl of extraction buffer.
5. Grind the mycelium for several minutes with a conical grinder, see Step 1 in the previous method.
6. Add 150 µl of 3 *M* sodium acetate, mix briefly, and incubate for about 10 min at −20°C.
7. Spin in a microfuge for 15 min (15,000 × g, room temperature).
8. Transfer the supernatant to a new 1.5-ml microfuge tube, and precipitate the genomic DNA by adding an equal volume of isopropanol.
9. Incubate for at least 5 min at room temperature.
10. Centrifuge as above.
11. Rinse the pellet in 70% ethanol. Invert the tube, and drain on a paper towel.
12. Air dry the DNA pellet, dissolve it in 50 µl of 1× TE, and store at +4 or −20°C.

b. TNS/PAS Procedure

This method, originally described by Slater[636] and de Graaff et al.[144] for the isolation of RNA, was modified for DNA isolation by Gruber.[244] Ground mycelia are incubated in a hot extraction buffer containing the detergents TNS (triisopropylnaphthalene sulfonic acid) and PAS (para-aminosalicylic acid) as well as phenol. After organic extraction with chloroform, the DNA is precipitated with ethanol.

Solutions:

Liquid nitrogen
TNS
PAS (sodium salt × 2 H_2O)
Extraction buffer: 1 *M* Tris-HCl, 1.25 *M* NaCl, 0.25 *M* EDTA, pH 8.0
Distilled water
Phenol (saturated with 1× TE; see Safety precautions)
Chloroform
96% ethanol
70% ethanol
1× TE: 10 m*M* Tris-HCl, 1 m*M* EDTA, pH 8.0

Method:

1. Dissolve 100 mg of TNS and 600 mg of PAS in 10 ml of distilled water in a 100-ml Erlenmeyer flask. Add 2.5 ml of extraction buffer and 7.5 ml of phenol. Preheat to 55°C.
2. Grind 2 g of frozen mycelia in liquid nitrogen using a mortar and pestle. Do not allow the powder to thaw.

3. Add ground mycelia to the preheated extraction solution. Incubate for 2 min at 55°C. Shake occasionally.
4. Add 5 ml of chloroform, and incubate for 2 min more at 55°C. Shake occasionally.
5. Transfer to a centrifuge tube, and spin for 10 min (3700 × g, room temperature) to separate the phases.
6. Transfer the aqueous phase (upper phase) to a new tube, and add 10 ml of phenol and chloroform (1:1). Shake gently for 2 min at room temperature.
7. Centrifuge as above.
8. Transfer the aqueous phase to a new tube, and add 10 ml of chloroform. Shake gently for 2 min.
9. Centrifuge as above.
10. Transfer the aqueous phase to a 30-ml centrifuge tube, and precipitate the genomic DNA by adding 0.03 vol of 3 M sodium acetate and 2.5 vol of 96% ethanol. Mix well, and incubate for 1 h or overnight at –20°C.
11. Centrifuge for 30 min (14,000 × g, 4°C).
12. Discard supernatant. Wash the DNA pellet twice with 70% ethanol, and centrifuge for 10 min (14,000 × g, 4°C).
13. Air dry the DNA pellet, dissolve it in 0.5 to 1 ml of 1× TE, and store at +4 or –20°C.

c. DNA Preparation via Protoplasts

The protocol given below (modified after Davis et al.[142]) is based on the preparation of protoplasts (also called spheroplasts in yeasts) by digesting the fungal cell wall components with a mixture of hydrolytic enzymes. This strategy facilitates subsequent DNA isolation considerably. Contaminating proteins are removed by SDS/potassium acetate precipitation and phenol/chloroform extraction. The protocol, though quite long, is especially efficient for yeast-like fungi with a large capsule (e.g., *Cryptococcus neoformans*).

Solutions:

Spheroplasting buffer: 1 M sorbitol, 0.1 M EDTA, 14 mM β-mercaptoethanol, pH 7.4
Zymolase: 15 mg/ml
0.5 M EDTA, pH 7.4
1 M Tris-HCl, pH 7.4
10% SDS
5 M potassium acetate, pH 5.2
Phenol and chloroform (1:1) (phenol is saturated with 1× TE; see Safety precautions)
Phenol and chloroform and isoamyl alcohol (25:24:1) (phenol is saturated with 1× TE)
Chloroform and isoamyl alcohol (24:1)

3 *M* sodium acetate, pH 5.2

96% ethanol

70% ethanol

RNase A: 10 mg/ml (to inactivate DNases, the RNase has to be preincubated for 15 min at 100°C)

1× TE: 10 m*M* Tris-HCl, 1 m*M* EDTA, pH 8.0

Method:

1. Prepare an overnight liquid culture, and harvest the yeast cells by centrifugation for 5 min (2700 × g, room temperature).
2. Wash cells twice with 1× TE, and centrifuge as above.
3. Resuspend cells in 3.2 ml of spheroplasting buffer, and add 100 μl of zymolase stock solution.
4. Incubate for 1 h at 37°C. Check formation of spheroplasts under microscope.
5. Centrifuge for 5 min (2300 × g, room temperature) to pellet the spheroplasts.
6. Resuspend spheroplasts in 3.2 ml of 1× TE, add 0.3 ml of 0.5 *M* EDTA and 0.3 ml of 1 *M* Tris-HCl, vortex briefly, add 160 μl of 10% SDS, and vortex again.
7. Lyse the spheroplasts by incubation for 30 min at 65°C (the suspension should become highly viscous at this stage).
8. Add 1 ml of 5 *M* potassium acetate, vortex briefly, and incubate for 1 h at 0°C.
9. Centrifuge for 25 min (20,000 × g, 4°C) to pellet cell membrane fragments, precipitated proteins, polysaccharides, and potassium dodecyl sulfate.
10. Transfer the supernatant to a new tube, add an equal volume of phenol and chloroform and isoamyl alcohol (25:24:1), and shake gently for 2 min.
11. Centrifuge for 15 min (2700 × g, room temperature) to separate the phases.
12. Transfer the aqueous phase (upper phase) to a new tube.
13. Repeat Steps 10 through 12 twice.
14. Add an equal volume of chloroform and isoamyl alcohol (24:1), and shake gently for 2 min.
15. Centrifuge as above, and repeat Steps 14 and 15 once.
16. Precipitate the DNA by adding 0.03 vol of 3 *M* sodium acetate and 2.5 vol of cold 96% ethanol. Mix well, and incubate for 1 h or overnight at −20°C.
17. Centrifuge for 30 min (14,000 × g, 4°C).
18. Wash the DNA pellet twice with 70% ethanol, and centrifuge for 10 min (14,000 × g, 4°C).

19. Air-dry the DNA pellet, and dissolve it in 6 ml of 1× TE.
20. Add 60 μl of RNase A stock solution, and incubate for 1 h at 37°C.
21. Add an equal volume of phenol and chloroform (1:1), and shake gently for 2 min.
22. Centrifuge for 15 min (2700 × g, room temperature) to separate the phases.
23. Transfer the aqueous phase to a new tube, add an equal volume of chloroform and isoamyl alcohol (24:1), and shake gently for 2 min.
24. Centrifuge as above.
25. Repeat Steps 16 through 18.
26. Air-dry the DNA pellet, dissolve it in 0.5 to 1 ml of 1× TE, and store at +4 or –20°C.

Steps 19 through 25 can be omitted if the presence of RNA is not considered as a problem.

C. QUANTITATION OF DNA

Two procedures are the most widely used for estimating DNA concentration. One method is based on the spectrophotometric measurement of UV absorbance at 260 nm. A major disadvantage of this approach is that RNA, oligonucleotides, proteins, and other contaminants interfere with the measurement. Moreover, microgram amounts of DNA are needed to ensure reliable readings. The second method is based on the UV-induced fluorescence emitted by ethidium bromide-DNA complexes. Here, a sample of unknown concentration is compared with known standards, usually after electrophoretic separation. The latter method is probably less accurate for purified DNA. It is, however, recommended if only nanogram amounts of DNA are available or if DNA samples are contaminated by, e.g., proteins or RNA.

An alternative procedure which is not treated here relies on the specific binding of Hoechst dye 33258 to DNA, but not to RNA (for details see Labarca and Paigen[383] and Riley et al.[580]). In this method, different dilutions of the DNA-dye mixture are prepared in microtiter plates alongside with standards, and the measurement is carried out in a fluorescence spectrophotometer.

1. Ethidium Bromide Staining

In this procedure, DNA samples are subjected to agarose gel electrophoresis and subsequently stained with ethidium bromide. The dye intercalates into the DNA double helix, and the intensity of fluorescence induced by UV light is proportional to the amount of DNA in the lane. Comparison to a set of standards, e.g., of lambda DNA, gives an estimate of the amount of DNA in an unknown sample. In contrast to the spectrophotometric method outlined below, this technique allows, at the same time, (1) DNA quantitation, (2) estimation of the extent of contamination by RNA (which usually runs ahead), and (3) evaluation of DNA quality (i.e., the extent of degradation). Several variations of this approach exist, one of which is given below. For alternative procedures, see Sambrook et al.[606] For details on setting up and running an agarose gel, see Section III.B.

Solutions:

Ethidium bromide solution: 10 mg/ml in water (see Safety precautions)
0.8% Agarose in electrophoresis buffer
Electrophoresis buffer: TAE, TBE, or TPE (see Section III.B)
Gel-loading buffer: 30% glycerol, 1% SDS, 0.25% bromophenol blue
DNA (e.g., phage lambda): different concentrations (e.g., from 0.01 to 0.2 µg/ µl) diluted in water or 1× TE

Method:

1. Mix an appropriate amount of the DNA sample (e.g., 5 µl) with 0.2 vol of gel-loading buffer. Mix 5 µl of each of a series of lambda DNA standards (this covers a range of 50 to 1000 ng of DNA if the concentrations given above are used) with 0.2 vol of gel-loading buffer.
2. Load samples alongside with standards onto a 0.8% agarose gel.
3. Electrophorese until the bromophenol blue dye front has migrated at least 2 cm.
4. Photograph the gel on a transilluminator using short-wavelength UV irradiation (see Safety precautions). Estimate the quantity of the DNA in the sample by comparing the intensity of the fluorescence with the standard DNAs.

2. Spectrophotometry

This technique measures the total amount of nucleic acids in a sample (including DNA, RNA, oligo-, and mononucleotides). Therefore, it is only useful for pure DNA preparations of a reasonable concentration.

Solutions:

1× TE: 10 mM Tris-HCl, 1 mM EDTA, pH 8.0 or distilled water

Method:

1. Dilute an aliquot of the DNA sample in 1× TE or distilled water (usually in a ratio of 1:100; e.g., 5 µl/500 µl) in a microcuvette.
2. Determine the optical density at 260, 280, and 320 nm against a blank (1× TE or water).
3. Calculate the DNA concentration in the sample using the formula 1.0 OD_{260} = 50 µg/ml (under standard conditions, i.e., 1 cm light path). The ratio of OD_{260} to OD_{280} provides some information about the purity of the

DNA sample. Pure DNA preparations show an OD_{260} to OD_{280} ratio between 1.8 and 2.0. Contamination with, e.g., proteins results in lower values. The OD_{320} should be close to zero.

III. DNA FINGERPRINTING BASED ON HYBRIDIZATION

The "classical" DNA fingerprint strategy relies on the hybridization of restriction-digested genomic DNA with labeled multiallelic, multilocus probes. A variety of steps are included (see also Figure 3, Chapter 2, Section II.B.1):

1. Complete digestion of genomic DNA with an appropriate restriction enzyme
2. Electrophoretic separation of the restriction fragments, usually performed on agarose gels
3. Denaturation and blotting of the separated DNA fragments onto a membrane (alternatively, drying of the gel matrix on a gel dryer)
4. Hybridization of the membrane (or the dried gel) to (non)radioactively labeled multilocus probes
5. Detection of hybridizing fragments (i.e., fingerprints) by autoradiography or by nonradioactive approaches

A. RESTRICTION OF DNA
1. Choice of the Enzyme

Four parameters should be considered when choosing an appropriate restriction enzyme for DNA fingerprinting experiments: length of recognition sequence, probe sequence, costs, and sensitivity to cytosine methylation of the target sequence.

Restriction enzymes with a four-base recognition sequence are most often used for DNA fingerprinting of human, plant, and animal genomes. Though the statistical probability of a particular four-base motif to occur in DNA is as high as 1 in 256 bases, the observed fragment distribution of eukaryotic genomic DNA cut by four-base enzymes and visualized by minisatellite or simple sequence probes usually ranks between very small and more than 20 kb,[343,837] yielding highly informative banding patterns. The larger fragments are generally more polymorphic than the smaller ones. The reason for the generation of fragments that are considerably larger than expected from the cutting frequencies of the enzyme is not yet clear. It could be the consequence of a clustering of different simple repetitive motifs, together with other types of repetitive DNA in regions of low sequence heterogeneity. For the analysis of less complex genomes, six-base cutters should also be tested and may work better in some cases. This is especially true for fungal species.[453,459]

An obvious requirement is that the chosen enzyme should cut outside of the repetitive target sequences. To ensure this, no enzymes should be used which recognize sequence elements contained within one of the hybridization probes.

A very important parameter for choosing an appropriate enzyme is its methylation sensitivity. A large number of restriction enzymes will not cut if a methylated cytosine is present in its target sequence.[498,606] To some extent, this is also true for *Hin*fI,[43,521] though this particular enzyme is highly informative with many probes and species and has been used frequently for fingerprint studies. Since DNA fingerprint analyses are undertaken in order to examine variability in DNA sequence rather than in DNA modification, one should refer to an enzyme which is definitely not cytosine methylation sensitive. We found *Taq*I and *Rsa*I, but also the six-cutter *Dra*I to be a good choice for fingerprinting plant genomes.

2. Reaction Components and Conditions

Digestion should be performed under the conditions recommended by the enzyme manufacturers. A ten-time concentrated buffer solution is usually supplied with the enzyme. Five to 10 µg of DNA are usually sufficient for fingerprint analyses, depending on the probe/species combination and also on genome size. Thus, fingerprinting very large genome sizes (such as many monocotyledonous plants) will require more DNA per lane.

To ensure complete digestion, at least 5 to 8 units (U) of enzyme per microgram of genomic DNA should be used and control digestions including lambda DNA in the sample may be performed. The appearance of lambda restriction bands of the expected sizes on the gel, superimposed on a smear of plant DNA fragments, indicates that the enzyme has worked properly. If this was not the case, inclusion of 100 µg/ml BSA or 4 mM spermidine in the restriction buffer may help. Complete digestion of lambda DNA is, however, only a hint and not a definite proof for complete restriction of genomic DNA. For unknown reasons, some target sites of a given restriction enzyme, even in purified genomic DNA, are less accessible than others (also if methylation is not involved), leading to so-called "hidden partials".[506] If such a phenomenon is suspected to have occurred in a particular case, a test series with increasing enzyme concentrations (up to 30 U/µg) should be performed. A "true" pattern remains unchanged throughout the different concentrations.

The desired reaction volume is adjusted with distilled water. If the digested DNA is directly applied to a gel, the total volume of the restriction assay must be adjusted according to the space limits exerted by the gel slots. In this respect, one has to take into account that restriction enzymes are usually supplied in 50% glycerol. Since glycerol concentrations exceeding 5% may cause altered target site specificity of some enzymes, the contribution of the enzyme solution to the total reaction volume should not exceed one tenth. If the DNA sample is too dilute, it may be concentrated in a SpeedVac™ centrifuge. Digestion may also be performed in a larger volume. DNA is then precipitated after digestion

by the addition of 0.1 vol of 3 *M* sodium acetate and 2 vol of ethanol, left at
−20°C for 30 min, spun in a microfuge (30 min, 12,000 ×g), washed with 70%
ethanol, spun again, drained, and redissolved in an appropriate volume of
electrophoresis buffer. Though somewhat more time consuming, digestion of
dilute DNA samples may lead to better results for less pure samples.

3. Setting Up Reactions

In the example outlined below, a total reaction volume of 40 μl, 5 μg DNA
for each sample, and an enzyme concentration of 10 U/μl are assumed. Restriction enzymes are sensitive and expensive. Do not take the enzyme out of the
freezer until it is actually needed. Always store it on ice. When digesting many
samples with the same enzyme, calculate the total volume of enzyme solution
you need, transfer it to a reaction tube, and put the original tube back into the
freezer. Always use a fresh pipette tip for dispensing enzyme from the original
tube.

Solutions:

10× concentrated buffer (according to the manufacturer)
Enzyme solution (usually 5 to 50 U/μl)
Genomic DNA
Distilled water

Method:

1. Pipette in the following order: (1) distilled water: 32− y μl; (2) 10×
   buffer: 4 μl, (3) genomic DNA (5 μg) y μl, (4) enzyme (40 U) 4 μl. If
   the required amount of water is the same for all samples, you may also
   prepare a "master mix" of water, buffer, and enzyme (which saves
   pipette tips and time). Distribute this mix into the reaction tubes, and then
   add the DNA.
2. Mix carefully, and centrifuge in a microfuge for a few seconds to collect
   the ingredients at the bottom of the tube.
3. Incubate for at least 3 h to overnight at the incubation temperature
   recommended by the supplier (37°C for most enzymes).
4. At the end of the incubation period, the sample can either be used directly
   for electrophoresis (see below), stored at −20°C, or ethanol precipitated
   as described above.

B. AGAROSE GEL ELECTROPHORESIS

Digestion of complex plant and fungal genomes with a four- or six-base-
specific restriction enzyme may yield several million restriction fragments
which have to be separated from each other. For classical DNA fingerprinting,

this is usually done by electrophoresis on horizontal agarose gels. For PCR-based fingerprint techniques, polyacrylamide gels are also used (see Section IV.C.1). In short, agarose is molten in electrophoresis buffer to yield a clear solution, cast into a gel mold, and allowed to solidify. DNA restriction fragments are applied to the gel, and a constant electric field is imposed. Under neutral or slightly alkaline conditions, DNA migrates toward the anode. Since the agarose matrix is acting as a molecular sieve with pore sizes depending on the agarose concentration, restriction fragments up to about 20 kb are separated according to size. After finishing electrophoresis, DNA in the gels is stained with ethidium bromide, photographed, and further processed as described in the forthcoming sections.

Since large restriction fragments are frequently more informative for DNA fingerprinting purposes than small ones, relatively low gel concentrations are usually applied (between 0.7 and 1.2%; see also Table 4 and Section IV.C). For example, a 0.7% agarose gel efficiently separates DNA fragments between 0.8 and 10 kb.[606] Three buffer systems are generally used, i.e., TAE, TBE, and TPE[606] (see below for buffer compositions).

The resolving powers of all three buffer systems are almost identical. Though TAE is the most commonly used buffer (mainly for historical reasons), it is probably the least advisable. TAE has a low buffering capacity as compared to TBE and TPE, and electrophoretic runs using TAE should not exceed 24 h without buffer recirculation or exchange. The higher buffering capacity of TBE allows to use it in a 0.5× concentration.[606] According to our own observations, nice fingerprint patterns and least "smiling" of bands are obtained using TPE.

As mentioned above, 5 to 10 μg of DNA per lane are usually sufficient for a fingerprint experiment. Applying too much DNA may actually result in inferior banding patterns and thus should be avoided. To ensure that approximately the same amount of DNA is present in all lanes to be compared, we recommend that a control electrophoresis with a small part of the digested samples is performed on a minigel. The DNA amounts intended for the analytical gel(s) can then be adjusted accordingly.

At least two or three lanes should be loaded with molecular weight markers, which are mixtures of restriction fragments of known size (0.2 to 2 μg, depending on gel size). After electrophoresis, these markers are visualized by staining, and their positions documented by photography. Some fingerprint probes (e.g., the M13 repeat) cross-hybridize with phage lambda DNA, which allows visualization of phage lambda-derived marker fragments on the autoradiogram. An alternative strategy is to rehybridize a fingerprint gel or blot to labeled marker DNA. In this case, very small amounts of marker (e.g., 100 pg per lane) can also be included in the sample lanes. Rehybridization to a marker-complementary probe then results in reliable in-lane markers. Although this strategy requires an additional hybridization step, it circumvents the often encountered problem of between-lane variation in gels. For nonradioactive detection methods, labeled markers (e.g., with digoxigenin) are commercially available.

Solutions:

Electrophoresis buffers (one of the three following):

1× TAE buffer: 40 mM Tris-acetate, 1 mM EDTA, pH 8.0 (adjust pH with glacial acetic acid)

0.5× TBE buffer: 45 mM Tris-borate, 1 mM EDTA, pH 8.0 (adjust pH with boric acid)

1× TPE buffer: 90 mM Tris-phosphate, 2 mM EDTA, pH 8.0 (adjust pH with 85% phosphoric acid)

Note: prepare electrophoresis buffers as 10 or 50× concentrated stock solutions and dilute prior to use.

Loading buffer: 0.25% bromophenol blue, 0.25% xylene cyanol, 30% glycerol in electrophoresis buffer or water

Molecular weight marker: DNA restriction fragments of known size (commercially available)

Agarose: e.g., 0.7% in electrophoresis buffer

Staining solution: 1 μg/ml ethidium bromide in electrophoresis buffer or water (see Safety precautions)

Method:

1. Suspend agarose (electrophoresis grade, e.g., Seakem LE) at the desired concentration in an appropriate amount of electrophoresis buffer in a bottle or flask (e.g., 2.8 g of agarose per 400 ml of electrophoresis buffer yields a 0.7% gel). Flasks should be covered, and bottles should be loosely capped. Do not use aluminum foil!

2. Boil the suspension in a microwave oven for 2 to 4 × 2 min. Swirl the bottle in between. Continue until the agarose has dissolved. A good check for complete dissolution is the complete disappearance of lens-like particles.

3. Let the molten agarose cool to 60°C, stirring helps to prevent uneven cooling. In the meantime, seal the open edges of the plastic tray supplied with the electrophoresis apparatus using tape. Insert a slot-forming comb. Band resolution is, to some extent, dependent on the shape of the teeth of the comb; sharp teeth yield sharp bands, but also allow less volume to be applied. Check that the teeth are not too close to the bottom of the gel mold. Fine holes in the bottom of a slot might allow your samples to escape in an undesired direction.

4. Make sure that the gel mold is in a horizontal position. For some electrophoresis apparatuses, the gel is best poured with the mold already in place. Carefully pour the agarose into the gel mold, remove small air

bubbles with a pipette, and let the agarose solidify [usually 1 h at room temperature or faster in a cold room (4°C) or by turning on the water cooler system of the electrophoretic apparatus].

5. When the agarose is solid, carefully remove the comb and the tape, and insert the gel mold into an electrophoresis apparatus filled with buffer. Electrophoresis runs best if there is not too much buffer on top (about 5 mm). Remove air bubbles from the slots. Connect the apparatus to a power supply, and check whether it is working correctly (before applying the samples!).

6. Add 0.2 vol of loading buffer to the DNA samples, mix, and centrifuge for a few seconds in a microfuge in order to collect the samples at the bottom of the tubes. The loading buffer adds color and provides a higher density to the samples, thus allowing their convenient application to the slots. Moreover, the dyes are moving toward the anode when voltage is applied (bromophenol blue runs about twice as fast as xylene cyanol), and this gives you an idea how far the electrophoresis has proceeded. Bromophenol blue usually runs close to the front (at the same rate as linear DNA of 300 bp in 0.5× TBE).[606]

7. Slowly load the samples into the submerged slots. Alternatively, samples may also be loaded into the dry slots (in this case, no loading buffer is needed). Buffer is then cautiously poured to the gel surface, and the gel is run for 20 min before more buffer is added.

8. Turn on power supply, and start the electrophoresis. Running conditions are usually 1 to 2 V/cm (i.e., distance between the electrodes) for 24 to 48 h. Longer runs give better resolution of large fragments. To minimize diffusion, the gels should be preferably run in the cold (4°C) or with water cooling.

9. After the run is completed, remove the gel from the apparatus, and stain for 15 to 60 min (depending on the gel thickness) in a tray with staining solution. Then rinse the gel briefly in water, place it on a UV transilluminator, and photograph under UV light (302 nm). Use an orange or red filter, which only allows the fluorescence of the ethidium bromide-DNA complex to enter the camera. For photographic documentation, place a transparent or fluorescent ruler alongside the gel to align marker sizes in the gel with fragment sizes in forthcoming autoradiograms.

C. GEL DRYING

For "Southern type" hybridizations, gel electrophoresis is usually followed by the transfer of the DNA fragments to a membrane support (Section III.D). With some kinds of probes (e.g., short, radiolabeled simple sequence oligonucleotides), however, hybridization can also be successfully carried out in a dried agarose gel matrix.[8,471,609,690] The gels are dried on a commercially available gel dryer. Once dried, the agarose is ultrathin and convenient to store. During subsequent hybridization and washing steps, dried agarose gels rehydrate only partially and are surprisingly stable and easy to handle.

For oligonucleotide hybridization, the drying method has several advantages over the conventional blotting techniques: it is faster, prehybridization steps can be omitted, a "100% transfer efficiency" (i.e., no transfer at all) is obtained, and signal strength is about five times higher than using blot hybridization.[471,690] Several authors have claimed that in-gel hybridization is also convenient with nonradioactive probes.[548,819,838] While this view is not supported by our own experience, at least in case of digoxigenated probes and colorigenic or chemiluminescent signal detection,[49] we nevertheless favor and recommend in-gel hybridization for DNA fingerprinting with radiolabeled simple sequence oligonucleotides.

Solutions:

Denaturation buffer: 0.5 M NaOH, 0.15 M NaCl
Neutralization buffer: 0.5 M Tris-HCl, 0.15 M NaCl, pH 7.0

Method:

1. After photographic documentation, transfer the gel to a tray filled with several volumes of denaturation buffer. Incubate for 30 to 45 min at room temperature under agitation.
2. Decant denaturation buffer (which can be reused). Rinse the gel twice with distilled water to remove excess NaOH, and incubate in several volumes of neutralization buffer for 1 h. The neutralization buffer can be reused several times, provided that its pH is controlled and readjusted.
3. Transfer the gel on two sheets of filter paper (e.g., Whatman 3MM) soaked with neutralization buffer, and cover the gel with plastic wrap. Place the gel onto a commercial gel dryer. The plastic wrap should face the rubber gasket. Insert a washing bottle and a cooling device between the gel dryer and the vacuum pump.
4. Apply vacuum (without heat) for 1 to several hours until the gel appears flat. Then turn on the heater to 60°C, and dry for 1 h more. Turn off vacuum, and check whether the filter paper is completely dry. If it is still damp, continue for another 30 min. The total duration of the process depends on the thickness of the original gel and the power of the vacuum pump.
5. Dried gels should be handled with care to avoid breakage and may be stored on their filter support at room temperature for years. For up to several weeks, partially rehydrated gels may also be stored in high salt solutions (6× SSC, e.g., between two subsequent hybridization experiments). Longer storage under aqueous conditions, however, results in successive rehydration and destruction of the gel. Regenerated gels (see Section III.G.3) should therefore be dried when not in use.

Note: It is also possible to dry the gel immediately after staining. Denaturation and neutralization of DNA are then performed within the dried gel. In this case, the incubation times for both denaturation and neutralization step can be reduced to 15 min each.

D. SOUTHERN TRANSFER

Whereas in-gel hybridization has several advantages when radiolabeled oligonucleotides are used as probes,[690] Southern blot hybridization seems preferable to us for larger probes (e.g., nick-translated or random-primed minisatellite probes) as well as for nonradioactive approaches. To generate a Southern blot, electrophoretically separated DNA samples are denatured (i.e., made single stranded) within the gel and transferred ("blotted") onto a membrane where they are bound to the surface. The original technique described by Southern[652] makes use of a high salt buffer, which transfers the DNA to a nitrocellulose filter by capillary forces. In the course of almost 20 years since its initial description, several variations on the theme have been developed, including alternative transfer buffers, driving forces (e.g., electrophoretic blotting, vacuum blotting), and types of membranes (e.g., diazobenzyloxymethyl paper, nylon membranes, charged membranes, and hydrophobic membranes). Nylon membranes are now generally used instead of nitrocellulose, mainly because of their high physical resistance. However, not all types of nylon membranes may be suitable for DNA fingerprinting. In a comparative study, Stacy and Jakobsen[658] found Hybond Nfp™ (a membrane recently developed by Amersham International for fingerprint purposes) and GeneScreen™ (NEN) to be optimally suited for DNA fingerprinting with a ^{32}P-labeled human minisatellite probe. Among the different membranes we tested for DNA fingerprinting with digoxigenated oligonucleotides [i.e., neutral nylon (Nytran™, Schleicher & Schuell; Hybond N™ and Hybond Nfp™, Amersham International), positively charged nylon (Boehringer Mannheim Biochemicals), nitrocellulose (BA-85, Schleicher & Schuell) and hydrophobic polyvinylidene difluoride (PVDF) membranes (Millipore)] the two Hybond membranes performed best. The choice of membrane is especially crucial for chemiluminescent signal detection (see Section III.H.3), since, e.g., PVDF and nitrocellulose membranes are known to quench light and, hence, also luminescence.[37]

Though this variant is somewhat slow, we still find capillary transfer in high salt buffer preferable to other methods in terms of simplicity, reliability, and efficiency of transfer. After blotting (usually overnight), the DNA is irreversibly fixed to the membrane either by heat treatment (2 h at 80°C, vacuum is necessary for nitrocellulose, but not for nylon membranes) or UV cross-linking (for nylon membranes only). Since the intensity and duration of UV treatment have to be carefully optimized and controlled in order to obtain reproducible results,[37] we prefer heat treatment.

Solutions:

Denaturation buffer: 0.5 *M* NaOH, 1.5 *M* NaCl
Neutralization buffer: 1.0 *M* Tris-HCl, 3.0 *M* NaCl, pH 7.0

Transfer buffer: 20× SSC: 3.0 M NaCl, 0.3 M sodium citrate, pH 7.0
6× SSC, 5× SSC, 2× SSC: prepare by diluting 20× SSC stock solution with
distilled water

Method (see Figure 12):

1. After photographic documentation, transfer the gel to a tray filled with
    several volumes of denaturation buffer. Incubate for 30 to 45 min at room
    temperature under agitation.
2. Decant denaturation buffer (which can be reused). Rinse the gel twice
    with distilled water to remove excess NaOH, and incubate in neutraliza-
    tion buffer for 1 h. Neutralization buffer can be reused several times,
    provided that its pH is controlled and readjusted.
3. While the gel is being neutralized, fill another tray with transfer buffer
    (20× SSC), and cover it with a glass plate so that it forms a bridge. Cut
    a sheet of filter paper (e.g., Whatman 3MM) to the same width, but the
    double length of the gel. Wet the filter paper with 20× SSC, and put it
    onto the glass plate so that its free ends are hanging down into the 20×
    SSC solution. Cut a nylon membrane to a size 5 mm wider and longer
    than the gel, wet it by floating on distilled water, and incubate it for 10
    min in 20× SSC. Do not touch the membrane with your fingers; use either
    gloves or forceps.
4. With the help of two sharp-edged plexiglass plates, take the gel out of
    the neutralization buffer, and put it in an inverse orientation (slot
    openings pointing downward) onto the filter on the glass plate. Re-
    move air bubbles between filter and gel by gently rolling a glass rod
    or a pipette over the gel. Surrounding the gel with either plastic wrap
    or used X-ray films prevents the direct flow of transfer buffer from
    the reservoir to the stack of paper towels to be placed on top of the
    device.
5. Place the wet nylon membrane centrally onto the gel. Remove air bubbles
    between gel and membrane (see above).
6. Wet two more pieces of filter paper (cut to the same size as the gel)
    in 20× SSC, and put them on top of the membrane. Remove air bub-
    bles.
7. Place a stack of paper towels on top, followed by a glass plate, and a
    weight of about 500 g. Too heavy weights cause early compression of the
    gel and prevent transfer.
8. Let transfer proceed overnight.
9. Remove the wet paper towels and the filter papers. Label the positions
    of slots and gel edges on the membrane using a pencil or waterproof ink.
    Alternatively, labeling can also be done before blotting.
10. Peel the membrane from the gel, rinse it for a few minutes in 6× SSC to
    remove residual agarose (which may cause hybridization background),
    and let it air dry on a sheet of filter paper.

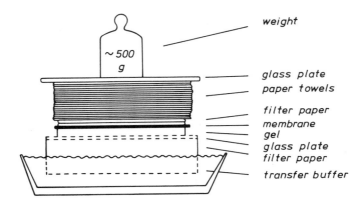

FIGURE 12. Schematic drawing of a typical Southern blot set up. See text for details.

11. Place the membrane between two sheets of filter paper, and bake it for 2 h at 80°C in an oven. DNA is irreversibly fixed by this step.

12. Store membranes between sheets of filter paper or in aluminum foil at room temperature under dry conditions.

Variation of Steps 3 through 6 (time saving):

3a. Decant neutralization buffer. Soak five pieces of filter paper (cut to the same size as the gel) in transfer buffer, and put these on top of the gel which is resting on a glass plate.

4a. Turn the whole stack upside down, and place it on the benchtop.

5a. Place the wet membrane onto the gel. Remove air bubbles.

6a. Place another stack of five filter papers, soaked in transfer buffer, on top of the membrane. Remove air bubbles.

For increasing the transfer efficiency of large DNA fragments (>10 kb), the DNA has to be partially hydrolyzed before transfer. This is either done by partial depurination in 0.25 M HCl (10 to 20 min, depending on gel thickness) or by short UV treatment (1 to 3 min on a transilluminator). We have omitted this step since it is usually not necessary for DNA fingerprint analyses and may also destroy the DNA if applied inappropriately.

Inverting the gel has two advantages: (1) the DNA side of the blot will have the same orientation as is shown on the photograph of the gel and (2) the lower surface of the gel (which is then in contact with the filter) is much smoother than the upper surface, allowing a more intimate contact between both and avoiding hybridization background (the surface structure of the gel is sometimes visible on autoradiograms).

E. GENERATION OF PROBES

There are several possibilities to obtain suitable probes for DNA fingerprinting. The easiest way is to purchase probes from commercial companies.

Simple sequence oligonucleotide probes may be synthesized with an automated DNA synthesizer (if available). Minisatellite probes can be cloned and multiplied in *Escherichia coli*. The most cumbersome way is to identify and isolate new probes from genomic libraries.

1. Commercially Available Probes

A variety of commonly used DNA fingerprint probes is commercially available, including the wild-type M13 phage,[726] the human minisatellites 33.15 and 33.6,[315] and a set of oligonucleotides complementary to simple repetitive sequences.[8] Some of these probes are subject to patent applications and are therefore only available through certain companies. The reader should refer to the manufacturers for commercial inquiries. For example, the "Jeffreys" probes (e.g., 33.5, 33.6, 33.15) are available through Cellmark Diagnostics, while the "Epplen" probes [e.g., $(CAC)_5$, $(GACA)_4$, $(GTG)_5$] are manufactured by Fresenius AG. Both companies also offer a variety of nonradioactively labeled probes.

An alternative worth considering is that many companies offer the synthesis of customer-designed oligonucleotides at a relatively modest price. From most manufacturers, these oligonucleotides are also available in a biotinylated or digoxigenated form. The costs per base vary between $2 and $10, depending on the company, length and amount of the oligonucleotide, and on the extent of purification.

2. Chemical Synthesis of Oligonucleotides

A variety of methods are available to synthesize DNA in the laboratory. The phosphoramidite method of oligonucleotide synthesis is the chemistry of choice because of inherently high coupling efficiencies and stability of the starting materials. Recently, this process has been simplified by the development of automated DNA synthesizers (e.g., Applied Biosystems, Beckman Instruments, Cruachem, Eppendorf, Millipore), which are capable of producing large quantities of any desired oligonucleotide. However, DNA synthesizers are relatively expensive (from $22,000 to $575,000) and therefore advisable only in situations where several laboratories can share the costs.

Since detailed instruction manuals are usually supplied by the manufacturers of DNA synthesizers, no protocol is given here. However, oligonucleotides may be purchased at different states of purity (crude preparations are generally cheaper). After cleaving it from the solid support matrix with ammonium hydroxide, a newly synthesized oligonucleotide still carries the protecting groups which have to be removed. A deprotection protocol as described by Cruachem[132] is given below. The final product can be used as a hybridization probe or PCR primer.

Solutions:

Concentrated ammonium hydroxide
Triethylamine

1 M MgCl$_2$
96% ethanol

Method:

1. Mix 1.5 ml of concentrated ammonium hydroxide and 50 µl of triethylamine in a 2-ml screw-cap microfuge tube.
2. Transfer the oligonucleotide from the column into the microfuge tube, and close the cap tightly.
3. Transfer the tube into a water bath or heating block preheated to 80°C, and incubate for 30 min.
4. Place the tube on ice.
5. Dry the oligonucleotide under vacuum, and resuspend it in 50 to 200 µl of water.
6. Add 1 M MgCl$_2$ to a final concentration of 10 mM. Mix gently, and precipitate the oligonucleotide by adding 5 vol of 96% ethanol. Keep at –20 or –70°C for 30 min to overnight.
7. Spin for 10 min in a microfuge (14,000 × g, room temperature). Wash the pellet with 96% ethanol. Centrifuge as above, dry the pellet, and dissolve it in 250 µl of water.
8. Determine the oligonucleotide concentration spectrophotometrically (see above; Section II.C); 1.0 OD$_{260}$ corresponds to approximately 20 µg/ml under standard conditions.

3. Isolation of Fingerprint Probes from Genomic Libraries

The isolation of genomic fingerprint probes requires the construction of a genomic DNA library. This is basically done by isolating DNA from the organism of interest, fragmenting the DNA by shearing or endonuclease digestion, ligating the individual fragments into cloning vectors (e.g., plasmids, bacteriophages, or cosmids), and transforming suitable host cells (usually *E. coli* strains) with these constructs. In order to screen for a clone of interest, the library (i.e., thousands of individual colonies carrying different fragments) is plated onto agar plates, blotted onto membranes, and hybridized to a suitable (non)radioactively labeled probe. If clones harboring potential DNA fingerprint probes are looked for, minisatellites from other organisms or any kind of short tandem repeat may be used as a screening probe. "Positive" clones (i.e., clones yielding a hybridization signal) are then isolated, and the inserted DNA fragment is excised (or amplified by PCR) and hybridized to restriction-digested genomic DNA of the organism of origin. A fingerprint pattern will occur if the experiment has been successful.

Since a considerable collection of useful DNA fingerprinting probes is already available and the description of molecular cloning is beyond the scope of this book, no cloning protocols are given here. For comprehensive treatments of this topic, the reader should refer to the commonly used molecular biology manuals (e.g., Berger and Kimmel,[44] Brown,[75] Davis et al.[140] and Sambrook et al.[606]).

4. Isolation and Processing of the M13 Probe

Double-stranded M13 DNA is one of the major sources for probes used in DNA fingerprinting. The exact genetic constitution (e.g., wild type, M13mp8, M13mp9, M13mp18, or M13mp19) is not relevant for this purpose. Double-stranded M13 is also termed replicating form (RF), in contrast to the infectious single-stranded form. M13 DNA is available from many manufacturers and can be used immediately upon purchase. However, it can also be used to produce more M13 DNA by transfection into *E. coli* and replication in the bacterium, eventually resulting in a large and cheap supply of probe. Since several handbooks give full details on transformation and transfection,[75,606] only some simple protocols and strategies are outlined here.

Competent cells (i.e., cells able to take up foreign DNA) of a suitable *E. coli* host strain (e.g., JM101, JM107, and JM109) are needed for the transfection. Cells can either be purchased in a competent state or, alternatively, be rendered competent by following one of several protocols (one is given below). Competent cells are grown on agar plates after transfection with M13 DNA. A resulting plaque is then used for producing large quantities of infected *E. coli* cells. From these, M13 DNA can be isolated, while the supernatant (containing single-stranded M13 DNA) may be kept as an M13 stock for future needs. For hybridization purposes, a minor part of the M13 genome is usually isolated, which consists of the tandem repeats (located in the protein III coding region; see Figure 5, Chapter 2).

The procedures described in this section can also be applied for plasmid transformation into *E. coli* and isolation of plasmid DNA. In that case, a selection marker (antibiotic) is needed to select for transformed bacteria, i.e., those containing a plasmid.

a. Preparation of Competent Cells

Solutions and material:

LB medium (autoclaved):[606] 10 g/l bacto-tryptone, 5 g/l bacto-yeast extract, 10 g/l NaCl, pH 7.0
Suitable *E. coli* strain
0.1 M CaCl$_2$ (filter sterilized, not older than 14 days)
LB plate: LB medium containing 15 g/l agar, freshly prepared

Method:

1. Prepare a pure culture of the *E. coli* strain on a plate by streaking out cells or prepare an overnight liquid culture (in LB medium).

2. Dilute the overnight culture 1:100 (e.g., 100 µl in 10 ml LB broth) or pick cells from a single colony on a plate, and transfer these into 10 ml of LB medium. For one transfection 1 ml of culture (after dilution) is sufficient.

3. Grow the cells at 37°C with vigorous shaking until the correct cell density and growth phase is reached. Monitor growth by checking the OD_{600} (optical density at 600 nm in a spectrophotometer). This absorption value should be approximately 0.3 and is attained 1.5 to 3 h after growth initiation, either from an overnight culture or from a colony on a plate, respectively.

4. Transfer the cells (e.g., 1 ml) into a cold tube or microfuge tube, and cool for 5 min.

5. Centrifuge the cells for 10 min at 4°C or for 30 s at room temperature in a microfuge (not longer!), and decant the supernatant carefully.

6. Resuspend the pellet in ice-cold 0.1 M $CaCl_2$ in 0.2 to 0.5× the original volume (Step 4), and store on ice for 30 min.

7. Repeat Step 5.

8. Resuspend the cells in ice-cold 0.1 M $CaCl_2$ in 0.04 to 0.1× the original volume (Step 4). These cells are kept on ice and will stay competent for 30 min to 24 h. Competent cells can also be stored at –80°C as a glycerol stock (10% glycerol).

b. *Transfection of Competent Cells with M13 Phage*

Solutions and material:

LB medium and plates, see previous protocol
Top agar: LB medium containing 7 g agar per liter
Competent *E. coli* cells
E. coli stock
M13 DNA, double-stranded

Method:

1. Prepare an overnight liquid culture of *E. coli* (from stock) in LB medium at 37°C with shaking.

2. Prepare sterile tubes with 3 ml melted top agar, and store at 47°C.

3. Add 10 to 100 pg M13 DNA (in 1 to 25 µl) to 100 µl competent cells, and keep on ice for 30 min. Then apply a heat shock of 37°C for 5 min or of 42°C for 1.5 min. The *E. coli* cells will take up the foreign M13 DNA.

4. Transfer back on ice, and add 400 µl of LB medium.

5. Transfer the transfected cells (in different quantities: 1 to 100 µl) to tubes with melted top agar, and add 200 µl of overnight culture cells. Mix the contents, and pour the melted agar immediately onto an LB agar plate.

6. Let the top agar solidify, invert the plates after 5 min, and incubate overnight at 37°C. Plaques will be visible as clear spots (no *E. coli* cell growth) on a background of bacterial growth.

Note: When *E. coli* is instead transformed with plasmids, no top agar with overnight culture cells is needed; the transformed cells are streaked out directly on agar plates.

c. Amplification of M13 Phage

Solutions and material:

LB medium, see previous protocol
Overnight culture of *E. coli*
E. coli transfected with M13

Method:

1. Mix 100 µl of an overnight culture of *E. coli* cells into 2 ml of LB medium in a sterile culture tube.
2. Touch the surface of a single plaque (resulting from the transfection, see above) with a sterile toothpick or applicator stick, and add the toothpick to the *E. coli* cells or wash the stick in the medium. Incubate the infected cells for 4 to 5 h at 37°C under vigorous shaking.
3. Transfer 1 to 2 ml of the culture to a sterile Eppendorf tube, and centrifuge for 5 min (12,000 × g, room temperature). The supernatant contains the single-stranded M13 DNA and can be kept as a phage stock at 4°C for long periods. This stock can be used for further amplification of M13 DNA (addition of 5 to 10 µl of phage stock to an overnight culture of *E. coli*).

d. M13 DNA Isolation

Here, we describe a simple small-scale preparation. Several preparations can be carried out in parallel to yield sufficient amounts of DNA (500 ng/ml) for numerous labeling procedures. For large-scale preparations see Sambrook et al.[606] The protocol given here is a modification of a technique for plasmid isolation. It may therefore also be used for isolating minisatellite or other probes cloned in plasmids or in M13.

Solutions and material:

Solution I: 50 mM glucose, 10 mM EDTA, 25 mM Tris HCl, pH 8.0; autoclave, and store at 4°C
Solution II: 0.2 M NaOH (freshly diluted from 10 M stock), 1% SDS; this solution should be prepared freshly
Solution III: 60 ml 5 M potassium acetate, 11.5 ml glacial acetic acid, 28.5 ml H$_2$O; this solution should be ice cold
Phenol and chloroform (1:1 ratio; see Safety precautions)
RNase: 10 mg/ml
TE buffer: 10 mM Tris-HCl, 1 mM EDTA, pH 8.0

E. coli cells infected with M13
96% ethanol

Method:

1.   Grow infected *E. coli* cells, transfer into an Eppendorf tube, and pellet the cells by centrifuging as described above.
2.   Resuspend the pellet in 100 µl of ice-cold solution I by vortexing. Tubes should be kept on ice.
3.   Add 200 µl of solution II, and mix (do not vortex) tubes by inverting. Keep on ice.
4.   Add 150 µl of solution III, and vortex. Keep on ice for 5 min.
5.   Centrifuge the tubes for 5 min (12,000 × g, preferably at 4°C), and transfer the supernatant to fresh tubes.
6.   Add an equal volume of phenol and chloroform, and vortex. Centrifuge the tubes for 2 min (12,000 × g, room temperature).
7.   Transfer the supernatant to fresh tubes, add 2 vol of ice-cold ethanol, and vortex. Precipitated DNA may be collected after 2 min at room temperature by centrifugation (5 min, 12,000 × g), but it can also be stored overnight at −20°C.
8.   Wash the pellets with 0.5 ml ice-cold 70% ethanol, and centrifuge (5 min, 12,000 × g).
9.   Decant the supernatant, and use a Pasteur pipette to remove as much ethanol as possible. Dry the pellets at room temperature or at 37°C for 10 min.
10. Dissolve the DNA in 25 µl of TE with RNase added (10 µg/ml).

Note: To counteract later problems when digesting this DNA, two precautions can be taken.

1.   Add phenol and chloroform in two steps instead of as a mixture. First add phenol (0.5 vol), vortex, and leave for 2 to 5 min, then add chloroform (0.5 vol), vortex, and centrifuge the tubes for 2 min (12,000 × g, room temperature).
2.   Include an extra step (6a) after the phenol-chloroform step (6): transfer supernatant to fresh tubes, and add an equal volume of chloroform and isoamylalcohol (24:1). Vortex tubes, and centrifuge for 2 min (12,000 × g, room temperature).

e. *Isolation of a 780-bp M13 Fragment*

For hybridization-based fingerprinting, either intact, full-length M13 DNA or the repeat-containing sequence of the M13 phage may be used as a probe. We shall describe here how a 780-bp fragment harboring this sequence can be isolated.

Solutions and material:

Restriction enzymes *Cla*I and *Bsm*I
Restriction enzyme buffers (see Section III.A)
Agarose minigel: 0.8% agarose in TBE buffer (see Section III.B)
Ethidium bromide staining solution (see Section III.B)
TBE buffer (TAE if the Geneclean® kit is used)
Geneclean® kit (BIO 101, Inc.) (optional)

Method:

1. Digest M13 DNA with *Cla*I and *Bsm*I, 5U/μg DNA, in the appropriate buffers (preferably the *Cla*I buffer). First, the *Bsm*I is added, and the DNA is digested at 65°C for 4 h or overnight. Then the sample is allowed to cool, and *Cla*I is added for further digestion at 37°C.
2. Separate the digested M13 DNA on a 0.8% agarose gel together with a molecular size standard, and stain with ethidium bromide. Three fragments should be visible. Cut out the 780-bp fragment with a razor blade.
3. Isolate the DNA fragment from the agarose by electroelution[606] or use, e.g., the Geneclean® kit.
4. Check the concentration of the 780-bp fragment on a 0.8% agarose gel. The fragment is now ready for use and can be labeled by, e.g., random priming (Section III.F.3).

A 780-bp fragment, isolated as above, has been used for most of the published M13 fingerprinting studies. However, a 282-bp fragment, containing only one repeat sequence, is also useful.[726] This smaller fragment is isolated by digestion with *Cla*I and *Hae*III.

F. LABELING OF PROBES

To detect a hybridization event, the probe which binds specifically to its target DNA sequence has to be labeled. While this is traditionally done by introducing a radioactive isotope (commonly ^{32}P) into the probe sequence and detecting hybridization signals by autoradiography, substantial progress has been made in recent years concerning the development of nonradioactive labeling and detection procedures (see Section III.H.3).

A number of techniques based on enzymatic reactions have been developed to introduce various kinds of labels into a probe (for an overview see Sambrook et al.[606]). Among these, nick translation[579] and random priming[191] are the most useful for the generation of uniformly labeled DNA probes, e.g., from cloned minisatellites. There are also several alternative methods especially designed for labeling M13 for fingerprint purposes.[40,682,764] Alternatively, RNA probes may be prepared by transcription of cloned minisatellite templates in the presence of labeled nucleotides.[98] 5′ end labeling with bacteriophage T4 polynucleotide kinase and a γ [^{32}P] nucleotide as a phosphate donor is an appropriate technique for generating radioactive oligonucleotide probes. Kits are commercially available for all these methods.

1. End Labeling

For radioactive analysis, oligonucleotide probes are end labeled with polynucleotide kinase using γ [^{32}P] ATP as a phosphate donor[8] and purified by ion exchange chromatography on Whatman DE-52. For nonradioactive analysis based on digoxigenin or biotin, end-labeled simple sequence oligonucleotides may either be purchased (e.g., from Fresenius AG) or prepared with the help of a commercial kit (e.g., from Boehringer Mannheim Biochemicals).

Simple sequence oligonucleotides are usually supplied in a dried or lyophilized form by the manufacturer. They are dissolved in distilled water or in TE buffer. For radioactive labeling, we recommend adjusting the concentration to about 3.3 pmol/μl. Oligonucleotides of this concentration can be used directly for the labeling reaction. For storage of probes which are often used, it is preferable to pipette aliquots of the oligonucleotide solution into reaction tubes (3 μl = 10 pmol per tube), dry them (e.g., in a SpeedVac™ centrifuge), and store them at 4°C. Frequent freezing/thawing circles which could damage the oligonucleotide are thus circumvented.

Solutions and materials:

Oligonucleotide probe: 10 pmol (dried or dissolved in 3.5 μl distilled water or TE buffer)
10× kinase buffer: 670 mM Tris-HCl, pH 8.0, 100 mM MgCl$_2$, 100 mM dithiothreitol. This buffer is usually supplied by the manufacturer of the polynucleotide kinase.
γ[^{32}P] ATP: stabilized aqueous solution (5000 Ci/mmol, 10 mCi/ml), e.g., Amersham PB.10218 (see Safety precautions)
T4 polynucleotide kinase
0.5 M EDTA, pH 8.0
TE buffer: 10 mM Tris-HCl, 1 mM EDTA, pH 8.0
10× TE buffer: 100 mM Tris-HCl, 10 mM EDTA, pH 8.0
0.2 M NaCl in TE buffer
0.5 M NaCl in TE buffer
Whatman DE-52 cellulose: To prepare DE-52 equilibrated in TE buffer, suspend dry DE-52 in 10× TE buffer, and let it swell overnight. Let it settle, and change buffer. Repeat several times until the pH of the suspension reaches 8 to 8.5. Then equilibrate the DE-52 material several times with 1× TE buffer, and store it in this buffer at 4°C until use.

Method (see Safety precautions):

1. Dissolve 10 pmol of oligonucleotide in 3.5 μl of distilled water in a reaction tube. Add 1 μl of 10× kinase buffer, followed by 5 μl (50 μCi)

of γ [^{32}P] ATP. Draw up radioactive solutions slowly to avoid contamination of the pipette, and discard the tip into the radioactive waste before releasing the pressure.

2. Close the tube, mix cautiously, and spin for a few seconds in a microfuge.

3. Add 2 to 4 U of polynucleotide kinase (usually 0.5 µl, depending on the manufacturer) directly to the mixture, and incubate for 30 min on ice (some suppliers also recommend incubation at 37°C).

4. In the meantime, prepare DE-52 columns. Use either Pasteur pipettes or 1-ml pipette tips plugged with glass wool or, more conveniently, disposable plastic columns (e.g., Econocolumns™ or Poly-Prep, Bio-Rad Laboratories). Fill the columns with 0.2 to 0.4 ml of DE-52 equilibrated in TE buffer. Wash with several volumes of TE buffer, close the outlet (e.g., with a plastic stopper), and store until use.

5. Stop labeling reaction by adding 1 µl of 0.5 *M* EDTA, and also add 90 µl of TE buffer to aid subsequent handling.

6. Remove the plastic stopper from the column, put a small Erlenmeyer flask below the outlet, and apply the labeled oligonucleotide solution to the column. Wash with 4 ml of TE buffer to elute unincorporated γ [^{32}P] ATP. When the washing solution has reached the top of the DE-52 cellulose, perform a second wash with 4 ml of 0.2 *M* NaCl in TE buffer.

7. Replace the Erlenmeyer flask by a 50-ml Falcon™ tube, and discard the flow-through washing solutions into the radioactive waste. Elute the oligonucleotide from the DE-52 cellulose with 2× 0.5 ml of 0.5 *M* NaCl in TE buffer. Discard the column into the radioactive waste.

8. Store the labeled oligonucleotide at –20°C until use.

2. Nick Translation

Labeled deoxynucleotides may be incorporated into double-stranded DNA by nick translation.[579] Low concentrations of DNase I are used to introduce "nicks" (i.e., single-strand breaks) within the DNA fragment to be labeled. DNA polymerase I from *E. coli* recognizes such nicks. By virtue of its combined 5′ to 3′ exonuclease and polymerase activity, this enzyme replaces the preexisting deoxynucleotides in the 3′ direction, resulting in a shift of the nick. Inclusion of highly radioactive deoxynucleotides (e.g., α [^{32}P] dCTP) in the reaction results in the generation of efficiently labeled double-stranded DNA molecules. Usually, more than 60% of the labeled deoxynucleotides are incorporated during nick translation, and DNA probes with a high specific activity (10^8 cpm/µg) are generated this way.

Nick translation kits are commercially available from several companies. It is generally convenient to follow the user's manual included in the kit. Using a kit is easier, but also more expensive per single labeling reaction. The protocol given below may be followed if no kit is available.

Solutions:

DNA sample (0.5 to 1.0 µg) dissolved in 1× TE or water

10× nick translation buffer: 0.5 M Tris-HCl, pH 7.5, 0.1 M MgSO$_4$, 0.1 M dithiothreitol, 500 µg/ml BSA (fraction V)

Pancreatic DNase I solution: 10 ng/ml in 1× nick translation buffer and 50% glycerol

DNA polymerase I: 5 U/µl

Unlabeled dNTPs: 1 mM dCTP, 1 mM dGTP, 1 mM dTTP (if dATP is used as a labeled nucleotide)

α [^{32}P] dATP: 10 µCi/µl (specific activity >3000 Ci/mmol; any labeled nucleotide can be used if the unlabeled dNTP mix is made up appropriately)

Distilled water

Stop solution: 1% SDS, 10 mM EDTA, 0.25% xylene cyanol, 0.25% bromophenol blue

Method:

1. Mix the following components on ice:
   2.5 µl 10× nick translation buffer
   3.0 µl DNA (0.5 to 1 µg) dissolved in 1× TE or water
   3.0 µl Unlabeled dNTPs
   5.0 µl α [^{32}P] dATP (=50 µCi)
   2.0 µl DNase I (10 ng/ml)
   2.0 µl *E. coli* DNA polymerase I (5 U/µl)
   7.5 µl Distilled water
   The final volume is 25.0 µl
2. Mix gently, and incubate at 14 to 16°C for 60 min.
3. Stop the reaction by adding 1 vol of stop solution.
4. Separate the radiolabeled DNA from the unincorporated dNTPs as described below (Section III.F.4).

3. Random Priming

The random priming labeling procedure was developed by Feinberg and Vogelstein.[191,192] Oligonucleotides of random sequence (usually a population of synthetic hexamers or octamers) are used as primers for DNA synthesis on single-stranded template DNA. The synthesis of the complementary strand in the presence of labeled deoxyribonucleotide triphosphates is catalyzed by the DNA polymerase I Klenow fragment which lacks the 5' to 3' exonuclease activity. Newly incorporated nucleotides are therefore not removed. The method allows the generation of probes with very high specific activity (>10^9 cpm/µg), requiring only a low amount of input DNA (25 to 100 ng). More than 70% of the labeled deoxyribonucleotide molecules are usually incorporated, and the resulting DNA probes are often longer than those obtained after nick translation.

Random priming kits are commercially available from different companies. It is generally convenient to follow the user's manual included in the kit. Using a kit is easier, but also more expensive per single labeling reaction. The protocol given below may be followed if no kit is available.

Solutions:

DNA sample: 10 ng/μl in 1× TE or water

10× labeling buffer: 0.25 M Tris-HCl, pH 8.0, 30 mM MgCl$_2$, 0.4% β-mercaptoethanol

Unlabeled dNTPs: 1 mM dCTP, 1 mM dGTP, 1 mM dTTP (if dATP is used as a labeled nucleotide)

α[^{32}P] dATP: 10 μCi/μl (specific activity >3000 Ci/mmol; any labeled nucleotide can be used if the unlabeled dNTP mix is made up appropriately)

Primer solution: commercially available random hexanucleotides (e.g., Pharmacia LKB, Amersham International, Boehringer Mannheim Biochemicals)

BSA (fraction V): 2 mg/ml

DNA polymerase I Klenow fragment: 0.5 U/μl

Distilled water

Stop solution: 1% SDS, 10 mM EDTA, 0.25% xylene cyanol, 0.25% bromophenol blue

Method:

1. Mix on ice:
   3.0 μl DNA sample (30 ng)
   16.0 μl Distilled water

2. Denature the DNA sample by heating to 100°C for 10 min, and place immediately on ice.

3. Add on ice:
   5.0 μl 10× labeling buffer
   4.0 μl dCTP
   4.0 μl dGTP
   4.0 μl dTTP
   5.0 μl Primer solution
   2.0 μl BSA
   5.0 μl α[^{32}P] dATP (=50 μCi)
   2.0 μl DNA polymerase I Klenow fragment (0.5 U/μl)
   The final volume is 50.0 μl

4. Mix gently, and incubate at room temperature for 3 to 4 h.

5. Stop the reaction by adding 1 vol of stop solution.

6. Separate the radiolabeled DNA from the unincorporated dNTPs as described below (Section III.F.4).

4. Removal of Unincorporated dNTPs

In order to avoid unspecific background hybridization and to protect the experimenter from unnecessary exposure to radioactivity, unincorporated labeled nucleotides should be separated from the probe. This is usually done by gel filtration through a Sephadex G-50 or Bio-Gel P-60 column. These procedures are suitable for the separation of DNA probes larger than 80 bp from mononucleotides. According to Sambrook et al.,[606] the spin-column variant of Sephadex G-50 gel filtration (see below) will also separate oligonucleotides as small as 16 bases from smaller molecules. For end-labeled simple sequence oligonucleotide probes, however, we prefer ion exchange chromatography (e.g., on Whatman DE-52; see Section III.F.1).

a. *Chromatography on Sephadex G-50*

This technique is based on gel filtration to separate molecules according to size.[606] DNA molecules larger than about 80 bp are excluded from the pores of the Sephadex beads, run in the void volume, and pass the column very fast. Small molecules, such as the mononucleotides, enter the pores and are retained in the column.

Solutions:

Sterile distilled water
1× TE: 10 m*M* Tris-HCl, 1 m*M* EDTA, pH 8.0
1× TEN: 10 m*M* Tris-HCl, 1 m*M* EDTA, 100 m*M* NaCl, pH 8.0

Sephadex G-50: Pretreat as follows:

1. Add Sephadex G-50 powder (medium or fine) to sterile water, allow to swell overnight at room temperature, and wash several times with sterile water. Ten g of Sephadex G-50 powder result in about 150 ml of swollen resin.
2. Equilibrate the resin in 1× TE (several changes), autoclave, and store at room temperature or at 4°C.
3. Prior to use, equilibrate the resin in 1× TEN (several changes).

Dye marker: 0.25% xylene cyanol, 0.25% bromophenol blue in water

Method:

1. Prepare a Sephadex G-50 column in a disposable 1-ml pipette tip, a syringe, or a Pasteur pipette plugged with a small amount of sterile glass wool. Using a Pasteur pipette, fill the column to about 80% of the available volume. Avoid trapping air bubbles.
2. Wash the column several times with 1× TEN.
3. Apply the labeled DNA sample mixed with the dye marker solution (in a volume of 100 µl or less) to the column.

4. Add 1× TEN to the column. Follow the separation of the two dyes. The size of the labeled DNA probe determines its speed, e.g., the 780-bp M13 probe runs in front, close to the xylene cyanol (greenish-blue dye).
5. Collect the probe into a microfuge tube. Measure the radioactivity of (a) the eluted DNA sample and (b) the Sephadex material using a hand monitor. The proportion of incorporated radioactivity can be roughly estimated from these values and should be greater than 50%. Discard the column into the radioactive waste.
6. Store the radiolabeled DNA probe at –20°C until use.

Instead of monitoring the position of the DNA probe with the help of dyes, the leading (DNA) peak of radioactivity may also be identified by collecting individual 200-μl fractions into microfuge tubes and measuring them with a hand monitor.[606]

b. Spin-Column Technique

The spin-column technique is also based on gel filtration through Sephadex or Bio-Gel columns. However, packing and running of the column are accomplished by centrifugation rather than by gravity.[606] In the variant given below, the chromatography column is prepared in a 1.5-ml microfuge tube. It may also be prepared in a disposable 1-ml syringe as described by Sambrook et al.[606] The method is fast and simple, but involves a slightly higher contamination risk than ordinary gel filtration. Therefore, a microfuge should be used which is designated for radioactive experiments only.

Solutions:

1× TEN: 10 m*M* Tris-HCl, 1 m*M* EDTA, 100 m*M* NaCl, pH 8.0
Sephadex G-50: Pretreat as described above

Method:

1. Using a hot needle, punch a small hole into the bottom of a decapped 1.5-ml microfuge tube.
2. Plug the bottom of the microfuge tube with sterile glass wool.
3. Place the microfuge tube on the top of another decapped microfuge tube. Add 800 μl of Sephadex G-50 equilibrated in 1× TEN.
4. Spin in a suitable microfuge (2 min, 735 × g).
5. Change the lower microfuge tube, add 500 μl 1× TEN, and repeat Step 4.
6. Change the lower microfuge tube, and gently apply the labeled DNA sample to the upper tube.
7. Repeat Step 4.
8. Remove the labeled DNA probe from the lower tube to a capped reaction tube. Evaluate amount of incorporated radioactivity as in previous protocol.
9. Store the radiolabeled DNA probe at –20°C until use.

5. Nonradioactive Labeling Procedures

Nonradioactive fingerprint probes can either be purchased commercially or, alternatively, labeled with the help of a commercially available kit for nick translation, random priming, or end labeling (e.g., Amersham International, Stratagene, Boehringer Mannheim Biochemicals). Generally, it is advisable to follow the protocols provided by the manufacturer. In our hands, the Boehringer Mannheim Biochemicals kits work very well.

G. HYBRIDIZATION

In the course of the hybridization reaction, the labeled, single-stranded DNA probe is expected to bind to complementary DNA sequences on the membrane (or dried gel), thereby revealing a more or less complex banding pattern of restriction fragments. However, single-stranded DNA generally tends to bind to membranes (otherwise Southern blotting would not work), and this unspecific binding of the probe would result in (non)radioactive signal generation all over the membrane. To prevent this, membranes are preincubated (i.e., prehybridized) in a buffer containing a variety of high molecular weight blocking agents (e.g., PVP-40, Ficoll, BSA, nonfat dry milk) and detergents (e.g., sodium dodecyl sulfate, or SDS). In most general hybridization protocols, single-stranded DNA from unrelated organisms is also included in the hybridization buffer to block the membrane from unspecific binding of the probe. In case of DNA fingerprinting with minisatellite and simple sequence probes one has to keep in mind that the target sequences of these probes are ubiquitously occurring in eukaryotic genomes. Consequently, inclusion of "nonspecific" eukaryotic DNA in the buffer might lead to an increase instead of a decrease of background. To that end, DNA should either be omitted completely from the hybridization buffer (see, e.g., Vassart et al.[726] and Westneat et al.[772]) or prokaryotic DNA (e.g., from *E. coli*) which typically lacks tandem-repetitive sequences should be used for blocking.[8,609] The prehybridization step is not necessary for in-gel hybridization with oligonucleotides, since dried gels have little tendency to bind single-stranded DNA probes unspecifically.

After prehybridization, the actual hybridization is performed in a buffer of similar composition, but containing the probe. The results of hybridization are strongly influenced by the stringency which has been applied, i.e., what percentage of base mis-pairing has been allowed to occur between probe and target (no mismatch = 100% stringency). Hybridization stringency depends on a variety of parameters such as the GC content of the probe-target complex, probe concentration, buffer composition (e.g., salt concentration and the inclusion of formamide), and, last but not least, temperature (for an overview see Meinkoth and Wahl[449]). For example, stringency may be increased by lowering the salt concentration or by increasing the hybridization or washing temperature. For practical purposes of DNA fingerprinting, it is not necessary to determine the exact stringencies of the hybridization reaction. However, all hybridization and washing conditions used with a given probe should be kept perfectly constant from experiment to experiment for reliable comparison of results.

Hybridization may either be performed in sealed plastic bags in a (shaking) water bath (or other kind of thermostate) or in glass tubes in a roller bottle oven. Though the equipment needed for oven hybridization is somewhat more expensive, we strongly recommend that an oven is used, at least for radioactive probes. Radioactivity is effectively shielded by the glass walls of the tube. The first posthybridization washing steps can be performed within the tube, thus avoiding the high contamination risk associated with removing radioactive probes and membranes from sealed plastic bags.

After hybridization, unbound probe is washed off the membrane (or the gel). The hybridization stringency is also influenced by the washing steps, i.e., by salt concentration and temperature of washing solutions.

1. Hybridization with Oligonucleotide Probes

The most commonly used procedure for DNA fingerprinting with simple sequence oligonucleotide probes has been introduced by Epplen and co-workers.[8,609] It is based on a protocol originally described for allele-specific oligonucleotide hybridization.[471,690] Here, we give a radioactive as well as a nonradioactive (digoxigenin-based) variant of the general protocol. The annealing temperatures (T_m) of the oligonucleotides are calculated according to a rule of thumb put forward by Thein and Wallace:[690] 2°C for each AT pair and 4°C for each GC pair, respectively (assuming a salt concentration of 1 M in the hybridization buffer and an oligonucleotide length of about 16 bp). Hybridization is carried out at $T_m -5$°C, for example, at 35°C for (GATA)$_4$ and 43°C for (GACA)$_4$. According to Thein and Wallace[690] and Miyada and Wallace,[471] these conditions result in 100% stringency, i.e., no mismatches are allowed. Though this might not hold true for all oligonucleotides containing simple repeat motifs,[234] we found hybridization results to be reliably reproducible if the conditions were kept constant between experiments.

a. Gel Hybridization with Radioactive Probes

Solutions:

Hybridization buffer: 5× SSPE, 5× Denhardt's solution, 0.1% SDS, 10 µg/ml fragmented and denatured *E. coli* DNA; sterilize by filtration; stock solutions which facilitate preparation of this buffer are given below
Probe: [^{32}P]-labeled oligonucleotide, add to an appropriate amount of hybridization buffer at a concentration of 0.5 pmol/ml
6× SSC (washing solution): 0.9 M NaCl, 0.09 M sodium citrate, pH 7.0

Stock solutions:

20× SSPE: 3 M NaCl, 0.2 M sodium phosphate buffer, pH 7.4, 0.02 M EDTA
100× Denhardt's solution: 2% PVP-40, 2% BSA, 2% Ficoll. Sterilize by filtration, and store in aliquots at –20°C.

20% SDS

2.5 mg/ml *E. coli* DNA: Dissolve in 10 mM Tris-HCl, 1 mM EDTA, pH 8.0; store in aliquots at –20°C, and denature by heating (5 min, 100°C) prior to addition to the hybridization buffer.

Method (see Safety precautions for work with radioisotopes):

1. Remove plastic wrap cautiously, and soak the dried gel in a tray filled with distilled water. After a few minutes, the backing filter paper detaches. Remaining pieces of filter paper may cause background and should be carefully wiped off the gel using gloves.
2. Transfer the gel into a new tray filled with 6× SSC, and incubate for 5 min.
3. Wind the gel onto a 10-ml disposable pipette, transfer it into a hybridization tube filled with 6× SSC, and unroll it to the inner wall of the tube. Pour off the 6× SSC, and replace with 10 ml of hybridization buffer including the labeled probe.
4. Hybridize for 3 h to overnight at T_m –5°C. Tubes should be carefully closed to avoid contamination and/or loss of probe. Better to control tightness twice than never; do this after a few minutes and after 1 h (when heat has built up inside the tube).
5. After hybridization, carefully decant the probe into a 50-ml Falcon™ tube. The probe may be reused several times. Store at –20°C.
6. Fill the hybridization tube up to one half with 6× SSC, close it, and wash off most of the unbound probe by shaking. Decant the washing solution to the radioactive waste, remove the gel from the tube (this should be done wearing gloves), and transfer it to a tray filled with 6× SSC.
7. Wash the gel in this tray for 3 × 30 min in 6× SSC at room temperature.
8. Transfer the gel to another tray containing 6× SSC prewarmed to hybridization temperature. Wash for 1 to 2 min (stringent "hot wash"[471]). According to our observations, neither the exact duration nor the temperature of this hot wash is too critical, and similar fingerprint patterns occur over a wide range of washing temperatures.
9. Transfer the gel to 6× SSC at room temperature. It is now ready for autoradiography (see Section III.H.1).

Note: The gels should not run dry during either step of hybridization and washing. Otherwise, severe background problems may be encountered.

b. Blot Hybridization with Nonradioactive Probes

Solutions:

For blot hybridization, blocking solution has to be prepared in addition to the solutions given above for gel hybridization. Hybridization buffer without the probe is used for prehybridization.

Blocking solution: 0.5% blocking reagent (Boehringer Mannheim Biochemicals) in 6× SSC. Dissolve by heating; prepare in advance, and allow to cool to room temperature.

Method:

1. Incubate the membrane in blocking solution in a tray at room temperature for 1 h.
2. Transfer the membrane to a hybridization tube as described above (DNA side facing inward), add 10 ml of hybridization buffer excluding the probe, and prehybridize overnight at T_m –5°C.
3. Decant prehybridization buffer, and replace by 10 ml of hybridization buffer (including the probe at a concentration of 5 pmol/ml) or, alternatively, just add the probe to the prehybridization buffer.
4–9. Proceed as described above for gel hybridization with radioactive probes. The membrane is now ready for nonradioactive signal detection (see Section III.H.3).

2. Hybridization with M13 and Other Minisatellite Probes

Two protocols for the hybridization of radioactively labeled minisatellite (and other genomic DNA) probes are described below: (1) a standard protocol according to Westneat at al.[772] and (2) a two-step procedure[750] based on the Westneat protocol, specifically designed for the use of M13 as a DNA probe.

a. Standard Protocol

Solutions and materials:

Radiolabeled DNA probe of the wild-type M13 phage
6× SSC: 0.9 M NaCl, 0.09 M sodium citrate, pH 7.0
(Pre)hybridization buffer: 7% SDS, 0.263 M Na_2HPO_4, pH 7.2, 1 mM EDTA, 1% BSA (fraction V)
Washing solution: 2× SSC, 0.1% SDS (dilute from stock solutions)

Method (see Safety precautions for work with radioisotopes):

1. Prewet the membrane with 6× SSC.
2. Wind the membrane onto a 10-ml disposable pipette, transfer it into a hybridization tube filled with 6× SSC, and unroll it to the inner wall of the tube (DNA side facing inward). Pour off the 6× SSC, and fill the tube with 10 ml of prehybridization buffer. Avoid trapping of air bubbles between the membrane and the tube wall! In this and later steps, hybridization tubes should be carefully closed to avoid contamination and/or loss of probe.

3. Prehybridize for 2 to 3 h at 60°C in a hybridization oven.
4. Denature the radiolabeled wild-type M13 probe by boiling for 10 min in a screw-cap microfuge tube, and transfer it immediately to ice.
5. Add the denatured radiolabeled probe (30 ng/10 ml of hybridization buffer) to the hybridization tube.
6. Hybridize overnight at 60°C in a hybridization oven.
7. Remove the hybridization solution, and rinse twice with washing solution.
8. Transfer the membrane to a tray, and incubate in three changes of washing solution:
   15 min at room temperature
   15 min at 60°C
   15 min at room temperature
   Do not allow the membrane to dry at any stage during washing!
9. Transfer membrane to filter paper, drain off excess liquid, and cover the damp membrane in plastic wrap. It is now ready for autoradiography (see below, Section III.H.1).

b. Two-Step Protocol

M13 DNA, when applied as a hybridization probe for DNA fingerprinting, is usually labeled by random priming. However, weak signals are often observed due to the small size of the hybridizing tandem repeat region (a few hundred basepairs out of 7 kb of phage DNA). One way to overcome this problem is to isolate and eventually subclone the repeat region itself. Alternatively, a two-step hybridization procedure may be used, which was introduced by Weihe et al.[750] In the first step, unlabeled single-stranded M13 DNA is hybridized to the genomic blot. The tandem repeat region of this unlabeled probe will bind to complementary minisatellites in the genomic DNA, while the remainder of the molecule remains single-stranded. These extended single-stranded regions then serve as target regions which are recognized in a second round of hybridization with a radiolabeled M13 probe (see Figure 13). Hybridization and washing buffers are according to Westneat et al.[772]

Solutions and materials:

Unlabeled single-stranded M13 DNA
Radiolabeled M13 DNA probe
6× SSC: 0.9 M NaCl, 0.09 M sodium citrate, pH 7.0
(Pre)hybridization buffer: 7% SDS, 0.263 M Na_2HPO_4, pH 7.2, 1 mM EDTA, 1% BSA (fraction V)
Washing solution: 2× SSC, 0.1% SDS (dilute from stock solutions)

Method (see Safety precautions for work with radioisotopes):

Steps 1 through 3 are performed as above (standard protocol).

1. Step:

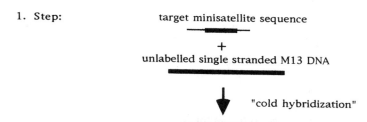

target minisatellite sequence

+

unlabelled single stranded M13 DNA

"cold hybridization"

one unlabelled M13 molecule hybridized to target sequence

2. Step:

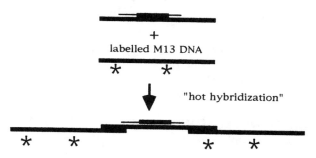

target minisatellite sequence; with increased size

+

labelled M13 DNA

"hot hybridization"

two labelled M13 molecules hybridized to target sequence

FIGURE 13. Principle of the two-step hybridization procedure. In the first step ("cold" hybridization), one unlabeled single-stranded M13 DNA molecule hybridizes to the target minisatellite sequence. In the second step ("hot" hybridization), two ^{32}P-labeled M13 DNA molecules hybridize to the overhanging ends of the first unlabeled M13 DNA molecule, resulting in an increased signal intensity.

4. ("Cold" hybridization): Add 10 ng of single-stranded cold (i.e., unlabeled) M13 DNA per ml hybridization solution. Hybridize overnight at 60°C.
5. Remove the cold hybridization probe, transfer the membrane to a tray, and incubate in three changes of washing solution:
   15 min at room temperature
   15 min at 60°C
   15 min at room temperature
6. ("Hot" hybridization): Return the filter to a clean hybridization tube as described in Step 2 (standard protocol), and add 10 ml of hybridization buffer.

Then continue with Steps 4 through 9 as above (standard protocol).

3. Stripping of Hybridized Gels and Membranes

After signal detection and documentation, gels and membranes may be reused with other probes for several times (up to ten times in our hands). The former probe has to be stripped off. Since it is imperative to prevent the membranes/gels from drying before removing the probe (which might bind irreversibly upon drying), stripping should be done as soon as possible. Several options exist to achieve this, and each membrane manufacturer provides their own protocol. We generally use one of three different procedures:

1. Wash gel/membrane in 5 mM EDTA at 60°C (2 to 4 × 15 min). For membranes, the temperature can also be elevated to 100°C.
2. Repeat the denaturation/neutralization step (see Sections III.C and III.D), i.e., wash gels/membranes in denaturation buffer (2 × 15 min), followed by neutralization buffer (30 min) at room temperature.
3. For membranes only: Boil a solution of 0.1× SSC and 1% SDS, add membrane, and let cool under swirling.

It is advisable to check for completeness of probe removal by using a hand monitor or reexposure. After removing the probe, stripped membranes should be stored under dry conditions at room temperature, whereas gels should be redried (see Section III.C).

H. SIGNAL DETECTION
1. Autoradiography

Radioactive as well as chemiluminescent signals (see Section III.H.3) are usually detected by exposing the gel/membrane to an X-ray film. For reasons of safety and to avoid artifacts, this is best done in X-ray cassettes. Alternatively, gel/membrane and X-ray film can be sandwiched between glass plates and inserted into light-proof plastic bags. Before applying the film, signal strength should be evaluated using a hand monitor. With some experience, the appropriate exposure time (between several hours and several days) can be roughly deduced from the amount of radioactivity as noted on the monitor. If signals are weak, different strategies may be followed to enhance signal intensity. First, different types of X-ray film are available. For example, Kodak XAR is about three times more sensitive than Kodak X-omat S. Second, intensifying screens (usually consisting of calcium wolframate) may be included in the cassettes. At low temperatures, these screens emit photons upon receiving radioactive β-particles, thereby increasing signal strength severalfold.[606] As with X-ray films, screens with different degrees of intensification are available. Two disadvantages of using intensifying screens are (1) a –80°C freezer is needed for exposure and (2) bands on the autoradiogram appear less sharp. Third, preflashing of X-ray films under carefully optimized conditions leads to a further enhancement of sensitivity. The preflashed film side should face the intensifying screen. It should also be mentioned that single- as well as double-coated X-ray films are available. Single-coated films (e.g., Amersham β-max or Kodak BIOMAX) give sharper bands, but are about twice as expensive as double-coated films (e.g., Amersham Hyperfilm).

Method:

1. After the final washing step, place gel/membrane on a sheet of plastic wrap, and drain excess liquid with filter paper.

2. Cover the gel/membrane with plastic wrap. Remove large air bubbles. Inclusion of a piece of tape between the upper and lower sheets of plastic wrap facilitates future unpacking before reusing the gel/membrane.

3. Evaluate signal strength using a hand monitor.

4. Transfer the gel into an X-ray cassette (with or without intensifying screens, depending on signal strength), go to the darkroom, and place a sheet of X-ray film between gel/membrane and intensifying screen. If strong signals are measured, it is conceivable to perform the exposure at room temperature without screens. In this case, exposure time has to be prolonged by a factor of about three.

5. If screens were used, transfer the cassette into a –80°C freezer.

6. After several hours to several days (depending on signal strength), remove the cassette from the freezer, and let it warm up to room temperature.

7. Open the cassette in the darkroom, and develop the X-ray film as follows: 5 min in X-ray developer (single-coated film may require more time) 30 s in water containing a few drops of acetic acid, 3 min in X-ray fixer (until the film turns transparent), 30 min in running water. The film is then allowed to dry. Handle wet X-ray films carefully since they are very sensitive to scraping. X-ray developing and fixing reagents are commercially available. Automatic X-ray processing machines are also available, but generally not required for a smaller laboratory.

2. Phosphorimaging

The technique of phosphorimaging was originally developed for medical radiography and has recently been applied to the quantitation of radioactive samples and to autoradiography in molecular biology laboratories.[333] Phosphors are chemical substances which emit visible light after induction by short-wave radiation. In contrast to fluorescence, phosphorescence persists after the induction ceases. When a membrane is exposed to a phosphorimaging screen, the pattern of radioactive signals is stored in the screen. Upon excitation by light of a certain wavelength in a phosphorimager apparatus, the stored pattern is "released" and immediately transferred to a computer.

The phosphorimaging technology offers several advantages over the traditional method using X-ray films and intensifying screens: (1) storage phosphors are 10 to 250 times more sensitive to incident radiation than X-ray film, resulting in greatly reduced exposure times. For example, samples which require overnight exposure to X-ray film can be accurately imaged after 1 h of exposure to a storage phosphor screen. Since maximum sensitivity is obtained at room temperature, –80°C facilities are not required. (2) Multiple samples can be exposed simultaneously and independently from the detection instrument. (3) Storage phosphor screens are quantitatively accurate over five orders of magnitude, as compared to only two orders of magnitude for X-ray film. Multiple exposures to compensate for the limited dynamic range of X-ray films are therefore not required. (4) Neither special treatments or chemicals nor a darkroom are needed since the reading of the image from a storage phosphor screen is carried out in a phosphorimager. A storage screen is reusable up to about 1000 times.

Phosphorimaging systems are presently supplied by two companies (Molecular Dynamics and Fuji). The complete package consisting of a Phosphor-Imager, computer hard- and software, standard phosphor screens, and an Image-Eraser costs between $48,000 and $84,000. Phosphor storage imaging plates, available from Fuji, Eastman Kodak, and Molecular Dynamics, cost between $900 and $2300 depending on screen size. Though the technology seems very promising, its widespread use will probably be limited by its high expense. Nevertheless, phosphorimaging may become the method of choice in situations where several laboratories can share the costs.

3. Nonradioactive Assays

Avoiding radioactivity and introducing nonradioactive detection methods has several inherent advantages: (1) neither health risks nor waste problems are encountered, (2) the longer stability of probes implicates a more convenient planning of experiments, and (3) no special laboratory facilities are required. The first generation of nonradioactive hybridization techniques used biotinylated nucleotide analogs. An enzyme (alkaline phosphatase or horseradish peroxidase) was attached to the biotin residues via a streptavidin bridge. A colorigenic substrate was added, and its enzymatic processing led to the formation of a dye precipitate at the hybridization site. Biotinylated probes were successfully used for fingerprint analyses (e.g., Pena et al.[548]). However, high background-to-signal ratios were sometimes encountered with biotin. One of the approaches to solve this problem was the development of labeling/detection procedures based on digoxigenated nucleotide derivates. The difference of the digoxigenin vs. biotin strategy is that digoxigenated probes bound to target DNA are recognized by a specific antibody which is conjugated to the enzyme. Background problems appeared to be less using these kinds of probes, possibly because of the limited distribution of digoxigenin in plants (the compound is confined to *Digitalis* species). This approach has also been successfully used for DNA fingerprinting.[757,837,838]

The next important step forward concerned the type of enzymatic substrate. Assays based on colorigenic substrates still suffered from at least two drawbacks which hampered their general use in hybridization experiments: (1) reuse of membranes is limited due to insoluble dye precipitates, which are only incompletely removed even by the use of organic solvents (such as dimethylformamide); and (2) the dye precipitate is formed on the membrane itself, thereby allowing only one "exposure time" per experiment.

TABLE 3
Advantages and Disadvantages of Radioactive Vs. Digoxigenin-Based
Oligonucleotide Fingerprint Detection Methods

Criterion	NBT/BCI P	AMPPD	^{32}P
Laboratory requirements	Normal	Normal	Isotope facilities
Health risk	Low	Low	Medium
Waste problem	Low	Low	High
Probe stability	1 year	1 year	2–4 weeks
Signal development	Medium	Fast	Slow
Signal-to-background ratio	Medium	High	High
Sensitivity	Medium	High	High
In-gel hybridization possible	Limited	No	Yes
Reuse of membranes/gels	Limited	Yes	Yes

With the advent of chemiluminogenic substrates based on substituted 1,2-dioxetanes, these disadvantages were largely overcome[37,49,67,170,501] (see Table 3). In the chemiluminescence-based detection system, a light-generating reaction is initiated by adding the substrate. The alkaline phosphatase cleaves off a protecting phosphate group within its substrate molecule. A destabilized dioxetane anion is formed, which rapidly decomposes into a chemiluminescence-generating compound. The generated light can be easily detected by exposure on X-ray or Polaroid films.

For nonradioactive oligonucleotide fingerprinting, we routinely use digoxigenin-labeled synthetic oligonucleotide probes and AMPPD [3-(2′-spiroadamantane)-4-methoxy-4-(3″-phosphoryloxy)phenyl-1,2-dioxetane, disodium salt] as a chemiluminescent substrate.[49] This approach circumvents the two major drawbacks of colorigenic detection mentioned above. Thus, no residual dye precipitates are formed, membranes are easily stripped off the probe as well as off the AMPPD substrate, and the fast signal development (1 min of exposure may be sufficient for strong signals) in combination with the remarkable stability of the chemiluminescence (up to 2 weeks[37,49]) allows one to obtain a practically unlimited number of exposures on X-ray films.

The chemiluminescent and colorigenic detection protocols given here are based on the procedures of Bierwerth et al.[49] and Zischler et al.,[837] respectively. The reader may also refer to the protocol supplied by Boehringer Mannheim

Biochemicals together with their DIG detection kit. A comparison of [32]P- and AMPPD-detected oligonucleotide fingerprints using the same membrane is shown in Figure 14. Membranes are used instead of dried gels. Though in-gel hybridization with colorigenic probes has been reported,[548,819,838] attempts to create chemiluminescent signals after digoxigenin-based in-gel hybridization were not successful in our hands.

Solutions:

Blocking solution: 0.5% blocking reagent (Boehringer Mannheim Biochemicals) in 6× SSC, dissolve by heating
Solution A: 0.9 M NaCl, 0.1 M Tris-HCl, pH 7.5
Solution B: 0.9 M NaCl, 0.1 M Tris-HCl, pH 9.5, 0.05 M MgCl$_2$; has to be sterilized by filtration or made up from sterile stock solutions; insoluble precipitates will form upon autoclaving
Antibody solution: Polyclonal sheep antidigoxigenin antibody conjugated to alkaline phosphatase (Boehringer Mannheim Biochemicals); stock solution, diluted 1:5000 in solution A (final concentration 0.15 U/ml)

For chemiluminescent detection only:

AMPPD solution: Dilute stock solution (10 mg/ml) 1:100 in solution B. Diluted solutions can be reused several times, but within a few weeks only. To lower costs, the AMPPD concentration may be further decreased according to the expected signal strength.

For colorigenic detection only:

Staining agarose: Prepare stock solutions of nitroblue tetrazolium salt (NBT, 75 mg/ml in 70% dimethylformamide) and of 5-bromo-4-chloro-3-indolyl phosphate (BCIP, 50 mg/ml in 100% dimethylformamide). Dilute to concentrations of 0.33 and 0.17 mg/ml, respectively, in 0.6% molten agarose cooled to 50°C.
Agarose: 0.6% agarose without dye (cooled to 50°C)
TE buffer: 10 mM Tris-HCl, 1 mM EDTA, pH 8.0

Method (chemiluminescent detection):

1. When hybridization to a digoxigenated probe and washing steps are completed, incubate membrane for 1 h at room temperature in blocking solution in a tray. This prevents unspecific binding of the antibody to the membrane.
2. Rinse membrane shortly in solution A.
3. Transfer membrane to a hybridization tube (DNA side facing inward), add antibody solution (5 ml/100 cm²), and incubate for 15 min at room

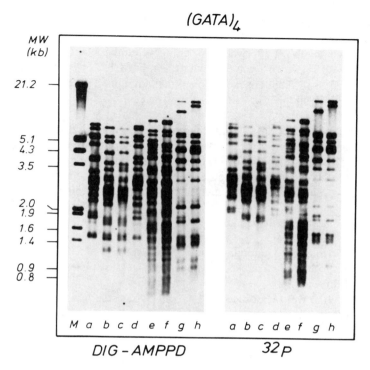

FIGURE 14. Comparison of nonradioactive and radioactive detection methods for oligonucleotide fingerprinting of (a through d) different chickpea accessions, (e, f) two individuals of
the wild tomato species *Lycopersicon peruvianum*, and (g, h) two varieties of cultivated tomato.
Genomic DNA was digested with *Taq*I, and two agarose gels were run with identical sets of
samples. One gel was blotted onto a nylon membrane and hybridized to DIG-(GATA)$_4$ (left panel),
the other gel was dried and hybridized to ^{32}P-(GATA)$_4$ (right panel). Digoxigenin-dependent
hybridization signals were detected by an antibody/alkaline phosphatase conjugate followed by
treatment with a chemiluminogenic substrate (AMPPD). Radioactive signals were detected by
autoradiography. Digoxigenated size markers (kb) are indicated in lane M.

       temperature under constant rotation in a hybridization oven. Alterna-
       tively, spread membrane onto a glass plate (DNA side facing upward),
       pipette the antibody solution on the surface, and incubate for 15 min.

4.   Discard antibody solution. Fill hybridization tube with solution A, and
       transfer membrane to a tray. Remove excess antibody by three washes in
       solution A on a rotary shaker (15 min each).

5.   Incubate membrane in solution B (2 × 15 min). Then spread membrane
       onto a glass or plexiglass plate (DNA side facing upward).

6.   Pipette AMPPD solution onto the membrane (5 to 10 ml/100 cm^2), and
       incubate for 5 min under gentle agitation.

7.   Decant the AMPPD solution (which can be reused several times), briefly
       drain the membrane on filter paper, cover it in plastic wrap, and incubate
       at 37°C for 15 min.

8. Document results by exposure to Kodak XAR (or equivalent) X-ray film at room temperature. Signal strength increases during the first 24 h, and slowly levels off afterwards. When X-ray films are developed within the first 48 h (which we recommend), 1 min to 1 h of exposure time are usually sufficient depending on signal strength. This allows to take an indefinite number of different exposures.

Note: After a few days, bands tend to show some equalizing effect (i.e., strong bands appear weaker and weak bands appear relatively stronger than before). This effect is probably due to substrate limitations as well as to the catalytic rather than stoichiometric means of signal generation.[49] It can be reverted by reincubation of the membrane with the substrate.

Method (colorigenic detection):

1. Perform Steps 1 through 3 as described above.
2. Prepare staining agarose, and pour it onto a glass plate. Allow the gel to solidify.
3. Place the wet membrane onto the gel (DNA side facing downward).
4. Overlay the membrane with molten 0.6% agarose cooled down to 50°C. Allow the gel to solidify. Incubate overnight in a dark, moist chamber at room temperature.
5. Stop color reaction by transferring the membrane into TE buffer.
6. Document results by photographing wet membranes in transmitting light.

Note: Due to the many steps involved in signal detection, the nonradioactive methods are more prone to unspecific background formation than those using ^{32}P. Background is often due to residual antibodies on the membrane. Thus, we recommend that membranes be washed very carefully after the antibody incubation step (each membrane in a separate tray). Increasing washing times, inclusion of 0.1% SDS in the first washing step, and centrifugation of the antibody solution in a microfuge prior to use[501] might also be helpful.

IV. PCR-BASED DNA FINGERPRINTING

A. PCR WITH ARBITRARY PRIMERS

In this section, we shall describe various methods used in PCR with arbitrary primers. Protocols for generating fragments with RAPD analysis, arbitrarily primed PCR (AP-PCR), and DNA amplification fingerprinting (DAF) are given. The discussion is focused on the RAPD technique, since this is the most widely used variant of arbitrarily primed PCR, but many of the remarks are also valid for other PCR-based fingerprinting techniques.

A major problem with PCR is the possible contamination by unwanted template DNA, e.g., from aerosols, skin, or hair of the researcher or from product carryover of a previous reaction. Several precautions can be taken to minimize this problem: (1) PCR should be carried out in a separate room, using a separate pipette set and high-purity water to prepare the reaction mixes, then DNA is added in yet another room with another pipette set (positive displacement pipettes also help to avoid contamination); (2) it is advisable to divide reaction components (buffer, dNTPs, mineral oil) into small aliquots (0.5 to 1 ml); (3) clean gloves should be used to handle the pipette tips and to prepare the reaction mixes; (4) the reaction mix (without polymerase and target DNA) may be irradiated with short-wave UV light prior to PCR to destroy contaminating DNA; and (5) a negative control reaction (no DNA added) should be included in each run.

Contamination is an especially serious problem when PCR is used for the detection of small quantities of DNA (such as bacteria on food) or for the analysis of partially degraded human DNA, where, e.g., skin particles of the researcher are a serious cause of contamination. With plants or fungi, contamination is usually less of a problem, and the precautions mentioned above should be sufficient.

1. Standard RAPD Protocol

Solutions:

10× buffer: 100 mM Tris-HCl, pH 8.3, 500 mM KCl, 20 mM MgCl$_2$, 0.01% gelatin (a *Taq* polymerase buffer is often supplied by the manufacturer of the enzyme and may contain additional ingredients)
dNTP stock: 2 mM dATP, 2 mM dCTP, 2 mM dGTP, 2 mM dTTP
Primer: 50 pmol/µl
Template DNA: 5 ng/µl
Polymerase: 5 U/µl Ampli*Taq* (Perkin Elmer)

Method:

1. Prepare a reaction mix for each primer ("master mix"), sufficient for all samples plus one negative control to which water is added instead of DNA. For the setup of the master mix, calculate 5 µl of buffer, 2.5 µl of dNTP stock, 0.2 µl of primer, 0.2 µl of polymerase, and 37.1 µl of H$_2$O for a volume of 45 µl per sample. The final concentrations are 10 mM Tris-HCl, 50 mM KCl, 2 mM MgCl$_2$, 0.001% gelatin, 100 µM dNTPs, and 0.2 µM primer. Use a specially designated PCR pipette set. Master mixes for only a few samples are preferably made by using primers diluted to 5 instead of 50 pmol/µl, as this is easier to pipette.
2. Divide the master mix into labeled reaction vials or into a microtiter plate. Add 5 µl of the template DNA solution or 5 µl of water (negative control).
3. Mix the contents, and centrifuge the vials briefly. Microtiter plates can be centrifuged in specially equipped centrifuges, but omitting this step does not seem to influence the reaction.
4. Overlay the reaction solution with two or three drops of mineral oil to prevent evaporation.

5. Put the vials or the microtiter plate into a thermocycler, and start the desired program. We use the following program (but see remarks on program in text):

   4 min 94°C (initial denaturing step)

   45 cycles consisting of:

   15 s 94°C (denaturing, minimum ramp time)

   45 s 36°C (annealing, minimum ramp time)

   90 s 72°C (elongation, ramp time 1.5 min)

   4 min 72°C (final elongation step), followed by cooling (if available).

6. After amplification, the vials can be stored at 4°C for a couple of days or at –20°C for a longer period if necessary. The samples are then electrophoresed on a 1.4% agarose gel with TBE, TAE, or TPE as a buffer (see Sections III.B and IV.C) and detected by staining with ethidium bromide. A suitable molecular weight marker is ϕX174 phage DNA, digested with *Hae*III.

A reaction volume of 50 µl is used in the RAPD method described above. If the DNA yield is sufficient and if the detection method is sensitive enough, smaller reaction volumes (e.g., 25 µl) are advised to save on the expensive polymerase.

2. Influence of Reaction Conditions and Components

Optimizing RAPD patterns is laborious since many reaction components as well as any part of the PCR program can be changed with quite unpredictable effects. Several papers describe how optimization can be achieved.[7,484,784] The significance of several reaction components influencing the quality of RAPD patterns was tested in a specially designed study by Wolff et al.[796] using a fractional factorial design. The brand of polymerase and thermal cycler, as well as annealing temperature and primer, were found to have a major impact on banding pattern quality. However, strict predictions on how to choose optimal combinations of reaction conditions could not be made. In the following subsections, the influence of several reaction parameters on RAPD results is discussed. Some results are not only valid for PCR with arbitrary primers, but also for PCR with simple sequence and minisatellite core sequence primers.

a. Primers

Primers can be purchased from several manufacturers [e.g., Operon Technologies, Inc., UBC (University of British Columbia), or Pharmacia LKB] or obtained by using a DNA synthesizer. Primer concentrations are generally optimal between 0.1 and 2.0 µ*M*.[784] In most species, the majority of RAPD primers result in fragment patterns with 6 to 12 fragments, while a few primers fail to amplify DNA. Fritsch et al.[202] studied the "amplification strength" of 480 different 10-mer primers in three plant species. The authors showed that although there were considerable differences among species; some general guidelines for primer design could be given. The G+C content had the highest prediction value; a high G+C content was positively correlated with primer strength. A positive, though less strong, influence on primer strength was also exerted by the G+C content in the four 3'-most bases. These observations should be tested on a broader range of species, since, e.g., in chrysanthemum, all 67 primers used so far have resulted in good fragment patterns,[793a] including primers that failed to amplify fragments in the three species studied by Fritsch et al.[202]

b. Polymerase

A large range of brands and types of polymerases are available for PCR, with *Taq* polymerase (Promega) and Ampli*Taq* (Perkin Elmer) probably being the most frequently used. Super*Taq* (HT Biotechnology) is one of the cheapest available enzymes, yet it gives satisfying results. All these polymerases lack 3' to 5' exonuclease (proofreading) activity, but instead have a 5' to 3' exonuclease activity. "Proofreading" allows the enzyme to check for correct basepair matching and, if necessary, to replace a false with the correct nucleotide. Misincorporation of nucleotides, as studied by denaturing gradient gel electrophoresis (DGGE), is, however, a rare event ($<10^{-5}$),[183] and a proofreading activity can therefore be omitted. Truncated polymerases have been designed which do not have a 5' to 3' exonuclease activity, e.g., Delta*Taq* (U.S. Biochemical), Klen*Taq*1 (AB Peptides), and the Stoffel fragment (Perkin Elmer). The Stoffel fragment lacks 289 amino acids in the N-terminal region, exhibits a higher degree of thermostability than full-length *Taq* polymerases, and tolerates higher Mg^{2+} concentrations. However, due to a slower processivity,[642] twice as much of the enzyme is required per reaction. Some scientists prefer the Stoffel fragment since banding patterns appear to be more polymorphic as well as more reproducible than those obtained with full-length polymerases.[642]

Different polymerases often give rise to different RAPD products,[642,784] see Figure 15. Therefore, the initial choice of polymerase is important; switching to another type of enzyme is likely to render comparisons with previous experiments impossible. In particular, the modified polymerases, such as the Stoffel fragment, result in totally different banding patterns.[642,793a] A comparison of Ampli*Taq* and the Stoffel fragment using identical primers and reaction conditions resulted in completely different RAPD fragments that also mapped, with one exception, to different genomic positions.[642] These results suggest that the number of polymorphic RAPD markers can be increased not only by using additional primers, but also by using another polymerase.

c. Thermocycler

A wide variety of thermocyclers are commercially available. Most of these have 40 to 48 wells (sample positions) in a block, others have 96 wells or are adapted for microtiter plates. The machines can be programmed to various extents. Many applications demand that the transition time between temperatures (also called ramp time or slope) can be programmed. Some new models use capillary tubes which allow very short runs (20 s per cycle) due to the

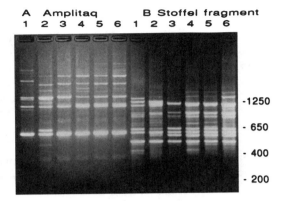

FIGURE 15. RAPD fragment patterns obtained from chrysanthemum cultivars with two different polymerases, (A) Ampli*Taq* (Perkin Elmer) and (B) Stoffel fragment (Perkin Elmer). The primer is 5'-ACC CCC GAA G -3'.Amplification products were separated on an agarose gel and stained with ethidium bromide. Lanes 1 to 6 are *Dendranthema grandiflorum* cultivars CH53, CH54, CH61, CH62, CH63, and CH64, respectively. Size markers (bp) are indicated.

optimal temperature exchange between thermoblock and reaction mix. Thin-walled tubes also give a better temperature exchange. Some thermocyclers also have a heated lid, which obviates the need for mineral oil.

The temperature regime should be consistent across all wells. This is, however, not always the case.[549] To test for position effects, fragment profiles from an identical template DNA sample, amplified in several wells, may be compared to each other. Generally, position effects seem to be negligible, as suggested by the consistent fragment patterns obtained in a study of vegetatively propagated plant material, analyzed with a total of 37 RAPD primers.[793a] Similar results were obtained by Devos and Gale[153] who found that annealing temperatures differing by 2°C did not lead to differences in the obtained fragment patterns.

Running the same program on different thermocyclers, however, may result in different amplification patterns.[424] This phenomenon is most likely caused by different temperature profiles in the reaction tubes. However, we performed identical RAPD reactions using three different thermocyclers (Perkin Elmer 480, MJ Research PTC100, and Pharmacia LKB) side by side and obtained the same fragment profiles with all three machines.[793a]

d. Temperature Profile

Originally, a temperature profile of 1 min at 94°C, 1 min at 36°C, and 2 min at 72°C was suggested for RAPD analysis.[782] Our protocol given above uses shorter time intervals at each step and generally works well for thermocyclers with an optimal temperature transfer between block and reaction tubes. Even more condensed programs may work in some machines. Yu and Pauls[825] optimized reaction times using an MJ Research PTC100 thermocycler and found that 5 s at 94°C, 30 s at 36°C, and 60 s at 72°C gave better results than more time-requiring programs. Shorter periods at 94°C, of course, prolong the lifetime of the polymerase.

Transition times between steps should not be too short, this is especially important for the ramp time between the annealing (36°C) and elongation temperature (72°C). For example, we found that 55 s on a Perkin Elmer 480 thermocycler gives unreliable results, whereas a ramp time of 90 s results in reproducible fragment patterns. Often, primer and target site only have 80 to 90% homology. If the temperature is raised too fast, the primer-template complex may denature again before the polymerase has elongated the DNA strand to a length which is sufficient to resist 72°C.

Another point of attention is the annealing temperature. According to Innis et al.,[308] the melting temperature (T_m) of primer and annealing site can be calculated from a general rule of thumb (2°C for each AT- and 4°C for each GC-pair, respectively). However, this rule[690] was developed for oligonucleotide hybridization in solutions containing 1 M NaCl and may not be applicable for calculating reliable annealing temperatures in PCR reactions. Alternatively, computer programs based on various different algorithus (e.g., Oligo)[599] are available for calculating melting temperature of primers. Usually, the annealing temperature is set to 5°C below T_m. For arbitrary primers, an annealing temperature of 36°C may be chosen as a preliminary.[784] However, amplification may also occur at higher annealing temperatures. In chrysanthemum, even primers with as little as 50% GC content may result in useful patterns at an annealing temperature of 40°C, and primers with 80 or 90% GC still amplify DNA at 44°C. We have also observed that primers with a high GC content frequently result in too many bands (i.e., a smear) with an annealing temperature of 36°C. Raising the annealing temperature to 40 or 42°C results in fewer and more distinct bands.[793a] Good results have been reported also with considerably higher temperatures, e.g., 48°C in combination with a buffer containing 1% Triton-X-100 and 0.1% gelatin.[397a]

RAPD experiments are usually performed with 45 cycles. However, e.g., 35 cycles have also been recommended in some studies.[484] According to our own experience with chrysanthemum, significantly higher yields of DNA are obtained after 45 cycles than after 40 or fewer cycles.[793a]

e. Template Concentration

Optimization of template concentration is extremely important for obtaining good RAPD patterns. Initially, a template DNA range from 5 up to 500 ng may be tried, with checking for changes in fragment patterns and background levels. Negative controls (no template) should also be included. Usually, RAPD patterns are affected by template concentration mostly in the very low concentration range.[784,793a] In some cases, deviating patterns or lack of bands were also encountered in the higher DNA concentration range,[484] perhaps due to impurities in the DNA or residues of CsCl or other compounds used for purifying the DNA.[595]

Primer 5'- CGG CCC CTG T -3'

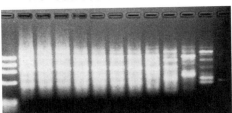

Primer 5'- TGG TCC TGC G -3'

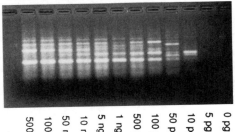

FIGURE 16. RAPD fragment patterns of a chrysanthemum cultivar obtained with a range of different template DNA concentrations. Two different primers were used as indicated. Amplification products were separated on an agarose gel and stained with ethidium bromide.

Between these extremes in the concentration range, there is usually a rather wide "window" where patterns are stable (see Figure 16). Koller et al.[367] report that patterns changed only quantitatively within a 200-fold range of DNA template concentrations. However, more narrow windows have also been reported.[153] A template concentration of 10 to 50 ng/50 μl-reaction volume is often considered optimal, although suggestions of 400 ng[361] have been made.

f. *Mg²⁺ Concentration*

Many studies have demonstrated a marked influence of Mg^{2+} concentration on the obtained RAPD patterns. While strong and reproducible bands are obtained over a wide range of Mg^{2+} concentrations, a change in concentration often results in a qualitative change of fragment patterns[784,796] (see also Figure 17). We suggest that Mg^{2+} concentrations are tested from about 1.5 up to 10 mM. Two mM generally seems to be a good starting point. The chelating effect of EDTA may affect Mg^{2+} concentrations if the DNA is dissolved in TE.

g. *Reproducibility and Quality of Banding Patterns*

RAPDs have often been criticized for low reproducibility, although we believe that this problem is now mostly solved. However, as any PCR-based technique, the RAPD method requires reaction conditions to be kept highly

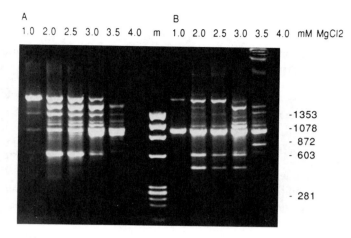

A
1.0 2.0 2.5 3.0 3.5 4.0 m B 1.0 2.0 2.5 3.0 3.5 4.0 mM MgCl2

-1353
-1078
- 872
- 603

- 281

FIGURE 17. RAPD fragment patterns obtained with a range of Mg^{2+} concentrations with primer 5'-GCC TGT CGA T-3' and two chrysanthemum cultivars, (A) CH53 and (B) CH54. Amplification products were separated on an agarose gel and stained with ethidium bromide. Size markers (bp) are indicated.

constant. Sometimes, certain chemicals such as dimethylsulfoxide (DMSO), Tween, BSA, or gelatin are added to make the reaction more specific or more efficient. Optimally, samples to be compared should be amplified simultaneously. Penner et al.[549] studied the reproducibility of RAPD analysis among six different laboratories and found considerable variation. Using the same protocol, but different machines, scientists were often unable to amplify DNA with many of the selected primers. On the contrary, allowing each scientist to use his/her own optimized protocol increased the mean reproducibility to 77% over all five primers (in a range of 36 to 100% for the different primers). The thermocyclers appeared to be the most important source of variation.

To counteract high background levels (smears), a "hot start" is sometimes advised, since it prevents the interaction of primer, template, and polymerase before reaching the denaturing temperature.[183] A part of the mix ($MgCl_2$, Tris-HCl, KCl, gelatin, dNTPs, primer) is divided over the reaction vials, and a wax germ (Perkin Elmer) is added to each vial. The wax is melted by placing the vials in the thermocycler or in an oven heated to 80°C. Upon returning to room temperature, a hard wax layer is formed. The remainder of the reaction mix containing the polymerase is then added together with the DNA template, and the program is started. The primers cannot anneal with the template DNA until the denaturing temperature is reached, i.e., after the wax layer has melted. This strategy may help to reduce background in particular cases. However, we have not seen any effects of a hot start on the quality of RAPD patterns[793a] and neither did Caetano-Anollés et al.[90] for DAF.

Another point of worry is the more or less frequent appearance of (often weak) bands in the negative control reaction (no DNA added), so-called "ghost bands". The origin and cause of these amplified fragment(s) is not known, although the condition is probably aggravated by the high number of PCR cycles commonly used in the RAPD method. Self-priming of partially annealed primer-dimers has been suggested.[308] Contamination of solutions by foreign DNA is probably not responsible, since repeated amplification of the negative control using the same solutions never result in identical banding patterns. However, it has been suggested that extremely low amounts of DNA from *E. coli* or the plasmid used to multiply and produce the enzyme may be involved.[63] Anyway, we do not believe there is any reason for alarm as long as the (weak) patterns in the negative control are different each time and contain no bands similar to the major patterns obtained with template DNA. We have never observed unexpected bands superimposed on or replacing the expected patterns when template DNA was added to the reaction mix. Generally, it is agreed that adequate quantities of template DNA prevent the occurrence of the ghost bands.[251,389]

Because competition between different target sites is involved in the arbitrarily primed PCR process, it is theoretically possible that the presence of a particular RAPD fragment is dependent on the presence or absence of other amplifiable target sequences. In other words, the genetic background may partly determine whether a particular DNA fragment is amplified or not. When tested experimentally by Heun and Helentjaris[284] (see Chapter 5, Section IV.B), the presence or absence of RAPD fragments in maize F_1 hybrids and their dominance behavior appeared to be reliable and not dependent on the genetic background. On the other hand, there are also studies that suggest that the genetic background may influence RAPD patterns in, e.g., somatic hybrids of potato.[30]

h. Modifications

Several modifications of the RAPD technique have been described. The use of two different primers (primer pairs) instead of a single primer may result in patterns that are quite different from the patterns generated by each primer alone.[87,768,784] Generally, only a few fragments are identical to those obtained with the same primers used singly, whereas several additional fragments appear. This nonadditive behavior may be explained by the competitive nature of primer target site selection. Fragments with annealing sites for two different primers at the ends do not form hairpin structures, and thus primers will not be outcompeted by internal hairpin formation, as was suggested[89,90] to be the case with fragments having identical primers at both ends. Using primers pairwise increases the number of polymorphisms that may be generated from a limited number of primers. However, fragments generated with single and with paired primers should not be combined in the same database since identical markers could occur twice. To our opinion, the use of additional single primers will generally give more new information than using paired primers.

Another modification is the digestion of DNA with restriction enzymes before or after the amplification (pre- or postdigestion). Such digestion can either result in the generation of less complex patterns that are easier to evaluate[752] or in the generation of polymorphisms between previously indistinguishable samples.[91,238] This strategy is usually pursued to increase the number of available polymorphic markers,[91] or to obtain codominant markers.[201] However, it detracts from some of the advantages of RAPDs, namely speed, low cost, and user friendliness.

3. RAPD Fragments Used as Probes

Southern blots are sometimes hybridized with probes derived from RAPD fragments. This approach is, e.g., used to establish whether the RAPD fragment of interest represents single copy, mid-repetitive, or highly repetitive DNA sequences. The fragment contains single or low copy sequences if only one or a few bands appear after hybridization to a genomic Southern blot with digested DNA. On the other hand, many bands or a smear indicate that mid- or highly repetitive sequences were amplified. In general, RAPD fragments have been shown to represent low as well as high copy DNA with no obvious preference.[348,432,572,573,784]

Amplification products corresponding to single copy sequences may be used as restriction fragment lengthpolymor phism (RFLP) probes and transformed into codominant markers, thereby circumventing the traditional cloning process[361,432,572,573] (see Chapter 5, Section IV.B). RFLP markers, derived from RAPD fragments, may have the additional advantage to reside in genomic regions, where no traditional RFLP markers are available.

Another reason for using RAPD fragments as probes is to test whether seemingly identical (i.e., comigrating) RAPD fragments from two or more individuals are indeed homologous[348,693,696,779] (see Chapter 5, Section III.A.2). This is extremely important if RAPD fragment profiles are used in studies of genetic relatedness based on the supposition that fragments with the same molecular weight are identical.

RAPD fragments to be isolated are excised from or punched out of the agarose or polyacrylamide gel, e.g., with a sterile Pasteur pipette. The fragment in the gel plug can be directly reamplified using the original primer and a small part of the plug.[89,779] However, this gel plug always contains some additional DNA molecules besides the desired fragment; smaller DNA molecules may be slowed or larger molecules may be dragged along with other fragments during gel electrophoresis. Though being low copy in the amplified DNA, these contaminating DNA sequences may be high copy in the genome and give a smear after hybridization. Purifying the reamplified product by agarose electrophoresis, followed by recovery from the gel and treatment with, e.g., the Geneclean® kit or a Sephadex G-50 spin column[779] will reduce these background problems.

Sometimes reamplification of RAPD fragments is unsuccessful; often one or several smaller fragments are amplified instead of, or in addition to, the original fragment.[696] This internal priming may be explained by the presence of imperfect priming sites within the original fragment to which the primer will anneal in the presence of a high concentration of template DNA, i.e., the originally amplified fragment. Performing reamplification with a series of dilutions may solve this problem.

Six-cutter restriction enzymes are advised for the digestion of genomic DNA, since in this type of experiment, four cutters often yield DNA fragments too small for efficient separation on an agarose gel. The gel can be blotted on any type of membrane (follow the directions given for Southern transfer in Section III.D). The hybridization and washing conditions are different from those in current DNA fingerprint protocols, but standard RFLP hybridization protocols will work.[606] A shorthand version is given below. For general information on probe labeling and hybridization steps, see Sections III.F–H.

Solutions:

Hybridization buffer: 1.5% SDS, 1 M NaCl, 10% dextran sulfate
Salmon sperm DNA: 10 mg/ml
Washing buffer I: 2× SSC, 0.1% SDS
Washing buffer II: 0.5× SSC, 0.1% SDS
Probe (RAPD fragment): 12.5 ng
Labeled nucleotide: α [^{32}P] dCTP (20 µCi)

Method:

1. Boil salmon sperm DNA for 5 min.
2. Add salmon sperm DNA to hybridization buffer (1 ml/100 ml).
3. Prehybridize the membranes for several hours at 65°C.
4. Label the probe by the random priming method (see Section III.F.3).
5. Boil the probe for 10 min, cool immediately on ice, then add the probe to the hybridization buffer.
6. Hybridize overnight at 65°C.
7. Wash membranes: 5 min at room temperature in washing buffer I; 5 minutes at 65°C in washing buffer I; 2 × 30 min at 65°C in washing buffer II.
8. Expose membranes to X-ray film as described in Section III.H.1.

An example illustrating the use of RAPD fragments as probes for genomic DNA as well as for RAPD DNA blots of chrysanthemum is given in Figure 18. This study comprised 14 cultivars (one individual of each) of the hybrid species chrysanthemum (*Dendranthema grandiflora*) and 14 individuals from 10 different species (Figure 18A) related to chrysanthemum.[793a] Three Southern blots, derived from (1) a RAPD gel of the 14 cultivars, (2) a RAPD gel of the 14 species, and (3) a gel with digested genomic DNA of 6 different chrysanthemum cultivars, were used for the hybridization experiments.

Seven RAPD fragments were isolated and used as probes in subsequent hybridization experiments with blots of the two RAPD gels to test the homology of comigrating fragments. Usually, two to three strong bands and four to eight weak bands were observed with each probe, indicating that several amplification products had hybridized in addition to the fragment from which the probe was derived (Figure 18B). One probe revealed polymorphisms among the 14 cultivars, whereas two probes showed pronounced quantitative variation in fragment intensity. Comparison of the different species demonstrated strong differences in hybridization signal intensity and also often in fragment size. This indicates that many different-sized fragments may hybridize to a given probe and that there is often only a low degree of homology between similar-sized fragments.

The same seven probes were then hybridized to a blot of restriction-digested genomic DNA, revealing two to six bands per lane and no polymorphisms between individuals (Figure 18C). The occurrence of multiple bands could be expected also from low copy number probes, since chrysanthemum is polyploid and may have gene duplications. The absence of polymorphism is somewhat surprising, since a very high proportion (79%) of clones from a *Pst*I genomic library had previously proved polymorphic.[797]

4. Standard DAF Protocol

The DAF method is presented here according to the standard protocols developed by Caetano-Anollés et al.[86,87,89]

Solutions:

10× buffer: 100 mM Tris-HCl, pH 8.3, 100 mM KCl, 60 mM MgCl$_2$
dNTP stock: 2 mM dATP, 2 mM dCTP, 2 mM dGTP, 2 mM dTTP
Primer: 50 pmol/μl
Template DNA: 5 ng/μl
Polymerase: 5 U/μl Ampli*Taq* or 5 U/μl Stoffel fragment (Perkin Elmer)

Method:

1. Prepare a reaction mix for each primer (master mix), sufficient for all samples plus one negative control to which water is added instead of DNA. For the set up of the master mix, calculate 2.5 μl of buffer, 2.5 μl of dNTP stock, 1.5 μl of primer, 0.5 μl of Ampli*Taq* (or 1.5 μl of Stoffel fragment), and 13 μl of H$_2$O (or 12 μl of H$_2$O) for a volume of 20 μl per sample. The final concentrations are 10 mM Tris-HCl, 10 mM KCl, 6 mM MgCl$_2$, 200 μM dNTPs, and 3 μM primer. Primers with a length of five and six nucleotides are used in a concentration of 30 and 15 μM, respectively.

2–6. Perform Steps 2 to 6 according to the RAPD protocol (Section IV.A.1), except for the thermocycler program, which is run for 35 cycles:

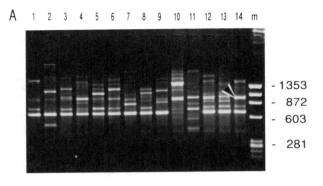

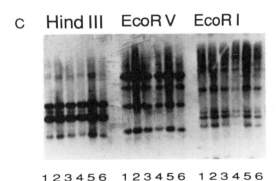

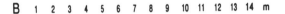

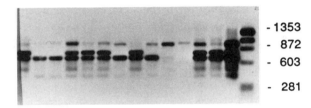

FIGURE 18. (A) RAPD fragment patterns obtained with representatives of different *Dendranthema* and *Chrysanthemum* species. Lanes 1 to 14 are *D. japonicum, D. boreale, D. yoshinaganthum, D. arcticum, D. pacificum, D. shiwogiku, D. indicum, C. zawadskii, C. nankingense, C. yezoense, D. weyrichii, D. × rubellum, C. wakaense,* and *D. grandiflorum*, respectively. The primer used was 5'-CGG CCC CGG T -3'. Amplification products were separated on an agarose gel and stained with ethidium bromide. Size markers (bp) are indicated. (B) Autoradiogram of the gel represented in (A) and hybridized with the marked reamplified ^{32}P-labeled fragment. (C) Autoradiogram of digested genomic DNA hybridized with the ^{32}P-labeled fragment marked in (A). Lanes 1 to 6 are chrysanthemum cultivars CH53, CH54, 9931, 9933, CH1, and CH6, respectively. The restriction enzymes used are indicated in the figure.

> 1 s at 96°C
> 1 s at 30°C

with a heating rate of 23°C/min and a cooling rate of 14°C/min.

7. Following amplification, the vials can be stored at 4°C for a couple of days or at −20°C for a longer period if necessary. Samples are electrophoresed on a polyacrylamide gel (with urea) and detected by staining with silver nitrate (see Section IV.C.2.b).

The main differences between the basic RAPD and DAF protocols are (1) higher primer concentrations in DAF, (2) wider range of primer lengths (e.g., very short primers are often used) in DAF, and (3) two-temperature cycles in DAF instead of three-temperature cycles in the other variants. The technique often results in highly complex banding patterns.

5. Standard AP-PCR Protocol

The AP-PCR method is presented here according to the standard protocol developed by Welsh and McClelland.[766]

Solutions:

10× buffer: Stratagene *Taq* polymerase buffer
1 *M* MgCl$_2$
dNTP stock: 2 m*M* dATP, 2 m*M* dCTP, 2 m*M* dGTP, 2 m*M* dTTP
Primer: 50 pmol/μl
Template DNA: 50 ng/μl
Taq polymerase
α[^{32}P] dCTP

Method:

1. Prepare a reaction mix for each primer (master mix), sufficient for all samples plus one negative control to which water is added instead of DNA. For the set up of the master mix, calculate 1 μl of buffer, adjusted with 1 *M* MgCl$_2$ to 4 m*M*, 1 μl of dNTP stock, 2 μl of primer, 0.025 U of *Taq* polymerase, and double-distilled water with a final volume of 9 to 9.5 μl, depending on the quantity of DNA to be added in Step 3. The final concentrations are 1× buffer with 4 m*M* MgCl$_2$, 0.2 m*M* dNTPs, and 10 μ*M* primer.
2. Partition the mix over the vials or the microtiter plate.
3. Add 0.5 to 1 μl of DNA (25 to 50 ng).
4. Start the program with two low-stringency cycles:
> 5 min at 94°C
> 5 min at 40°C
> 5 min at 72°C

5. Continue with ten high-stringency cycles:
    1 min at 94°C
    1 min at 60°C
    2 min at 72°C
6. Add to each reaction vial 90 µl of the following second master mix containing for each vial 2.25 U *Taq* polymerase, 1× buffer, 0.2 m*M* dNTPs, and 5 µCi α [^{32}P] dCTP. Run the high-stringency cycles described in Step 5 another 20 to 30 times.
7. Separate the amplified DNA on a 5% polyacrylamide and 50% urea gel in 1× TBE, and visualize by autoradiography (see also Sections III.H.1 and IV.C).

The main characteristics of the AP-PCR technique, in comparison to RAPDs and DAF, are (1) the amplification reaction is divided into three parts, each with different stringencies and concentrations of constituents. Two low-stringency cycles allow for mismatch annealing during the generation of the primary products. In later cycles, these products are further amplified under high-stringency conditions. Finally, labeled deoxynucleotides are added to allow for signal detection; (2) high primer concentrations are used in the first cycles and (3) primers (of variable length) originally designed for other purposes (e.g., sequencing primers) are chosen arbitrarily for these experiments.

B. PCR WITH SIMPLE SEQUENCE AND MINISATELLITE PRIMERS

As outlined in Section III, minisatellites and simple repetitive sequence motifs are usually used as hybridization probes for classical DNA fingerprinting. In the present section, we will describe the use of such tandem-repetitive DNA sequences as single primers in the PCR reaction. An example of a minisatellite-specific primer is the core sequence of the wild-type phage M13 (GAGGGTGGXGGXTCT, see Figure 5, Chapter 2). Examples of simple sequence-specific primers are the synthetic oligonucleotides (CA)$_8$, (CT)$_8$, (CAC)$_5$, (GTG)$_5$, (GACA)$_4$, and (GATA)$_4$.

The approach relies on the existence of inversely oriented tandem repeat arrays made up of the same or a very similar short motif. When such inverted repeat blocks are located close to each other (i.e., within about 5 kb), the inter-repeat region may be amplified by PCR with one tandem repeat-specific primer. Since the copy number of the repeated motif in the target DNA is usually higher than that of the primer, pairing may occur in different registers. This is true for both ends of a given amplicon. Therefore, the initial amplification products of a particular inter-repeat region are not necessarily fragments of distinct length, but rather comprise a population of fragments of slightly different size. However, in the course of 40 PCR cycles, internal priming is supposed to result in the successive shortening of a given fragment, with the shortest possible product (primed by the innermost repeat) predominating in the end. Cloning and sequencing of PCR products obtained by simple sequence- or minisatellite-primed PCR verified that each end is 100% complementary to the primer sequence and does not contain additional repeats.[452a,753a] In a variation of the technique described here, single fragments of defined length may be obtained when "anchored" primers are used, i.e., primers carrying one or more unique nucleotides in addition to the repeat.[835]

In contrast to arbitrary primers, a high homology between simple sequence-primers and their binding sites can be assumed. In initial studies on simple sequence-primed PCR,[460,500] the estimation of annealing temperatures was based on the Wallace rule for oligonucleotide hybridization, i.e., 2°C for each AT-pair, and 4°C for each GC-pair.[308,471] Accordingly, 50°C was used for the M13 core sequence, $(CA)_8$, $(CT)_8$, $(CAC)_5$, $(GTG)_5$, and $(GACA)_4$[460] (see Section IV.A, for the calculation of melting and annealing temperatures). Our recent studies, however, showed that considerably higher annealing temperatures may still yield distinct banding patterns.[753a] For example, a single $(CATA)_4$ primer used in conjunction with tomato template DNA yields a complex fingerprint by annealing at 40°C (which is the melting temperature according to the Wallace rule). If the annealing temperature is raised to 43°C or 46°C, some bands become weaker or disappear, whereas others become more prominent.[753a] A few bands are still present at 50°C, as is a band derived from a cloned fragment with 100% match between template and primer. A similar phenomenon is observed with more GC-rich primers such as $(GACA)_4$ and $(CAG)_5$. Here, banding patterns are still obtained by annealing at 60°C, which is 10°C above the Wallace temperature.

We interpret these results in the following way. Bands amplified by simple sequence primers can basically be generated by two mechanisms: (1) RAPD-like bands result from binding of the primer to cryptically simple sequences[687] in the template, allowing mismatch pairing during the first cycles. Depending on the extent of the mismatch, these bands do not occur at elevated annealing temperatures; (2) true inter-repeat bands result from 100% match priming to template regions containing inversely repeated perfect simple sequences. These bands are still generated using annealing temperatures which considerably exceed those calculated by the Wallace rule. However, there is some evidence from work on fungi that such bands may be less paymorphic than those generated at lower annealing temperatures.[452a]

The protocol given below follows the guidelines given by Meyer et al.[460] It usually results in reproducible PCR fingerprints. However, as is the case with RAPDs, many experimental parameters may be altered in order to optimize banding patterns for each particular study. This is especially true for choosing appropriate annealing temperatures (see above).

Solutions:

1× TE: 10 mM Tris-HCl, 1 mM EDTA, pH 8.0
Genomic DNA: 10 ng/µl
Primer: 10 ng/µl

Taq polymerase: 5 U/μl

10× *Taq* polymerase buffer: 100 m*M* Tris-HCl, pH 8.3, 500 m*M* KCl, 15 m*M* MgCl$_2$, 0.01% (w/v) gelatin

10× dNTP mix: 2 m*M* each of dATP, dCTP, dGTP, dTTP

50 m*M* magnesium acetate

Light mineral oil

1.4% Agarose in electrophoresis buffer (see Sections III.B and IV.C.1.a)

Electrophoresis buffer: TAE, TBE, or TPE (see Section III.B)

Gel-loading buffer: 30% glycerol, 0.25% xylene cyanol, 0.25% bromophenol blue

Distilled water

Method:

1. Dilute the genomic DNA to a concentration of 10 ng/μl.
2. Mix the following components in a 0.5-ml microfuge tube on ice:

    2.5 μl genomic DNA (10 ng/μl)

    5.0 μl 10× *Taq* polymerase buffer

    5.0 μl 10× dNTP mix

    3.0 μl 50 m*M* magnesium acetate

    3.0 μl primer (10 ng/μl)

    0.5 μl *Taq* polymerase (5 U/μl)

    31.0 μl distilled water

    The final volume is 50.0 μl

Note: For processing many samples, we recommend that a master mix is prepared consisting of water, 10× buffer, 10× dNTPs, magnesium acetate, primer, and *Taq* polymerase. This ensures the even distribution of all reaction components among the samples and reduces the number of pipetting steps.

3. Vortex briefly, and spin for 20 s in a microfuge.
4. Overlay the samples with two or three drops of light mineral oil.
5. Transfer the samples to a thermocycler that is programmed with 40 cycles of:

    20 s at 93°C (denaturation step)

    60 s at the annealing temperature

    20 s at 72°C (elongation step)

    Amplification is followed by a final extension period of 6 min at 72°C. Set the ramp time as short as possible.
6. When the reaction is completed, mix the amplification products with 0.2 vol of gel loading buffer, and electrophorese in a 1.4% agarose gel (see Sections III.B and IV.C.1.a). Stain with ethidium bromide, and photograph the gel using short-wavelength (302 nm) UV irradiation.

Comments:

1. A test series of several template concentrations should be performed to obtain optimal PCR fingerprint patterns.

2. Once optimized, amplification conditions should be kept as constant as possible. Whereas we found the technique less sensitive than RAPDs to changes in template concentration, the banding patterns are nevertheless influenced by Mg^{2+} and primer concentration as well as by the annealing temperature (see above). To maximize reproducibility, the same thermocycler, *Taq* polymerase, and standard program should always be used within a given set of experiments.

3. Ampli*Taq* DNA polymerase (Perkin Elmer) or *Taq* DNA polymerases (Gibco-BRL) are used by the authors, but other *Taq* polymerase will probably also do the job.

4. The agarose concentration depends on the size distribution of the amplified PCR products and should be adapted to each new species investigated.

5. Magnesium acetate is included in the reaction since it was found to increase the reproducibility of banding patterns in fungal DNA amplifications.[460] However, later studies on plant DNA showed that reliable results are also obtained using magnesium chloride only (1.5 to 3 m*M*).[753a]

The PCR fingerprinting approach described above applying classical DNA fingerprinting hybridization probes as single primers in the PCR was successfully used to detect DNA polymorphisms in fungal and plant genomes[406,459,460,500] (see also Heath et al.,[279] Zietkiewicz et al.,[835] and Chapter 5). Distinctive, polymorphic multifragment profiles were obtained from all fungal and plant species tested so far (see, e.g., Figure 19). The technique combines the advantages of DNA fingerprinting with highly polymorphic probes on one hand with those of the PCR technique on the other. To our experience, the quality of the DNA preparation has little influence on the amplified polymorphic bands. Thus, CsCl-purified DNA and crude miniprep DNA from the same fungal strain resulted in identical PCR fingerprint patterns.[460]

C. SEPARATION AND DETECTION OF PCR PRODUCTS
1. Separation Procedures
a. Agarose Gel Electrophoresis
Agarose gels can be prepared in different concentrations, e.g., 0.6 to 4%. For RAPD fragments, which are usually 300 to 2000 bp, a 1.4% agarose is recommended (Table 4). In addition to different brands of agarose, special separation media or mixtures of agarose and other chemicals are commercially available, e.g., Visigel™ (Stratagene) or MetaPhor™ (FMC Bioproducts). These are claimed to give a better resolution than agarose gels and/or a higher

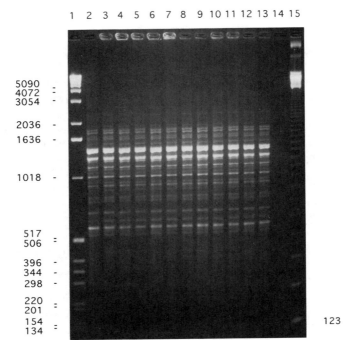

FIGURE 19. PCR-based DNA finger prints obtained by amplification of genomic DNA from *Cryptococcus neoformans* strain 101 using the M13 core sequence (GAGGGTGGXGGXTCT) as primer. To investigate the reproducibility of this fingerprint technique, genomic DNA was isolated every 4 weeks from subcultures of strain 101 over a period of 1 year (lanes 2 to 13). Size markers are in lanes 1 (1-kb ladder, GIBCO-BRL) and 15 (123-bp ladder, GIBCO-BRL). Control without DNA is in lane 14.

strength than polyacrylamide gels. However, these brands are also much more expensive than standard agarose. Electrophoresis of PCR-generated fragments is usually carried out in a 0.5 or 1× TBE buffer (1× solution: 90 mM Tris-borate, 2 mM EDTA). For preparing and running agarose gels, see Section III.B.

b. *Polyacrylamide Gel Electrophoresis*

Polyacrylamide (PAA) gels are generally preferred for achieving high reso-lution, especially in the low molecular weight range (50 to 1000 bp). Gels of different concentrations can be prepared covering a wide range of molecular weights, from 3.5% gels which are suitable for 100 to 2000 bp molecules up to 12% gels for 40 to 200 bp molecules (Table 4). Diffusion of small molecules is less pronounced than in agarose gels, and single basepair differ-ences may be scored. Arveiler and Porteous[23] advocate a mixture of 3% PAA

TABLE 4
Optimal Separation Ranges of Agarose and Polyacrylamide Gels According to Sambrook et al.[606]

Agarose		Polyacrylamide	
Percentage	mol wt region (kb)	Percentage	mol wt region (bp)
0.3	5–60	3.5	1000–2000
0.6	1–20	5.0	80–500
0.7	0.8–10	8.0	60–400
0.9	0.5–7	12.0	40–200
1.2	0.4–6	15.0	25–150
1.5	0.2–3	20.0	6–100
2.0	0.1–2		

and 0.25 to 1% agarose. The agarose will strengthen the gel and prevent smiling of bands.

Nondenaturing PAA gels are used for the separation of double-stranded PCR fragments, with 1× TBE as electrophoresis buffer and low voltages (1 to 8 V/cm). As opposed to agarose gels, nondenaturing PAA gels do not only separate fragments according to their molecular weight, but base composition and sequence may cause up to a 10% difference in running distance. In contrast, denaturing PAA gels are used for separation or purification of single-stranded DNA fragments, e.g., in a sequencing gel. Here, inclusion of urea or formamide in the gel suppresses base pairing, and DNA mobility is not influenced by base composition. Denaturing PAA gels containing urea are used in the DAF and AP-PCR procedures to enhance band resolution.[86,766]

He et al.[277] and Dweikat et al.[172] used DGGE[194] for detection of high numbers of polymorphisms between RAPD fragments which behaved either monomorphic or slightly polymorphic on agarose gels. The separation strategy of DGGE is based on the melting properties of DNA and allows discrimination of double-stranded DNA fragments that differ for one bp only. A DGGE gel contains a linear gradient of denaturants (urea or formamide). To denature the DNA, the sample is heated before application to the gel (see Chapter 7, Section I, for more information on DGGE).

PAA gels are prepared from a mixture of monoacrylamide and bisacrylamide. The ratio of these two chemicals influences the specific molecular structure (pore size) of the polymeric net. TEMED and ammonium persulfate are added to the solution. Ammonium persulfate initiates the polymerization, which is based on a free-radical mechanism. The formation of free radicals is base catalyzed, and TEMED usually functions as the base. Oxygen acts as an inhibitor, and, therefore, all oxygen must be removed from the solution by deaeration prior to polymerization. The protocol given below is for a nondenaturing 5% PAA gel (no urea) of 15 × 13 × 0.2 cm, but can easily be adapted to other sizes or other percentages.

Solutions and chemicals:

Acrylamide stock solution: 38% acrylamide and *N*,*N*′-methylene bisacrylamide
  (19:1 ratio) (see Safety precautions)
1× and 10× TBE buffer
TEMED (*N*,*N*,*N*′,*N*′-tetramethylethylenediamine)
Ammonium persulfate
Loading buffer: 0.25% bromophenol blue, 60 m*M* EDTA, 30% glycerol (or
  loading buffer used for agarose gels)

Method:

1. Clamp the two thoroughly cleaned glass plates of the electrophoresis
   apparatus with spacers in between, and seal the spacers with, e.g., 3%
   agarose or tape.
2. Mix 5.3 ml of acrylamide stock solution with 4 ml of 10× TBE and 30.4
   ml of H_2O, and deaerate under vacuum for 10 min. A sidearm flask or
   a hypodermic needle that is inserted through the rubber top of a small
   laboratory bottle can be used. The acrylamide stock solution should not
   be more than 2 to 4 weeks old.
3. Add 20 µl of TEMED and 280 µl of 10% ammonium persulfate solution
   (freshly made), and mix the solution gently, avoiding air bubbles.
4. Pour the solution between the plates, and put the slot former in place. A
   syringe may be helpful for thin gels. Air bubbles are avoided by tilting
   the gel mold and pouring the gel on one side.
5. Polymerization takes approximately 1 h. The gel can be stored overnight
   if kept humid (e.g., wrapped in tissues moistened with TBE buffer).
6. Remove the tape from the glass plates and the spacer on the bottom (if there
   is one). Insert the glass plates with the gel into the electrophoresis apparatus.
7. Fill the electrophoresis tanks with 1× TBE, remove the slot former, and
   clean the wells thoroughly using a pipette (this is especially important
   with urea-containing gels). Preelectrophorese the gel for 20 min before
   loading the samples. Wells have to be cleaned again before sample
   application (urea, if present, diffuses into the wells which prevents the
   samples from sinking to the bottom).
8. Mix 5 to 10 µl of the PCR reaction with 0.2 vol of loading buffer.
   Samples may be concentrated if volumes are too large: (1) by precipitation of the DNA and dissolving in a smaller volume or (2) by using a
   SpeedVac™. Use special narrow pipette tips or a Hamilton syringe to
   deposit the sample at the bottom of the well. Apply a suitable molecular
   weight marker to one or more of the lanes.
9. Run the gel at 15 to 100 V (1 to 8 V/cm) at room temperature.

For denaturing gels, the acrylamide solution is made 7 *M* in urea. Ready-to-
use urea-containing PAA solutions (Sequagel-6 and Sequagel-8) provided by

National Diagnostics save you the handling of poisonous acrylamide. Prices are comparable to making your own solutions.

After electrophoresis, the gel is removed from both glass plates and subsequently stained with ethidium bromide or silver nitrate (see below). The gel can also be stained while still attached to one glass plate, thus facilitating the handling especially of thin or low-percentage gels. In this case, one plate has to be pretreated with Bind-Silane and the other with Repel-Silane (Pharmacia LKB) following the protocol of the manufacturer, Sambrook et al.,[606] or Tegelström.[688] The gel sticks to the glass plate treated with the Bind-Silane and can be dried on this glass plate for a permanent (but very expensive) record. In the DAF technique, the gel is cast with a polyester backing (GelBond PAG Polyester film, FMC Bioproducts).[86]

2. Detection Procedures

A wide variety of techniques is available for the detection of PCR-generated fragments. We prefer the most simple and fast separation/detection system, i.e., agarose gel electrophoresis followed by ethidium bromide staining. The second option is silver staining of PAA gels. This method is highly sensitive, which is especially helpful when using very short primers (five to eight nucleotides). Both methods are described below. Specific requirements may influence the choice of method.[88] For mapping purposes, simple and unambiguously scored patterns are preferred as obtained with ethidium bromide staining of agarose gels. Genotype identification may, on the other hand, require fingerprints of high complexity. In this case, PAA electrophoresis/silver staining could be the method of choice.

For special applications demanding even higher levels of sensitivity and discrimination, radioactive detection methods may be the preferred alternative. The radioactive label can either be ^{32}P, which is highly sensitive but decays rapidly, or ^{35}S, which gives sharper bands but requires longer exposure times. RAPD fragments may be labeled in various ways: (1) radioactive nucleotides can be included in the last cycles of PCR, (2) one or both PCR primers can be end labeled with ^{32}P or ^{35}S (but also with fluorescent dyes;[14,90] see Chapter 7), and (3) the fragments can be end labeled after finishing the amplification reaction.[468] Labeled PCR products are then separated on an agarose or PAA gel and autoradiographed after gel drying.

For most applications using PCR-based fingerprinting, either ethidium bromide or silver staining will be suitable (Figure 20). While both methods are easy to perform and costs are similar, the use of PAA gels and silver staining is more laborious. Radioactive detection demands isotope facilities and greater skills. It is also more expensive than the other detection methods.

a. Ethidium Bromide Staining

DNA fragments can be stained with ethidium bromide whether separated on agarose or PAA gels. This compound is a powerful mutagen and carcinogenic (see Safety precautions). DNA amounts as small as about 10 ng per band can

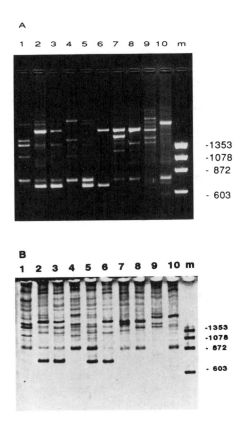

FIGURE 20. Influence of DNA fragment separation and detection methods on RAPD patterns generated with primer 5'-TGC TCA CTG A-3'. Lanes 1 to 8 are chrysanthemum cultivars CH53, CH54, 9928, 9929, 9930, 9931, CH4, and CH5; lane 9 is *Dendranthema japonicum*; and lane 10 is *D. boreale*. Lanes 3 to 5 are offspring from the cross CH54 × CH53. Amplification products were separated (A) on an agarose gel and stained with ethidium bromide and (B) on a 4% polyacrylamide gel and stained with silver nitrate. Size markers (bp) indicated.

be detected in agarose gels, thus providing sufficient sensitivity for most PCR experiments (except for very small fragments or minor amplification products).

Staining of agarose gels can be achieved in two ways: (1) by adding ethidium bromide to the somewhat cooled agarose solution as well as to the electrophoresis buffer or (2) by staining the gel after electrophoresis in buffer or water containing ethidium bromide.

Ethidium bromide is usually prepared as a stock solution (10 mg/ml) and stored in the refrigerator in the dark. For gels, buffers, or staining solutions, a final concentration of 0.5 to 1 μg/ml is needed. Staining takes 5 to 30 min (depending on the gel dimensions) in a tray on a rotary shaker. Rinsing the gel afterwards or destaining for 15 min in water or 1 mM MgSO$_4$ avoids spillage of ethidium bromide on the UV transilluminator and sometimes produces a higher contrast.[606] Ethidium bromide staining solutions can be reused several times, although the solution is light sensitive and should be changed regularly. Ethidium bromide waste should be disposed of in a legal and environmentally safe way.

Ethidium bromide stained gels can be photographed on a UV transilluminator using a Polaroid system with an orange filter (e.g., Kodak 22A Wratten) and Polaroid type 57 or 667 (3000 ASA) film. Exposures on black and white film can also be made using a standard 35-mm camera equipped with the same filter. As UV light is highly mutagenic and destructive, eyes and skin should be covered by glasses and protective clothing (see Safety precautions).

b. Silver Staining

Electrophoresed DNA fragments can also be detected with silver nitrate staining.[468,688] However, silver staining of agarose gels will not give better results than staining with ethidium bromide and, therefore, is not generally used.[688] PAA gels, on the other hand, are usually stained with silver, as this enhances sensitivity for about two orders of magnitude, thus enabling the detection of bands containing 10 to 30 pg DNA.[468] Consequently, RAPD, DAF, or AP-PCR patterns fragments generally exhibit more bands when visualized by silver staining of PAA gels as compared to ethidium bromide staining of agarose gels (see, e.g., Bassam et al.,[36] Caetano-Anollés et al.,[87] and Pang et al.[537]).

Several protocols for silver staining have been described, most of which require approximately 2 h.[36,56,226,468,688] One should be aware that silver does not only stain DNA, but also RNA and proteins. The presence of restriction enzymes, polymerase, and BSA, therefore, should be minimized.

Commercial kits for silver staining are available from several manufacturers (e.g., Bio-Rad Laboratories), but the technique is easily applied also with homemade solutions. Use high-quality (double-distilled) water and a separate container for each gel. For gels adhering to a glass plate, separate clean containers should be used for each different solution. Glass containers are easier to clean, but plastic containers will also work satisfactorily. Most solutions have to be freshly prepared; prepare the silver solution and the developer during the staining process. Silver solutions should be disposed of in a legal and environmentally safe way.

Solutions and chemicals:

0.1% CTAB
1 M NaOH, freshly prepared
25% Liquid ammonia stock solution
35% Formaldehyde stock solution (use in fume hood)
Glycerol
$AgNO_3$
Sodium carbonate

Method according to Tegelström:[688]

1. Rinse the gel in H_2O for 3 to 5 min.
2. Let the gel soak under gentle agitation for 30 min in 0.1% CTAB or in H_2O.
3. Incubate the gel in 0.3% ammonia for 15 min under gentle agitation (1.3 ml stock solution per 100 ml solution).
4. Prepare the silver solution in a flask with a magnetic stirrer: dissolve 0.2 g $AgNO_3$ in 125 ml H_2O, and add 0.5 ml $1M$ NaOH. The solution will turn cloudy and brownish. Add 0.5 to 0.6 ml of 25% ammonia to the silver solution drop by drop. When the solution has cleared, add another two drops of 25% ammonia.
5. Pour off the ammonia from the gel, and add the freshly prepared silver solution (Step 4). Incubate under gentle agitation for 20 min.
6. Prepare the developing solution (2% sodium carbonate, 0.02% formaldehyde). Sodium carbonate is first dissolved with intense stirring, then the formaldehyde (60 µl stock solution per 100 ml solution) is added.
7. Rinse the gel briefly in H_2O.
8. Add the developer to the gel. Staining takes 5 to 25 min under gentle agitation.
9. Stop the staining process by a quick rinse with water, and fix the gel in 1.5 to 3% glycerol for 30 min.
10. The gel can now be photographed and, if on a glass plate, dried.

V. EVALUATION OF FRAGMENT PATTERNS

In the previous sections of this chapter, we presented a variety of means to generate fingerprint patterns. The present section deals with the evaluation of these patterns, i.e., their **translation** into a biological meaning. Several commonly used approaches are described, and some inevitable problems and pitfalls are discussed. The majority of remarks will be of general significance for fingerprinting techniques; if they apply only to either hybridization-based or PCR-based fingerprinting, this will be stated.

A. FRAGMENT SIZING AND MATCHING

A fingerprint pattern obtained from a particular DNA sample is rarely informative on its own. Instead, patterns originating from different samples have to be compared to each other. To do this, individual bands within a lane are assigned to particular positions (which is often done by molecular weight marker-assisted sizing), and different lanes are screened for comigrating (i.e., matching) bands. The present section deals with problems and some general considerations associated with fragment sizing and matching.

The preciseness and accuracy of band scoring are critically dependent on several methodological parameters, including DNA quality, completeness of restriction, electrophoretic conditions, and means of signal detection. Gel electrophoresis deserves particular attention in this respect, since the mobility of DNA fragments is often somewhat uneven across a gel. One example is the so-called "smiling effect", resulting from lower electrophoretic mobility in the outermost lanes. Such mobility artifacts may be caused by irregularities in the electric field as well as by residual impurities in the DNA preparation (e.g., proteins), and the resulting band shifts can lead to severe misinterpretations of band matching. Variation caused by band shifts may be checked for by scoring replicate pairs of individual samples and by the inclusion of in-lane molecular weight standards (see below).

Whenever possible, samples to be compared to each other should be run in adjacent lanes. This is particularly important if high-precision comparisons are required (e.g., in paternity analyses). Large sample numbers, however, will make it necessary to determine band matching between lanes which are widely separated and even between lanes derived from different gels. In these cases, appropriate standards have to be included at least every four to five lanes on each gel. In most cases, molecular weight markers are used for this purpose. However, one of the investigated samples may also serve as a standard, especially if it contains a set of invariable bands present in all individuals. Another useful size standard is a restriction-enzyme digested DNA sample of the researcher's own blood — always available (see Figure 33, Chapter 5)!

For hybridization-based DNA fingerprinting, precision of sizing and matching may be improved by adding a small amount of molecular weight marker to each sample. If visualized by hybridization and, e.g., autoradiographic detection, such in-lane markers allow the convenient standardization of each lane, even when different gels are compared.[206] In-lane markers are particularly important in large PAA gels where the "smiling effect" is especially pronounced.[671]

Even if highly efficient standardization procedures are employed, the decision to regard two closely spaced bands as different or identical is always somewhat subjective and prone to error. Therefore, additional precautions should be taken to minimize the number of misinterpretations; (1) only the easily scorable part of a fingerprint pattern should be considered for the analysis. For example, the low molecular weight range of fingerprints generated by minisatellite hybridization often contains many closely spaced or comigrating bands. It is advisable to exclude this molecular weight range from the analysis also since small fingerprint fragments are generally less variable than larger fragments.[318,514] (2) Bands that cannot be accurately scored throughout all lanes to be compared should be excluded from the analysis; (3) comigrating fragments of different intensity should not be treated as identical if the intense band is more than twice as strong as the faint band (a factor of two might reflect homo- vs. heterozygous state).

Fingerprint patterns may either be evaluated by the eye and hand of the investigator or by automated methods (image analysis, see below). Fragment sizing and matching by eye is most often done by scoring the autoradiograms (or photographs) directly, usually with the help of a transparent ruler. Alternatively, bands may be copied onto an acetate overlay. This is especially helpful if in-lane molecular standards are visualized by a separate hybridization reaction. Usually, the presence or absence of a fragment at a particular mobility position is noted, although occasionally additional information may be obtained by noting fragment intensities.[149] In order to minimize errors, fingerprints should be evaluated by at least two independent investigators, and band positions which are subject to ambiguous interpretation should be excluded.

In recent years, a variety of computerized image analysis systems have appeared on the market which, among other specifications, were also designed for the evaluation of DNA fingerprints. Image analysis is usually based on a high-quality video camera or a scanning device for visible and UV light, radioactivity, or fluorescence-labeled DNA. Suitable software systems allow the interactive editing of the primary image on the computer screen, including background reduction, band sizing with the help of (in-lane or external) molecular weight standards, band matching, and comparison across gels. Primary images as well as processed data can be stored in the computer and used for later comparisons. Most importantly, image analysis allows the investigator to set intensity thresholds for the bands to be scored and mobility thresholds for recognizing a match between two bands.[220]

Alternative image analysis techniques have also been employed. For example, phosphorimaging is not only an efficient way of signal detection (Section III.H.2), but it may also be used for sizing of fragments and storage of data in a computer. The sonic digitizing technique, initially designed and successfully used for scoring sequencing gels, was tested for its applicability to fingerprints on autoradiograms by Galbraith et al.[206] Two microphones, positioned in the upper left and right of a lightbox, receive sounds from a pencil that clicks when a band is touched. The autoradiogram and lanes are standardized by clicking at the corners of the autoradiogram as well as top and bottom of the lanes. However, the technique was not recommended by the authors, since the digitization of fingerprint fragments on an autoradiogram gave unsatisfactorily high error rates. An elegant technique to visualize PCR-generated fragments is the use of fluorescence-labeled PCR primers[355,834] (see Chapter 7). With this technique, sizing of the PCR products can be done by real-time laser scanning with an automated DNA sequencing device (e.g., Applied Biosystems Model 362 Gene Scanner and Model 373 DNA sequencer combined with GENESCAN 672 software[14]). Three distinct fluorescent dyes are available for fragment detection, while an in-lane molecular marker may be labeled with a fourth dye. This technique may not only be used for the simultaneous detection of several PCR-amplified microsatellites,[355,834] but also for analyzing DAF and RAPD patterns.[14,90]

Taken together, accurate fragment sizing and matching is a difficult step of the DNA fingerprinting technology and precautions must be taken to avoid misinterpretations. In order to determine which level of accuracy is needed, the aim of the study must be considered. For example, deciding whether a suspect should be prosecuted for murder is very different from deciding whether a specific plant is the putative "father" of a set of seedlings. A decision also has to be made concerning the means of fragment evaluation. Though being costly, image analysis and direct acquisition of data into a computer are very helpful if large numbers of data have to be compared (such as in studies on population genetics and epidemiology). On the other hand, for comparing, e.g., small sets of offspring plants with their putative parents, pattern evaluation by eye is sufficient and more cost effective.

B. STATISTICAL PROCEDURES
1. Introduction

Once the positions and matches of fingerprint bands have been scored, the data are ready for interpretation. The three main application areas for PCR- and hybridization-based DNA fingerprinting are (1) the identification of genotypes, (2) the assessment of genetic diversity and/or relatedness, and (3) segregation and linkage analysis for genetic mapping. In some of these areas, qualitative evaluation of the fragment pattern by eye may be sufficient to give a quick answer to the investigated problem. In general, however, the data have to be analyzed quantitatively with the help of various statistical methods.

For several reasons, statistical treatment of multilocus fingerprint data is not easy, and the equations used in these analyses often yield only approximations. There are several main problems associated with the interpretation of fingerprint data: (1) it is usually not known whether a given band is in the homo- or heterozygous state (RAPD- and hybridization-based fingerprint bands are treated as dominant characters, although hybridization-based fingerprint bands are not genetically dominant); (2) since allelic pairs cannot be assigned to each other, it is not possible to calculate allele frequencies for specific loci; (3) it is not known whether comigrating bands actually represent homologous loci (see below); and (4) it is not known whether individual fingerprint bands represent independent characters. Instead, they may be actually linked or even allelic to each other (see below). As a consequence of these limitations, discussed in more detail below, many statistical analyses normally used in population genetics are not easily applicable to fingerprint data. However, if homozygous and heterozygous state of bands can be discerned, either directly or after segregation analysis of the offspring of the individuals under study, some general population genetic equations will be applicable.[272,753]

Some statistical procedures have been specifically developed for DNA fingerprint data. To that end, certain assumptions have been made. The first assumption is that comigrating bands are homologous. However, comigration of unrelated bands cannot be excluded. One main reason for "false matches" is the limited resolving power of the most oftenly used agarose gels (e.g., for high molecular weight bands >10 kb, a size difference of, e.g., 500 bp usually cannot be discriminated). Another reason is that small, unrelated fragments often comigrate incidentally. The difference in variability of the different size classes causes additional problems, since many equations rely on the assumption of uniform low frequencies of alleles and uniform mutation rates. The second assumption is that fingerprint bands are independent characters. However, bands treated as unrelated may in reality be alleles of the same locus or, alternatively, tightly linked or in linkage disequilibrium with each other. Consequently, their respective alleles are not transmitted independently, but instead as a haplotype. This is a problem, especially in estimating relatedness of individuals. Linkage disequilibrium is a deviation from the random association of alleles in a population caused by substructuring of a population or by high levels of inbreeding. The occurrence of linkage disequilibrium is especially problematic in population studies and paternal exclusion. Theoretically, problems associated with allelism and linkage may be approached by analyzing the offspring of a biparental cross for transmission frequencies, segregation, and linkage of fingerprint fragments (see Section V.B.5). Tightly linked fragments, whether detected by the same or a different probe, should then be counted only once in subsequent studies. In practice, however, it will usually not be possible to perform a cumbersome genetic analysis.

The majority of statistical procedures developed so far have been dedicated to the evaluation of hybridization-based human or bird fingerprints. From the comparatively few plant studies published so far, it appears that the level of variation and the numbers of alleles per locus is generally smaller in plants than in animals (see Chapter 2, Section II.B.4, for a discussion). While this may be partially explained by selection, inbreeding, or vegetative propagation, different mutation rates of these sequences in plants vs. animals may also be involved. Unfortunately, no data are yet available on mutation rates of tandem repeats in plants or fungi. If polymorphism and the number of alleles per locus in plants and fungi are actually different from those in many animal species, even more precautions should be taken when applying equations originally developed for human and animal fingerprints to plants and fungi.

2. Band Sharing, Similarity Index, and Genetic Distance

Once fingerprint patterns have been generated and scorable bands are assigned to specific positions in all lanes to be compared, different strategies may be followed to quantify the pairwise similarity of the genotypes represented in the different lanes. Most commonly, a "similarity index" (S) is calculated from band sharing data of each pair of fingerprints according to the formula:

$$S = 2 \times n_{ab}/(n_a + n_b) \tag{1}$$

where n_a and n_b represent the total number of bands present in lanes a and b, respectively, and n_{ab} is the number of bands which are shared by both lanes.[495,773] S can acquire any value between 0 and 1, where 0 means "no bands in common" and 1 means "patterns are identical". Another estimator is[76]

$$S' = (n_{ab}/2) \times (1/n_a + 1/n_b) \tag{2}$$

While S′ and S give identical values if $n_a = n_b$, S is a downward biased estimate if the band numbers are different. The similarity index was also denoted as "D" by some authors,[773] but this is somewhat confusing since D is often also used as an estimate of dissimilarity or genetic distance.

A variety of alternative measures for expressing the similarity levels of fingerprints have been proposed and used in various studies. For example, Jaccard's similarity coefficient may be used which only takes positive matches into account (both bands are present) according to the formula:[544]

$$J_{ij} = C_{ij} / \left(n_i + n_j - C_{ij}\right) \tag{3}$$

Here, C_{ij} is the number of positive matches between two individuals, while n_i and n_j is the total number of bands in individuals i and j, respectively. This measure of similarity has sometimes been used in RAPD studies.[544,663,730] Since the absence of a RAPD band may have several different causes, it was argued that using the mutual absence of bands may be improper for calculating the similarity. However, e.g., Stiles et al.[663] compared J and S values for their data set and found no major differences between the two estimates.

If it is possible to score the difference between a homozygous and a heterozygous band correctly from intensity differences or this information is deduced from segregation data, a genetic similarity index (GS) may be calculated as proposed by Mannen et al.[427] A faint band is supposed to be in a heterozygous state, whereas a strong band is in a homozygous state. For each particular band position, the similarity S_b is denoted as 1 if both individuals are either homozygous (two strong bands), heterozygous (two faint bands), or if no bands occur at this position. An S_b value of 0.5 is assigned if the combination consists of a faint and a strong band, or a faint band and no band. Finally, S_b is denoted as 0 if one individual is homozygous (strong band) and no band occurs in the other. For the pairwise comparison of two lanes, the sums of all $(S_b = 1)$ and $(S_b = 0.5)$ values are denoted as p and q, respectively. For N scored loci, the genetic similarity is then calculated as

$$GS = (2p + q) / 2N \tag{4}$$

The uniformity index (U) can be used for calculating the genetic uniformity and the level of inbreeding within inbred lines:[377]

$$U = (1/N) \sum_{i=1}^{N} v_i \tag{5}$$

where v_i is the frequency of the ith band, and N the number of different bands observed.

The similarity index for unrelated individuals is usually not zero, but is often (e.g., in humans) as high as 20% following hybridization with minisatellite probes. The level of this so-called "background band sharing" strongly depends on the combination of probe and species and has to be taken into account if relatedness is to be deduced from similarity indices. If x is the proportion of bands shared by unrelated individuals, then the probability that the offspring has a specific band inherited from one of the parents is[322]

$$P = \left[2x - 1 + (1-x)^{3/2} \right] \Big/ x \tag{6}$$

In human casework evaluation, for example, a conservative value of $x = 0.25$ is deliberately used. For plants, background band sharing after hybridization with minisatellite probes generally appears to be much higher (i.e., 0.3 to 0.5), even for obligately outcrossing species.[13,514,591,798]

If all bands represent statistically independent characters, the chance that n bands would all match in two randomly taken individuals of a population, assuming a background band sharing level of $\overline{S}$, is

$$P = \overline{S}^{\,\overline{n}} \tag{7}$$

This equation is, however, an overestimate because it does not take into account that two individuals do not always have the same number of bands. Alternatively, the probability of a random match can be estimated with[314]

$$P = \left(\overline{S}^2 + (1 - \overline{S})^2 \right)^{\overline{n}/\overline{S}} \tag{8}$$

This value is important as an estimate of how accurately, e.g., cultivars or clones are identified. In other words, it gives the likelihood of falsely identifying a cultivar or clone. It should be noted, however, that such estimates largely depend on the experimental setup. Meaningful comparisons can only be carried out within the same study using identical experimental conditions.

To compare S and n among species or populations, sampling variances can be calculated. If a sufficient number of bands per individual are scored, n and S behave like normally distributed variables. Therefore, the sampling variance of the number of bands per individual is

$$\mathrm{Var}(n) = \frac{k\left(\overline{n^2} - \overline{n}^2 \right)}{k-1} \tag{9}$$

where k is the sample size.[423] The sampling variance of the average number of bands, $\overline{n}$ is var(n)/k. Calculating the sampling variance for $\overline{S}$ is done by

$$\text{var}(\overline{S}) = \frac{\Sigma \left(S_{xy} - \overline{S}\right)^2}{a - 1} \tag{10}$$

where a is the number of pairwise comparisons, and S_{xy} is the similarity index of all pairs x and y.[643] This calculation is, however, only correct if all individuals are used only once in the pairwise comparisons, usually in comparison to one of the neighboring lanes. If instead all lanes are compared to all other lanes (which is usually done), the calculations become far more complicated because of nonindependence of S_{xy}.[422,423]

A special problem in genotype identification is caused by the fact that many plants and fungi reproduce vegetatively. However, Brookfield[68] showed that DNA fingerprinting may be effective in distinguishing whether individuals are members of the same clone or not, provided that the frequency of sexual reproduction is considerably greater than the mutation rate of the fingerprint loci.

3. Population Genetic Analyses
a. Calculation of Mutation Rate

The mutation rate (μ) of fingerprint bands in a given probe (or primer)/species combination is usually calculated from the observed proportion of nonparental bands in the offspring. This is certainly an underestimate because of the possible comigration of a mutant band with other bands.[322] The next equation corrects for this phenomenon by considering the background band sharing rate:

$$\mu_t = \mu_o / (1 - x) \tag{11}$$

where x is the fraction of background band sharing among nonrelated individuals, and μ_o is the observed mutation rate. Mutation rates may be as high as 5% in certain human minisatellite loci[318] and strongly interfere with, e.g., paternity testing. However, no reliable estimates for simple sequence and minisatellite mutation rates have yet been presented for plants and fungi.

b. Population Size

The effective population size (N_e) is the number of individuals that take part in reproduction and are effectively random mating. N_e can be roughly estimated from the mean similarity (S) of fingerprint patterns and the mutation rate (μ) derived from the proportion of nonparental bands in the progeny. However, this estimate is only valid if there are numerous alleles per locus occurring each at a low frequency. Lynch[423] gives two equations for the calculation of N_e. The first one:

$$N_e \approx \frac{4 - 3\overline{S}}{8\mu\overline{S}} \tag{12}$$

gives a slightly overestimated value, and the second one:

$$N_e \approx \frac{1-\overline{S}}{4\mu\overline{S}} \qquad (13)$$

gives an underestimation. The true value lies somewhere between the values as calculated from equations 12 and 13.

c. Heterozygosity

The level of heterozygosity should preferably be known, e.g., when judging the informativeness of a genetic marker. Moreover, the overall proportion of homozygous loci may yield valuable information about the breeding system of a species (e.g., selfing vs. outcrossing, apomixis etc.; see Chapter 5, Sections II and III). If the frequency q of each particular allele is low and similar for all alleles (as assumed by Jeffreys et al.[316]), then the mean homozygosity equals the estimated allele frequency q. Because this assumption is usually not fulfilled, this is a maximum estimate. Lynch[423] gives an equation to calculate the homozygosity from the population size and the mutation rate:

$$\text{Hom} \approx 1/(1+4N\mu) \qquad (14)$$

There are several other methods for estimating the average heterozygosity,[662] homozygosity, and identity by descent[422] for hybridization-based fingerprint data. However, allele frequencies of fingerprint bands must be known to solve these equations.

d. Inbreeding Coefficient

The inbreeding coefficient (F_{IS}) is a measure of the extent of inbreeding in a population due to selfing and/or mating between related individuals. F_{IS} can be calculated from the number of bands before (n_F) and after (n_o) one generation of random breeding, if the number of loci L is estimated from the mean similarity S and the mean number of bands $\overline{n}$:[423]

$$L \approx \overline{n}(4-\overline{S})/4(2-\overline{S}) \qquad (15)$$

$$F_{IS} \approx \frac{\overline{n}_F + \overline{n}_o}{L - \overline{n}_o} \qquad (16)$$

e. Genetic Diversity and Population Subdivision

Wright[807] developed the idea of population subdivision and gave equations for calculating estimates of the extent of substructuring within a population

(S_{ij}). By comparing two populations i and j, the index of between-population similarity $(\overline{S'_{ij}})$ is corrected by the within-population similarity $(\overline{S_i}$ and $\overline{S_j}$, for the populations i and j, respectively).

$$\overline{S}_{ij} = 1 + \overline{S}'_{ij} \frac{\overline{S}_i + \overline{S}_j}{2} \tag{17}$$

Lynch[422] showed how the variance of the measure for population subdivision and its significance can be calculated. The distance between populations i and j, i.e., D_{ij}, is derived from S_{ij}:

$$D_{ij} = 1 - S_{ij} \tag{18}$$

Though the estimators of similarity within (S_w, the mean of all S_i) and between populations (S_b, the mean of all S_{ij}) are all biased, an approximate measure of population differentiation F_{ST} can be calculated from these similarities:[422]

$$F'_{ST} = \frac{1 - S_b}{2 - S_w - S_b} \tag{19}$$

or

$$F''_{ST} = \frac{D_b}{D_w + D_b} \tag{20}$$

The calculated values of F_{ST} are downward biased. Standard errors can be calculated using a Taylor expansion.[422,423] Wolff et al.[798] used the population structure equation of Lynch[422] to calculate $F\text{¢}_{ST}$ values in a study on three *Plantago* species using minisatellite hybridization-based fingerprinting. It was shown that the inbreeding species *P. major* had a higher $F\text{¢}_{ST}$ value (0.74) than the outcrossing *P. lanceolata* (0.53).

The gene diversity statistics of Nei[494] were used by Dawson et al.[143] in a RAPD study of wild barley populations, *Hordeum spontaneum*. To partition the genetic diversity into a within- and a between-population component, heterozygosity was calculated within populations and within the total species, H_S and H_T, respectively. The heterozygosity was calculated as

$$H = 2p(1-p) \tag{21}$$

where p is the frequency of a given RAPD fragment. The population differentiation can be estimated with

$$F_{ST} = \frac{H_T - H_S}{H_T} \tag{22}$$

If H_T and H_S are taken as estimates of the total and within-population diversity, respectively, the between-population diversity D_{ST} can be calculated as

$$D_{ST} = H_T - H_S \tag{23}$$

The gene diversity, expressed relative to the total population diversity, is then

$$G_{ST} = D_{ST}/H_T \tag{24}$$

The proportion of the diversity within a population $(1 - G_{ST})$ compared to the between-population diversity G_{ST} was calculated to be 57% in the wild barley study.[143]

Russell et al.[598] performed a similar RAPD study on cocoa populations. The authors used Shannon's index of phenotypic diversity to calculate the diversity H within and between populations. The Shannon's index of diversity H is given by the following formula:

$$H = -\sum p_i \log_2 p_i \tag{25}$$

where p_i is the frequency of a given RAPD fragment.

Most recently, a computer program called AMOVA (analysis of molecular variance) was developed by Excoffier et al.[186] This program was successfully used in a RAPD study on buffalo grass, *Buchloë dactyloides*, to partition the variability among individuals within populations, among populations within regions, and among regions.[304]

f. Paternity Testing and Estimation of Relatedness

Fingerprint-derived bands are usually inherited in a Mendelian fashion. Consequently, one of the first applications of multilocus DNA fingerprinting in humans was the estimation of relatedness between parents and offspring, i.e., paternity testing. Since then, parentage analysis based on hybridization and PCR fingerprints has been applied to many other organisms. While these studies mainly focused on behavioral ecology such as the analysis of mating success (see, e.g., Burke[82]), paternity investigations are now increasingly performed in plants (see Chapter 5, Section III.B).

Paternity testing is comparatively simple. The PCR- or hybridization-derived fingerprint patterns of mother, father, and offspring are compared. All bands occurring in the offspring must either be derived from mother or father. While the occasional occurrence of nonparental bands may be a consequence of mutation (see Section V.B.3.a), a larger number of such bands will exclude paternity with a high level of probability. It appeared from studies in humans that the fraction of unassigned bands ranges between 0 and 0.2 for true fathers and between 0.4 and 1 for false fathers, respectively.[322] If data on mutation rates are available, the chance that multiple unassignable bands are attributable to mutation can be calculated.[83] In case of doubt, the use of additional probes may reveal the true father. It should be mentioned, however, that nonparental bands may also occur artifactually, especially with RAPDs.[577]

The experimental design is somewhat complicated if there are many putative fathers, as is often the case in natural populations. However, this is not a major problem if the effective population size is small and all putative fathers can be screened for the presence of bands indicative for paternity. Lewis and Snow[401] showed how high frequencies of recessive alleles may alleviate problems caused by the dominant nature of RAPD markers.

Hadrys et al.[252] determined paternity of large offspring clutches in dragonflies by pooling DNA of several offspring and performing a densitometric analysis of the resulting fragment patterns. The authors used "synthetic" offspring (i.e., artificially mixed DNAs) to test for the occurrence of artifacts. A good relationship was observed between the signal strength of specific fragments and the number of copies present in the synthetic offspring. This strategy seems to be applicable also to plant studies.

The similarity index S of two individuals is correlated with their genetic relatedness (r) (see Haig et al.[253]). However, S is always an overestimate of r because nonrelated individuals have a background band sharing level (x) above zero[569] (see also above). When third-, fourth-, or higher-order relationships within and between animal populations were investigated by DNA fingerprinting, it became evident that elevated levels of background band sharing render the use of similarity indices for estimating relatedness difficult.[421-423] Taking the band sharing of unrelated individuals into account, the expected similarity can be calculated as

$$E(S) = x + r(1 - x) \qquad (26)$$

In a theoretical study with 25 loci, the standard errors for estimating a first-, second-, third-, and fourth-order relationship were calculated as 14, 20, 35, and 53%, respectively, demonstrating that these estimates are inaccurate.[421] In the more distant relationships, the relatedness is strongly overestimated. The only reliable distinction can be made between a first-order relationship on one hand, and higher-order relationships on the other. Therefore, the application of DNA fingerprinting, at least with probes resulting in hypervariable patterns, appears to be limited in this respect. Chakraborty and Jin[103] recently demonstrated that relatedness between individuals as well as evolutionary relationships between populations can be approached by DNA fingerprinting data with a unified theory of allele sharing between individuals. The only critical parameters are the degree of heterozygosity per locus and the number of loci. Whereas these are seldom known in plant studies, hybridization-based fingerprinting should still yield good discrimination between, e.g., parent-offspring pairs on one hand and random pairs on the other, provided that the mean heterozygosity is at least 0.7 and that a minimum of ten loci are surveyed.

To test the relatedness data suggested by RAPDs, Tinker et al.[696] analyzed 27 inbred barley lines, with known kinship, using 19 RAPD markers. The resulting matrix of Roger's distances was compared to the matrix with known kinship coefficients r. These matrices were significantly different using the Mantel test, mainly due to a lack of correspondence for the more distant kinships (i.e., low kinship coefficient values).

For practical purposes, e.g., in plant breeding, it is often sufficient to obtain an approximate rank order of relatedness. For example, it is often desirable to select individual plants that are the least related to each other and, therefore, may give rise to the highest heterosis effects.

g. *Outcrossing Rates*

Determining outcrossing rates in plant populations by DNA fingerprinting is based on a strategy similar to parentage determination. The fragment pattern of the mother plant is compared to the fingerprints of its offspring. If nonmaternal bands occur in the offspring, these must have resulted from outcrossing or from mutation. The analysis of data in this comparatively simple way presupposes the existence of many polymorphic fingerprint loci in the population. If, however, variation is limited and many plants have the same genotype, the exclusion of selfing is not so straightforward.

There are several ways to estimate outcrossing rates for individual plants and for populations using codominant markers such as allozymes and RFLPs (e.g., Neale and Adams,[492] Shaw et al.,[625] Weir[753]). Fritsch and Rieseberg[201] used dominant RAPD markers to determine outcrossing rates in populations of the androdioecious plant species *Datisca glomerata* (see Chapter 5, Section II.A). The authors applied the multilocus maximum likelihood algorithm of Ritland,[581] which was modified to allow inclusion of dominant markers.

4. Phenetic and Phylogenetic Analysis

In recent years, several superb books have appeared that describe molecular phenetic and phylogenetic analyses adequately and in great detail, e.g., Hillis and Moritz[286] and Soltis et al.[648] Since the difference between: "phenetic" and "phylogenetic" is not universally agreed upon we would, however, like to start this section by quoting from Abbott et al.[2]: "**Phenetic patterns** are patterns of overall resemblance and difference among organisms based on many heritable characteristics. Often there are discontinuities so that the pattern reveals groupings with differing ranges of variation within groups and varying degrees of difference between them. These patterns are usually observed among the organisms or fossil organisms at a given time. **Phylogenetic patterns** show how the phenetic pattern changes with time; they form a branching tree" and

"... if we think of phenetic diversity as a two-dimensional pattern, phylogeny has a third dimension, time. A slice through such a three-dimensional tree gives a phenetic picture at a given time."

Several different statistical techniques are available for the analysis of phenetic and, to a lesser extent, phylogenetic patterns with DNA fingerprint data. For application of these techniques in plants and fungi, see Chapter 5, Sections II, III, and V.F.

a. Ordination Techniques

Among the ordination methods, principal component analysis (PCA) is probably the most well-known technique. This method makes use of a multidimensional solution of the observed relationships. PCA resolves complex relationships into interactions of fewer and simpler factors. Presence and absence of fragments in OTUs (operational taxonomic units) describe a hyperellipsoid in a multidimensional space. To simplify the description of these "clouds" of points, the (principal) axes through the hyperellipsoid are calculated. The successive principal axes, representing the first major axis, the second major axis, etc., account for the greatest, the second greatest, etc. amount of variation.[643] PCA has been used successfully to analyze DNA fingerprint data in several studies.[635] Advanced statistical software packages perform this type of analysis.

A related technique is the principal coordinate analysis (PCO). It was used by, e.g., Demeke et al.[149] to demonstrate relationships between *Brassica* species on the basis of RAPD data. In this technique, the data matrix is not derived from the presence and absence of all fragments in all OTUs, but rather from the distances (or similarities) between the OTUs. In a more recent study using PCO on RAPD data from various *Juniperus* species, the computer program PCO3D was introduced[4] (see also Chapter 5, Section II.B.2).

A third ordination method is MDS (nonmetric multidimensional scaling), which aims at preserving the rank order of magnitude of distances between the points, so that, for example, larger distances in the original model are larger also in the simplification. Though it has not been much applied yet for DNA fingerprint data, MDS appears to be quite useful.[218]

b. Construction of Dendrograms

The aim of reconstructing a phylogenetic tree or dendrogram is to give the best estimate of the past evolutionary history of a group of OTUs. Usually, these OTUs are related species, but recently the same kind of analysis has also been applied to analyze phenetic patterns among individuals, cultivars, or populations within a species. These techniques will be elaborated in the following sections. Basically, dendrograms may be obtained by three different strategies.[115,286,753] The first strategy comprises the distance matrix methods, also referred to as cluster analysis. These methods define a specific sequence of steps for constructing the best tree. In the first step, a data matrix of pairwise distances is created (e.g., by using D = 1 − S as distance value; S corresponds to the similarity index based on band sharing). The two OTUs with the minimal distance to each other are then grouped together, and the matrix is reduced by one row and one column. After this, a second round of clustering is performed, again grouping the two closest OTUs together. This process is reiterated until a single OTU remains (see, e.g., Clegg[115] and Swofford and Olsen[673]). The most elementary distance matrix algorithm is UPGMA (unweighted pair group method using arithmetic average). However, UPGMA assumes a rigid molecular clock, which means that evolution rates along all branches of the tree are more or less the same. A more sophisticated clustering method is the neighbor-joining method. In this algorithm, differences in evolution rates are corrected for to some extent.

In all distance matrix methods, distances between OTUs are based on the overall similarity between pairs of OTUs. No fragments are considered to be more informative than others. Only one optimal dendrogram is given, and there is no method for comparing or ranking suboptimal dendrograms. The clustering methods are computationally simple and can be used with large data sets. A variety of software packages are available for these types of analysis, e.g., SPSS, SAS, and PHYLIP (Felsenstein).

The second strategy uses parsimony methods and aims specifically at providing phylogenetic patterns. Here, a specific criterion (i.e., minimizing overall tree length) is defined for comparing alternative phylogenies, and specific algorithms are used to compute trees that are in maximum concordance with this criterion. Parsimony methods calculate how many steps (changes, from presence to absence or vice versa) are required for a particular tree to be the most likely one. In contrast to distance matrix-based methods, some characters (fragments) may be more important for inferring a tree than others. Parsimony methods can be biased by multiple substitutions and by variation in evolution rate.[115] Many data sets are too large to evaluate all possible trees. Therefore, heuristic methods are often used to find the best (i.e., most parsimonious) tree. Consequently, the resulting best tree(s) may be correct for some branches (local optimum), but may be less optimal for the tree as a whole (global optimum). Several software packages can perform parsimonious analyses very elegantly, and many options are available. For MS-DOS computers, Hennig86 (Farris) and PHYLIP (Felsenstein) are available. For MacIntosh computers, PAUP (Phylogenetic Analysis Using Parsimony; Swofford) is very useful, especially in combination with McClade (Maddison and Maddison). An MS-DOS version of PAUP will soon be available.

The third strategy relies on the maximum-likelihood method, using standard statistical methods for a probabilistic model of evolution. With this method all possible trees have to be evaluated, which means that large data sets must be handled.[115]

So far, fingerprint data usually have been analyzed with UPGMA and parsimony methods. A variety of reports, mainly based on RAPD data, have shown that the analysis of genetic relationships with the help of multilocus fingerprints is a useful tool for inferring phenetic and, to some extent, also phylogenetic relationships[257,340,341a,346,347,598,718-720,779] (see also Chapter 5, Sections II, III, and V.F). Addresses to obtain computer programs are given in Appendix 5.

c. Assumptions and Limitations

The reliability of phenetic and phylogenetic evidence obtained by any of the methods outlined above strongly depends on two assumptions: (1) the independence of characters within each OTU and (2) the homology of characters compared between different OTUs. As already outlined in a previous section, these assumptions are not necessarily met by DNA fingerprint data, no matter if they are derived from hybridization or PCR. Considering the first assumption, we rarely know whether bands treated as independent characters are actually linked or allelic to each other. Tight linkage between hybridization-based fingerprint bands has been observed in several studies and may result from clustering of different kinds of tandem repeats (see Chapter 2, Section II.C). If linked or allelic fragments are used as characters for estimating relatedness, the same character will be counted twice or more. The second assumption relies on the orthology of comigrating fragments. However, nonhomologous fragments of approximately similar size may comigrate, thereby appearing to be homologous to each other. It is hardly feasible, or even impossible, to test for homology of a large number of DNA fingerprint fragments. Some investigators have done this for comigrating RAPD fragments from different individuals or species (see Section IV.A.3), and Thormann and Osborn[693] showed that 20% of the comigrating bands in a between-species comparison were actually not homologous. However, there are only a few studies available yet which may not be fully representative.

Another complication occurs if a considerable amount of variation is already present within each OTU, e.g., a species. The analysis is then performed analogous to the use of allozyme frequencies for phylogenetic purposes. Correct estimation of band frequencies is needed, and the use of distances to design a tree is the only option. The problem of variation within an OTU may be circumvented by pooling strategies. For example, Yang and Quiros[818] found within-cultivar polymorphisms in celery using the RAPD technique. To perform a parsimony analysis (PAUP), the authors pooled the DNA of 5 to 20 plants for each cultivar. However, the pooling of individuals within an OTU is generally not the best strategy and should only be applied after a thorough analysis of levels of polymorphism.

To conclude, it can be stated that two main assumptions for constructing dendrograms (i.e., independence and homology of characters) are often not met by fingerprint data. Consequently, trees based on hybridization fingerprinting and RAPDs have to be interpreted with caution. DNA fingerprinting is probably not advisable as a phylogenetic tool at an interspecific level, especially since other techniques are available which are more suitable for such purposes (RFLP analysis, DNA sequencing; see Chapter 2, Section I, and Chapter 6). However, if cautiously interpreted, fingerprint data may well serve as one criterion among others to evaluate genetic relationships at an intraspecific level or between closely related species.

5. Linkage Analysis and Genetic Mapping

PCR- and hybridization-generated fingerprint bands are inherited to the offspring according to the Mendelian rules. Thus, bands which are polymorphic between two parents of a cross can be followed in a segregating progeny population and analyzed for linkage to each other as well as to other types of markers or characters (e.g., RFLPs, allozymes, phenotypic traits). Linkage analysis and genetic mapping are useful for a variety of applications: (1) for the evaluation of relatedness, e.g., in population or taxonomic studies (see previous section), it is important to know whether two markers are truly independent or actually linked to each other; (2) linked markers provide anchor points for the physical mapping of genomic regions of interest and allow the map-based cloning of genes by chromosome walking strategies; (3) the availability of a genetic linkage map and/or of genetic markers linked to a trait of interest provide invaluable support to classical breeding strategies. This is not only true for monogenic traits (e.g., many disease resistance traits), but also for quantitative trait loci (QTLs), i.e., genes that, when acting together, determine the expression of a quantitative rather than a qualitative character.

Molecular markers may assist plant breeding by a variety of means. Markers linked to a trait of interest can be used for marker-assisted selection at the seedling stage, i.e., selecting individuals from a segregating generation for further breeding. This is an especially useful strategy if the respective trait is hard to score or is only expressed in mature plants. Linked markers may also be used to speed up the introgression of specific genes or chromosome parts of a related wild species into a cultivated species by backcross breeding. In this breeding strategy, the screening for the presence of the donor-derived character intended to introgress, and screening for maximal similarity to the recipient line and for minimal similarity to the donor line is often laborious and time consuming. Moreover, a considerable number of backcross generations are usually required to remove that part of the donor chromosome that is physically linked to the introgressed character. A selection scheme was recently proposed[285] in which multilocus fingerprints serve as a criterion for "similarity" between backcross progeny and recurrent parent. Only those individuals that have the minimal number of marker alleles of the wild species, except for the desired chromosome segment, are used for the next backcross.

The first step to create a genetic map is the selection of a suitable mapping population, which may be either an F_2, a backcross population, or a set of recombinant inbred lines (see Reiter et al.[573] and Williams et al.[784] for criteria to choose a population suitable for RAPD mapping). Parents are usually selected according to maximum diversity, either concerning the general level of polymorphism or relative to a particular trait (e.g., high vs. low level of fungal resistance). Markers (i.e., hybridization probes, RAPD primers, etc.) are selected according to their ability to differentiate between the parental genotypes and are then tested for their allelic state in each individual of the segregating family.

How is linkage recognized? If two particular markers are located in close vicinity to each other (i.e., if they are physically linked on the same chromosome), their respective alleles are not assorted independently to the individual members of the mapping population. Instead, pairs of alleles preferentially occur in combination (linkage in coupling) or exclude each other (linkage in repulsion). Since the physical neighborhood of markers can be abolished by meiotic recombination, the extent of linkage depends on the recombination rate and, hence, on the distance between the two markers.

In practice, a simple chi-square analysis will already yield some information whether there is linkage or not between two markers.[643] However, to obtain recombination rates r and genetic map distances from linkage data, the influence of double recombinations has to be taken into account. This can be achieved by a variety of algorithms (see, e.g., Allard[11] and Ritter et al.[582]). The genetic distance on a map is expressed in centiMorgans (cM). For recombination rates smaller than 0.10, the map distance in cM is approximately r × 100 (i.e., a recombination rate of 0.01 = 1% recombination = 1 cM). The following example illustrates how linkage can be determined by chi-square analysis. We consider a backcross situation, where one parent is heterozygous for two marker loci (Aa and Bb) and the other parent is homozygous for the recessive alleles (aa and bb): AaBb × aabb. Assuming 85 offspring, we obtain the following numbers of observed and expected genotypes:

Marker B	Observed Marker A		Expected Marker A	
	Aa	aa	Aa	aa
Bb	38	4	20.26	21.74
bb	3	40	20.74	22.26

Is the observed genotype distribution in the offspring significantly different from the expected values? If we apply a chi-square test,

$$X^2 = \sum \left[(\text{observed value minus expected value})^2 / \text{expected value} \right]$$

we obtain a X^2 value of 59.32, $p < 0.001$. The probability that the null hypothesis (no linkage) is true is smaller than 0.001. This means that the difference between observed and expected genotype distribution is statistically significant, and that markers A and B are linked to each other. The recombination rate is

$$(4+3)/(38+4+3+40) = 0.085$$

and the map distance is approximately 8 cM.

Various computer programs are available that calculate recombination values and map distances. They also give a most likely order of markers within linkage groups and thus create genetic maps. LINKAGE-1 (Suiter)[669] is based on chi-square analysis and only allows the evaluation of pairwise (two-point) analysis of recombination values. GMENDEL (Knapp)[409] uses a log-likelihood method or G statistics. JOINMAP (Stam)[660] accepts data with different expected segregation ratios and integrates data from different populations. MAPMAKER (Lander)[387] performs multipoint analyses using maximum likelihood in F_2 and backcross families. A comprehensive application manual is usually provided together with the software of genetic mapping programs. Unfortunately, no program is yet available which is specifically designed for the evaluation of multilocus fingerprints (including RAPD bands) for linkage analysis. Addresses to obtain computer programs are given in Appendix 5.

Until recently, molecular marker maps were mainly derived from RFLP and allozyme data. However, theoretical studies as well as computer simulations have clearly shown that hybridization-based DNA fingerprint data may be useful for the detection of, e.g., major genes for quantitative traits in domestic animal species,[254] and statistical tests for linkage have been produced for this kind of data. A few applied studies have tested the usefulness of hybridization-based fingerprinting for linkage analysis and generating maps. For example, minisatellite-derived maps have been established for the human[765] and mouse genome,[339] and linkage studies have been performed in cattle[214,216] and sheep.[129] Linkage maps based on hybridization fingerprints have also been established in some fungi[593] (see Chapter 5, Section V.G), but not yet in plants.

The situation is quite different with PCR-generated fingerprints. The ease with which RAPD patterns are generated prompted a multitude of studies toward using RAPD-based linkage maps, especially in plants (see Chapter 5, Section IV.B). Finding markers which are linked to a particular monogenic trait do not necessarily require a saturated genetic map. Instead, one can make use of pairs of near-isogenic lines if available[432,540] or follow the strategy of bulked segregant analysis.[351,461] RAPD fragments have already been widely used in localizing useful agronomic traits (mainly resistance traits) and seem to have a great future there. To give a few examples, Paran and Michelmore[539] found eight RAPD markers associated with disease resistance in lettuce using MAPMAKER. Klein-Lankhorst et al.[361] identified a RAPD marker associated with a nematode resistance gene, located previously with RFLP markers. In sugar beet, RAPD markers were mapped together with a nematode resistance gene and a hypocotyl color gene using LINKAGE-1.[708] Yu and Pauls[826] used a maximum-likelihood method for calculating distances of RAPD markers associated with somatic embryogenesis in tetraploid alfalfa. For more information on these and similar studies, see Chapter 5, Section IV.B.

From the studies performed so far, it appears that multilocus markers in general and RAPD markers in particular are very useful for mapping monogenic traits as well as QTLs. However, multilocus strategies do not only implicate advantages, but also disadvantages for linkage analyses. The advantages include the following: (1) multiple loci are screened simultaneously, i.e., more information per marker is obtained; (2) higher levels of polymorphism are often encountered than with RFLPs and allozymes; and (3) especially in case of RAPDs, new markers are obtained easily and fast. On the other hand, there are several disadvantages: (1) scoring for presence or absence of a band is often more error prone than with RFLPs and allozymes[129] (see Section V.A); (2) it is not yet clear to what extent mapping data obtained from one cross can be transferred to other crosses; and (3) the transformation of a linked fragment to a locus-specific probe is not as straightforward as with RFLPs. To be used as reliable markers, RAPDs require the development of SCARs (see Chapter 7), while hybridization-based fingerprint fragments (due to their repetitive nature) have to be cloned and sequenced before they can be used as locus-specific starting points for, e.g., chromosome walking experiments. Another disadvantage is that (4) allelic pairs are usually not identified. Consequently, the presence of a band has to be treated as a dominant character, and the inability to differentiate between homo- and heterozygotes causes loss of information. For linkage analyses with dominant markers, the use of F_2 mapping populations is therefore discouraged, while backcross populations and recombinant inbred lines may be used.[784] Dominant markers are also not a problem if haploid mapping populations are available, which is the case in many fungi,[593] anther-derived plants, or haploid megagametophytes.[702]

Chapter 5

APPLICATIONS OF DNA FINGERPRINTING IN PLANTS AND FUNGI

Various methods of DNA analysis have been termed "DNA fingerprinting", however, we will here concentrate on those where commercially available multilocus DNA probes or polymerase chain reaction (PCR) primers have been utilized. These probes and primers are not species specific and will therefore usually work for many different species of plants and fungi once the technical difficulties have been mastered. For a closer description of the laboratory procedures and for evaluation of DNA fingerprints and calculations, see Chapter 4, and for a brief presentation of alternative methods used in analyses of genetic variation, see Chapters 2 and 6.

In this chapter, we shall demonstrate how DNA fingerprinting can be applied in plants and fungi, especially in the fields of identification (e.g., genotypes, strains, and cultivars), paternity analysis, estimation of genetic relatedness, and genome mapping. Due to space limitations, only some representative examples will be given. For a more comprehensive list of papers where these methods have been applied, see Appendices 1 to 4.

I. DEVELOPMENT OF METHODS AND INITIAL INVESTIGATIONS

A. MINISATELLITE DNA PROBES DETECT GENETIC VARIATION

DNA fingerprinting was first introduced to plant genome analysis in 1988. Initial experiments were performed using either the M13 repeat probe discovered by Vassart et al.[726] or the human minisatellite probes 33.6 and 33.15 developed by Jeffreys et al.[315,316] (see Chapter 2, Section II.B).

In the very first DNA fingerprint paper dealing with plants, Ryskov et al.[600] demonstrated DNA fragment pattern differences between two varieties of barley, *Hordeum vulgare*, following hybridization of *Hae*III digested DNA samples with the M13 probe.

Somatic stability of M13-regenerated DNA fingerprints was reported in another paper on plants in the same year by Rogstad et al.[588] In this study, identical DNA fragment profiles were found in samples derived from two separate branches of a large cottonwood tree, *Populus deltoides*, as well as in samples from a mature tree and its sucker plant. In contrast, DNA samples obtained from six different cottonwood trees showed extensive variation in fragment patterns. Variation was shown also to occur between the two trees investigated in each of *Polyalthia glauca* and North American quaking aspen, *Populus tremuloides*. On the other hand, two plants of an inbred tomato cultivar, *Lycopersicon esculentum*, exhibited completely identical fragment patterns.

In a third paper appearing that year, Dallas[137] was able to distinguish rice cultivars by hybridization to the human 33.6 minisatellite probe. The author obtained unique DNA fragment profiles for each of seven *Oryza sativa* cultivars and for two *O. glaberrima* cultivars. Since rice reproduces by selfing, levels of within-cultivar variation were expected to be very low. This expectation was confirmed by finding identical fragment profiles in 14 seedlings of an *O. sativa* cultivar. Perhaps most important in Dallas' study was the analysis of DNA fragment inheritance. A total of 14 F_2 plants from a mass crossing between two rice cultivars were scored for presence or absence of fingerprint bands. All fragments exhibited in the offspring except one could be related to one of the grandparental genotypes.

In the following year, Braithwaite and Manners[64] reported the first application of the human minisatellite probes 33.6 and 33.15 to detect polymorphisms between subgroups of the filamentous fungus *Colletotrichum gloeosporioides*. One year later, the M13 probe proved useful in distinguishing different *Fusarium* strains.[472]

B. OLIGONUCLEOTIDE PROBES ARE INTRODUCED

In 1989, synthetic oligonucleotide probes that recognize simple repetitive DNA sequences were introduced to plant DNA fingerprinting. Weising et al.[755] demonstrated different DNA fragment patterns between three barley cultivars by hybridization to a (GACA)$_4$ probe as well as between three accessions of chickpea, *Cicer arietinum,* by hybridization to a (GATA)$_4$ probe. Since then, numerous papers have been published on "oligonucleotide fingerprinting" in plants, suggesting that the number of informative simple sequence-complementary probes is considerable. Moreover, the level of variation detected is highly dependent on which probe is used.[761] Therefore, a careful matching between probe and material may enable us to focus our research at suitable level of genetic variability.

The first report on DNA fingerprinting in fungi using simple sequence probes demonstrated that a collection of diverse yeast species and strains, *Saccharomyces* spp., could be distinguished by a synthetic poly(GT) probe.[736]

For comprehensive reviews on early studies of simple sequence and minisatellite fingerprinting, see Nybom[511,512] and Weising et al.[759,760] for plants and Meyer et al.[455,459] for fungi.

C. PCR-BASED METHODS PROMPT INCREASED RESEARCH ACTIVITY

The introduction of PCR-based methods constituted a new milestone in the field of DNA fingerprinting (see Chapter 2, Section III). When sequence information is available, **specific** DNA primers may be tailored for amplifying areas of interest, sometimes resulting in high levels of polymorphism (see d'Ovidio et al.,[159] Ness et al.,[499] Rogowsky et al.,[586] Weining and Langridge,[752] Williams et al.,[786] Wolff et al.[797]and Xue et al.[814]). However, here we shall concentrate on research and application of PCR conducted with **arbitrarily** chosen primers, thereby foregoing the need for sequence data. Two such methods were published in 1990[766,782] and a third in 1991.[86,87]

Welsh and McClelland[766] used the arbitrarily primed (AP-PCR) method to distinguish various bacterial strains as well as three varieties of rice. The random amplified polymorphic DNA (RAPD) method of Williams et al.[782] demonstrated polymorphism between two lines each of the fungus *Neurospora crassa* and maize, *Zea mays*. Also, an analysis was conducted on 66 F_2 individuals segregating from a cross between two lines of soybean, *Glycine max* and *G. soya*. Eleven RAPD markers were mapped in the soybean genome relative to classical restriction fragment length polymorphism (RFLP) markers. The DNA amplification fingerprinting (DAF) method was introduced by Caetano-Anollés et al.[86,87] who amplified DNA from a wide variety of organisms, including the fungus *Candida albicans* and several plant species. Polymorphisms were detected between different cultivars of soybean and of several turf grasses, e.g., *Zoysia*.

Due mainly to the rapidity with which large numbers of samples can be processed, the PCR-based methods are increasing in popularity. Selected primers may reveal levels of genetic variability that are of the same order of magnitude as those commonly observed following hybridization-based DNA fingerprinting. Inheritance has been proved to be Mendelian, at least for the majority of detected PCR amplification products, thereby allowing genetic analyses. However, the obtained PCR products generally behave as dominant markers, in contrast to, e.g., the codominant RFLPs. For recent reviews on PCR-based methods using arbitrary primers, see also Caetano-Anollés and Bassam,[85] Meyer et al.,[459] Newbury and Ford-Lloyd,[502] Rafalski and Tingey,[562] Rafalski et al.,[563] Tingey and del Tufo,[694] Tingey et al.,[695] Waugh and Powell,[739] and Williams et al.[784]

II. WILD PLANT SPECIES

Several applications of DNA fingerprinting in population genetics, ecology, and systematics have been described in the literature. Some of these studies concerning wild plant species will be mentioned here. Corresponding studies carried out on mainly cultivated material are instead treated in Section III. Genome mapping and related areas are treated in Section IV. For additional information, see also Appendices 1 and 2.

A. DNA FINGERPRINTS DETECT GENETIC DIVERSITY AND RELATEDNESS IN PLANT POPULATIONS

Among all the various approaches taken up till now for estimation of genetic diversity and relatedness in plant populations, the DNA fingerprint methods appear extremely promising, as demonstrated in a number of recent publications dealing with a wide variety of plants.

1. Cross-Pollination vs. Selfing: The Influence of the Breeding System on Diversity

The existence of a close association between the breeding system and levels of genetic variation has been demonstrated previously with a variety of different methods. However, due to its unique ability to detect also very low levels of genetic variation, DNA fingerprinting may become one of the most useful tools for this kind of research. Several recently published population studies have already demonstrated how genetic diversity can be resolved to a much higher extent using DNA fingerprinting than with more traditional methods. Here, we shall give some examples featuring species with various levels of genetic variation.

DNA fingerprints appear to be individual specific, especially in cross-pollinated species as may be exemplified by the box elder, *Acer negundo*. This North American species is dioecious, i.e., male and female flowers occur on separate trees. Since there is no vegetative propagation, each tree is derived through genetic recombination and should therefore be genotypically unique. This assumption was tested by hybridizing the M13 probe to restriction-digested DNA of 21 different trees. Individual-specific fingerprints were actually obtained by Nybom and Rogstad.[514] Since box elder is insect pollinated, gene flow is expected to decrease with increasing geographic distance between plants. Such a relationship was to some extent corroborated by the DNA fingerprint data, showing mean bandsharing levels of 64 to 68% for comparisons among neighboring trees and 48 to 58% for comparisons among trees growing further apart.

Another dioecious species is the buffalograss, *Buchloe dactyloides*, which is native to the semi-arid regions of the Great Plains in North America. A study of population genetic variation was carried out with RAPDs on four populations by Huff et al.[304] Analysis of molecular variance (AMOVA[186]) was used to apportion the variance among individuals within populations, among populations within adaptive regions, and among the two regional ecotypes sampled. Every individual proved to be genetically distinct. Nevertheless, a measurable divergence was encountered among local populations and also among regional ecotypes. The seven primers used partitioned the variation quite differently, suggesting that representative results can be obtained only if a reasonably large number of primers are used.

RAPDs were used also when levels of genetic variation were analyzed between and within populations of two widely cultivated leguminous tree species native to Central America and Mexico, *Gliricidia sepium* and *G. maculata* by Chalmers et al.[104] Of these species, at least *G. sepium* is obligately outcrossing. Five trees from each of two populations of *G. maculata* and eight of *G. sepium* were sampled. Polymorphisms between *Gliricidia* genotypes were observed with 10 of the 11 PCR primers evaluated. The number of polymorphic loci varied between primers as well as between populations. Some populations were monomorphic with the majority of the primers, whereas other populations showed high levels of variation. Population-specific markers could be identified. The RAPD results were in close concordance with previous results on the variation for morphological characters in the same set of populations. When the genetic diversity estimated by RAPD data was partitioned into within- and between-population components, about 60% of the diversity found in *G. sepium* was shown to occur between populations. However, different primers yielded different results, some detecting variability mainly between populations and others instead within populations. This again, as in the *Buchloë* example above, illustrates the fact that numeric interpretations of RAPD data should be based on a substantial number of primers to avoid a serious bias in the results.

DNA fingerprint studies may be used also to analyze dispersal events. For example, substantial variation was found in a recently introduced population of the leguminous shrub, *Hippocrepis emerus (=Coronilla emerus),* in central Sweden by hybridization to a synthetic poly(GT) probe in studies by Lönn and co-workers.[417,418] In contrast, very low levels of within-population variation were found in native *H. emerus* in both Norway and Sweden, whereas a high degree of differentiation was found between regional populations. This pattern of variation probably reflects the species' history of population disjunction.

DNA fingerprinting has yielded valuable results not only for many outcrossing species, but also for selfing species. One example is the Chilean annual *Microseris pygmaea*, which is thought to have arisen by dispersal of only one seed from North America.[722] While isozymes revealed only low levels of variability in this species, hybridization of DNA samples with $(GATA)_4$ resulted in highly polymorphic fingerprints in a study by van Houten et al.[722] Most field-collected plants from one and the same population proved to have different fragment patterns. A study on the segregation of fragments in the offspring revealed some deviation from Mendelian distribution ratios as well as the occurrence of novel bands. To explain this, the authors suggested that the mutation rates of $(GATA)_n$ repeats may be unusually high in this particular species.

RAPDs were used in another study on *M. pygmaea* by van Heusden and Bachmann.[720] Ten strains from nine representative populations were analyzed with 24 primers. For each strain, DNA was extracted from 20 pooled offspring, obtained by selfing from a parental plant. The fingerprint data confirmed that coastal and inland populations form two distinct monophyletic groups. On the other hand, relatively weak differentiation was observed among populations within each of these groups, indicating some gene flow and recombination between neighboring populations. RAPDs were also used in population studies of two other selfing annuals in the genus *Microseris*: *M. bigelovii* and *M. elegans.*[718,719] Whereas the data suggested considerable gene flow for the former species, genetic recombination appears to be very scarce in the latter. The results from the RAPD study on *M. elegans* were found to be compatible with those obtained from $(GATA)_4$ hybridization as well as isozyme analyses.[718]

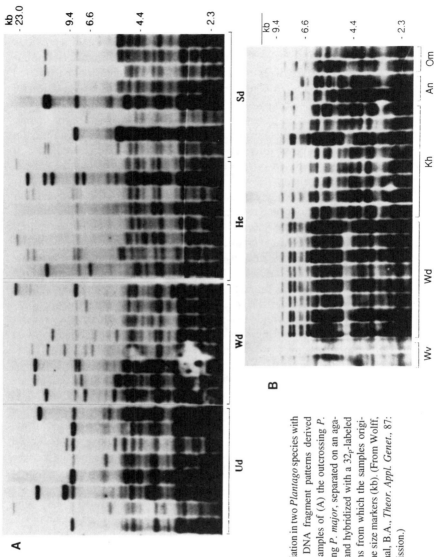

FIGURE 21. Genetic variation in two *Plantago* species with different breeding system. DNA fragment patterns derived from *TaqI*-digested DNA samples of (A) the outcrossing *P. lanceolata* and (B) the selfing *P. major*, separated on an agarose gel, Southern blotted, and hybridized with a 32$_P$-labeled M13 probe. The populations from which the samples originated are indicated, as are the size markers (kb). (From Wolff, K., Rogstad, S.H., and Schaal, B.A., *Theor. Appl. Genet.*, 87: 733–740, 1994. With permission.)

Whereas most examples have dealt with one species only, several species with different breeding systems have been studied in plantain, *Plantago,* by Wolff et al.[798] Within- and between-population variation in three *Plantago* species was investigated in a set of Dutch populations. The levels and patterns of variation detected by hybridization of DNA samples to the M13 probe were in close accordance with those expected from the different breeding systems of the species (Figure 21). The highly selfing *P. major* showed relatively little variation within but pronounced differentiation between populations. On the other hand, the obligately outcrossing *P. lanceolata* exhibited high variation within and moderate differentiation between populations. The fingerprints obtained for *P. coronopus,* with a mixed breeding system, showed levels of variation intermediate between *P. major* and *P. lanceolata.* The data obtained for intrapopulational variability by M13 fingerprinting corresponded, generally, to previous assessments of allozyme variation.[792]

Unexpectedly low levels of genetic variation were encountered in a DNA fingerprint study by Rogstad et al.[592] of the insect-pollinated North American paw-paw, *Asimina triloba,* when DNA samples were hybridized to the M13 probe. Wide-range samples of trees collected from the northwestern U.S. yielded an average bandsharing of 68 to 70%, while instead plants collected from within small areas yielded 75 to 84%. Even higher bandsharing values were obtained for narrow-range samples collected within the state of Missouri. Here, all 20 plants sampled within a plot of 100 m^2 exhibited identical DNA fingerprints. Moreover, fragment patterns from nine offspring plants proved to be identical with their seed parent. These data suggest that clonal propagation and/or high levels of inbreeding are responsible for the lack of genetic variation in *A. triloba* populations. However, apomixis, albeit not described previously for this genus, cannot be ruled out.

The perennial herb *Datisca glomerata* is androdioecious, i.e., individual plants are either strictly male or hermaphroditic. The maintenance of such a system requires a minimum twofold higher reproductive success of the male plants as compared with the hermaphrodites. An even greater advantage is required in partially self-fertilizing populations, as the gain in fitness through increased pollen production is least when few ovules are available for outcrossing. Thus, a high level of outcrossing in the hermaphrodites is a critical point in explaining the evolution of this very rare reproductive system. In a study using RAPDs, the offspring of 12 hermaphroditic plants and their seed parents collected in a population in California, were analyzed by Fritsch and Rieseberg.[201] A survey of 340 primers yielded 12 polymorphic loci. An outcrossing rate of 92% was demonstrated, thereby satisfying the theoretical requirements for the maintenance of androdioecy in this population. In a second Californian population, investigated with 19 polymorphic RAPD marker loci, an outcrossing rate of 65% was demonstrated in the hermaphrodites, which is still sufficiently high to account for the maintenance of androdioecy.

Hordeum spontaneum is thought to be the immediate evolutionary progenitor of cultivated barley. A RAPD analysis was performed by Dawson et al.[143]

in an attempt to quantify levels of genetic variation between and within 20 *H. spontaneum* populations in Israel. Distribution of variability between and within populations was found to vary considerably between loci. On the average, 57% of the genetic diversity was distributed within and 43% between populations. Individual estimates of within-population variability correlated reasonably well with those obtained in previous isozyme studies for the same populations. Cluster analysis UPGMA, as well as PCA and regression analyses, grouped the different populations in a pattern that suggested an association between RAPD data and ecogeographical factors such as the amount of rainfall and type of plant community.

2. Vegetative Propagation, Sexuality, and Apomixis: The Influence of the Reproductive System on Diversity

Many plant species form clones by various means of vegetative propagation. A species that is well known for its vigorous root suckering is the North American quaking aspen. Single clones of this wind-pollinated tree species may cover large areas. In a study by Rogstad et al.[591] on the genetic variability and clonal nature of quaking aspen, plants were collected from (1) several different locations in Colorado, (2) along a 4-km footpath at one of these locations, and (3) in two dense stands about 1 km apart. A surprisingly high number of different genotypes were encountered, with all plants sampled in different locations exhibiting different DNA fragment profiles. Average bandsharing values of 35 to 49% were calculated following digestion with one of the enzymes *Dra*I, *Hae*III, or *Hinf*I and hybridization to the M13 probe (Rogstad et al.[591] and unpublished results). However, according to the fingerprint data, the two dense stands comprised two and three clones, respectively, with the three different restriction enzymes yielding consistent results. More recently, RAPD analysis has also proven useful in the identification of aspen and poplar clones.[100,410]

The genetic diversity of stable vs. declining stands of another vegetatively propagating species, namely the reed *Phragmites australis*, was investigated at five lakes in Germany using two different DNA fingerprint methods: (1) hybridization of restriction-digested DNA samples to the $(GATA)_4$ probe and (2) PCR-based fingerprinting with $(GACA)_4$ as a primer.[379,500] With both techniques, plants from different lakes exhibited different DNA fingerprints. Sites within a lake sometimes consisted of different genotypes, whereas others appeared to be homogeneous.

Apomixis is the ability to set seed without prior fertilization.[24] However, pollination is frequently necessary to trigger the embryo development in a process called pseudogamy. Since many apomictic plant species also propagate vegetatively, clones can originate either from apomictically derived seeds or by some means of vegetative propagation. Genotypic distribution was investigated by Nybom and Schaal[516] in two North American *Rubus* species with DNA samples hybridized to the M13 probe. Both species exhibit vigorous vegetative propagation, but the way seed is produced differs. Whereas the black raspberry, *R. occidentalis,* reproduces sexually, the highbush blackberry, *R. pensilvanicus*, appears to be facultatively apomictic. Among the 20 shoots sampled from each species along a 600-m transect, only 5 different fragment patterns (i.e., genotypes) were found in *R. pensilvanicus,* as opposed to 15 in *R. occidentalis.* Moreover, *R. pensilvanicus* shoots with an identical genotype occurred up to 500 m apart and only 2 to 4 m in *R. occidentalis.* Most likely, the large distance between plants of identical genotype in *R. pensilvanicus* is due to dispersal of apomictically produced seed.

A similar study was conducted in Sweden, involving another pair of *Rubus* species, which both propagate vigorously by root suckers. However, the red raspberry, *R. idaeus*, is a sexual diploid, whereas the blackberry, *R. nessensis,* is an apomictic tetraploid. In order to assess the number of clones typically encountered within populations of these two species, 25 plants of each were collected from an approximate 150 × 50 m area in a brush forest. DNA samples were hybridized to the M13 probe as well as to a poly(GT) probe (Figure 22).[13] All 25 red raspberry plants exhibited individual-specific fingerprints, with a mean bandsharing of 45% for the M13 probe and 53% for the poly(GT) probe. The fact that vegetative propagation could not be demonstrated suggests that clones of red raspberry plants, if they exist, are quite restricted in this population. On the contrary, all six well-dispersed blackberry plants analyzed so far exhibit identical DNA fragment patterns and apparently belong to a single, large clone.

3. Rare Species

An example of complete genetic homogeneity indicated by allozyme data is provided by *Lactoris fernandeziana*.[66] This polygamodioecious (unisexual as well as bisexual flowers, with both sexes on the same plant) shrub is the only living representative of the primitive flowering plant family Lactoridaceae, which has a fossil pollen record dating 69 million years back. The species is found only on the island of Masatierra in the Juan Fernandez archipelago off the coast of Chile. A total of 27 plants from 15 populations, representing the entire geographic range, were recently analyzed with RAPDs by Brauner et al.[66]; 16 primers were used, resulting in a total of 106 bands per plant that could be scored reliably. Of these, 26 proved to be polymorphic. Only 11 of the primers yielded band differences between individuals, with an overall polymorphism of 1.6%. Genetic diversity in this species indeed appears to be very low, and the populations have few detectable differences. Preliminary RAPD data on nine species of two other rare genera (*Dendroseris* and *Robinsonia*) endemic to the Juan Fernandez indicate that the majority of these species are considerably more variable than *L. fernandeziana.*[66]

4. Gymnosperms

Exceedingly low levels of genetic variation were encountered by Mosseler et al.[478] in a RAPD investigation of red pine, *Pinus resinosa,* using 69 primers. Several populations in the main distributional area around the Great Lakes area

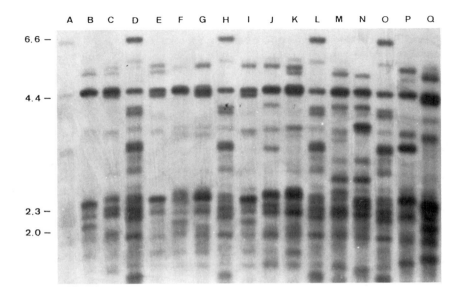

FIGURE 22. Genotypic distribution of the sexual raspberry, *Rubus idaeus*, vs. the apomictic blackberry, *R. nessensis*. DNA fragment patterns were derived from samples collected in the same location, DNA samples were digested with *Hae*III, separated on an agarose gel, Southern blotted, and hybridized with a ^{32}P-labeled M13 probe. Lanes B, C, E to G, I to K, M, N, P, and Q are *R. idaeus*; lanes A, D, H, L, and O are *R. nessensis*. Size markers (kb) are indicated.

in the U.S. and Canada, as well as some disjunct and long isolated populations in Newfoundland, showed almost identical fragment patterns. These results accord with a previously noted lack of allozyme variation, which has been explained by the species having passed through a genetic bottleneck during past glacial episodes of the Holocene. For comparison, three individuals each of the widely distributed white spruce, *Picea glauca*, and black spruce, *P. mariana*, were also analyzed. These exhibited high levels of RAPD variation as expected.

5. General Considerations

The different methods of DNA fingerprinting have proved to be very versatile tools in plant population studies as clearly demonstrated by the above-mentioned examples. Successful applications include e.g. detection of genetic variation within and between populations, characterization of clones, analysis of breeding system, and analysis of ecogeographical variation. These studies have taken benefit of the hitherto unparaleled ability of DNA fingerprinting to distinguish between even closely related genotypes.

Often, attempts are also made to quantify the fingerprint differences between genotypes and thereby to estimate levels of genetic relatedness. However, several factors may interfere with this kind of analysis, including comigration of nonallelic markers, linkage disequilibria between marker loci, inability to observe and evaluate smaller fragments, possible linkage of marker loci with loci under selection, and high and variable mutation rates. Some of the potential errors concerning evaluation of fragment patterns are treated in Chapter 4, Section V. When discussing the applicability of multilocus fingerprinting for plant population studies, one should also keep in mind that most limitations of the multilocus techniques (i.e., comigration of nonhomologous bands, linkage of bands, and band shifts)[421–423] are not only valid for DNA fingerprints generated by hybridization to multilocus probes, but also for DNA fingerprints generated by PCR amplification with arbitrary primers.[693]

B. DNA FINGERPRINTS YIELD INSIGHTS INTO PLANT TAXONOMY

Plant taxonomy research generally relies on a large number of characteristics to define taxa at different levels and to study their phylogenetic relationships. DNA fingerprinting may become a useful addition to this already existing reservoir of taxonomic tools, especially for problems involving the lower taxonomic levels as suggested by some examples mentioned below.

1. Apomictic Microspecies

Application of the species concept in apomictic plant genera has caused much controversy, with some authors recognizing vast numbers of so-called microspecies and other authors lumping these into larger aggregates.[24] In order to investigate variation between apomictic microspecies in *Rubus*, DNA samples from one plant each of several Swedish species were hybridized to the M13 probe as well as to a poly(GT) probe.[512] All species could be clearly distinguished from each other with each of the probes. Different fingerprints were obtained also in a PCR-based analysis of the same samples using either (GACA)$_4$ or the M13 repeat sequence as a primer (Figure 23).[513a]

In contrast, almost no within-species variation was detected with hybridization to the above-mentioned probes in a study by Kraft and Nybom.[370] of some Swedish and German *Rubus* species. Intraspecific variation among the Swedish populations was only found in *R. insularis*, where a geographically isolated population on the west coast deviated somewhat from the standard type. When Swedish and German populations of *R. infestus* were compared, only minor band differences were encountered, and none at all were found in *R. polyanthemus*. These results suggest that most of these microspecies have been derived through a single chromosome recombination event and that the resulting genotype, more or less unaltered, has subsequently spread over a large area.

Some results of the above-mentioned study are likely to affect future taxonomic treatment in *Rubus*. Thus, Swedish populations of *R. hartmanii* proved to be identical to a German population of *R. fuscus,* whereas Swedish populations of what was up till now regarded as *R. fuscus* exhibited a different fingerprint. Moreover, identical DNA fingerprints were found when two pre-

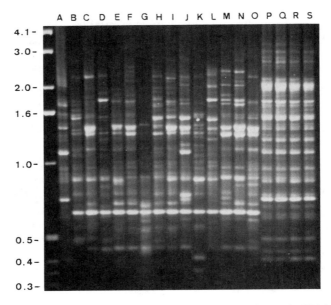

FIGURE 23. PCR-based DNA fingerprints obtained by amplification of genomic DNA from some apomictic blackberry species in Sweden using the M13 core sequence (GAGGGTGGXGGXTCT) as a primer. Amplification products were separated on an agarose gel and stained with ethidium bromide. Lanes A to O are one plant each of *Rubus scissus, R. sulcatus, R. lindebergii, R. grabowskii, R. hartmanii, R. vestitus, R. axillaris, R. infestus, R. polyanthemus, R. radula, R. divaricatus, R. wahlbergii, R. pedemontanus, R. pyramidalis,* and *R. insularis*; Lanes P to S show one plant from each of four different populations of *R. nessensis.* Size markers (kb) are indicated.

sumedly different species were compared, i.e., the Swedish *R. scheutzii* with the German *R. muenteri.*

Common dandelions, genus *Taraxacum*, comprise obligately apomictic triploids as well as some sexual diploids. To investigate whether apomictic microspecies are uniclonal or multiclonal, DNA fingerprints were obtained by van Heusden et al.[721] with the $(GATA)_4$ hybridization probe for some North and Central European species. One of these species, *T. hollandicum*, appears to be represented by the same clone in Czechoslovakia and France, as suggested by identical DNA fingerprints in the four populations sampled, whereas a slightly different fragment profile was encountered in a Dutch population. On the other hand, identical DNA fragment patterns were found in Dutch populations of *T. gelricum, T. maritimum,* and in two slightly dissimilar morphotypes of *T. palustre*, thereby suggesting that these three microspecies must have derived from the same clone. Two other morphotypes of *T. palustre* exhibited somewhat deviating fingerprints, most likely due to either somatic mutations or to genetic recombination between fairly similar parental genotypes. This investigation suggests that a taxonomic revision of the species complex would be appropriate, at least if additional probes yield consistent results.

2. Sexual Species

Introgression and hybrid speciation was studied with RAPDs in some American species of *Iris* by Arnold et al.[21] A total of 122 individual plants were analyzed, representing three and five populations, respectively, of the parental species *I. fulva* and *I. hexagona*; one population of the putative hybrid species *I. nelsonii;* and a hybrid population between *I. fulva* and *I. hexagona*. Three primers were used, resulting in several species-specific RAPD markers. One population each of the parental species showed some introgression, whereas the hybrid population exhibited intermediate frequencies for these markers. The *I. nelsonii* population also showed intermediate frequencies for some markers, whereas others appeared to have become fixed. In a later paper on the origin of *I. nelsonii,* a third species, *I. brevicaulis*, was also implicated.[20] Species-specific amplification products allowed estimation of the extent to which any of the three putative parental species were represented in each investigated individual in a population of *I. nelsonii* and also in two morphologically variable populations. Together with the analysis of the maternally inherited chloroplast DNA, these results suggested that *I. nelsonii* has been derived by pollination of *I. fulva* with the other two species, followed by recurrent backcrossing to the former species.

Another case study of hybrid speciation investigated with RAPDs even involved different genera, as ×*Margyracaena skottsbergii* from the Juan Fernandez islands was analyzed together with some putative parental species by Crawford et al.[130] Thirteen primers produced 18 consistent species-specific bands for *Acaena argentea* and 27 for *Margyricarpus digynus*, with all 45 bands present in the presumed hybrid ×*Margyracaena skottsbergii*. The RAPD analysis ruled out another putative parent, *A. ovalifolia*, since none of its 23 unique bands were found in ×*M. skottsbergii*.

Systematic relationships in *Juniperus* were investigated by Adams and Demeke[4] in an effort to examine the applicability of RAPD studies at different taxonomic levels. Between one and five plants were used per taxon in a material comprising 44 taxa and three sections. Similarity measures were computed for a total of 1624 RAPD bands, and principal coordinate analyses (PCO) were calculated on the entire material as well as on different subsets. It was concluded that RAPDs may be of taxonomic use ranging from sectional to varietal levels. Albeit nonhomologous band comparisons are sometimes scored as homologous, the statistical methods used are robust enough to offset such errors. In contrast, the authors advise against using RAPD data to construct parsimony based trees, since a few nonhomologous characters can drastically affect the tree shape. More powerful computer algorithms for PCO are now being developed to allow the elimination of bands that are polymorphic within species. This should render the RAPD method even more useful at the higher taxonomic levels.

3. Other Plant Groups

Even though most investigations published so far have dealt with sporophytic plants, DNA fingerprinting appears to be just as useful in other plant groups. Isolates of the red alga *Gelidium vagum* were investigated in a taxonomic study by Patwary et al.[544] using RAPDs. A total of 13 homozygous tetrasporic lines (obtained by selfing haploid gametophytes), as well as a clone of *G. latifolium*, were analyzed with 47 selected primers, revealing 322 loci of which 117 were polymorphic. A similarity matrix for all pairwise comparisons was calculated, and a dendrogram was constructed by application of UPGMA cluster analysis. As expected, the distance between different *G. vagum* isolates was much smaller than the distance between *G. latifolium* and *G. vagum*. It was suggested that the RAPD method may be useful in discriminating algal species as well as determining relationships within species.

Total DNA was isolated from species of the aquatic fern *Azolla* and its microsymbiont, the cyanobacterium *Anabaena azollae,* and analyzed with RAPDs by van Coppenolle et al.[716] The material comprised a total of 25 *Azolla* accessions, including all known species and encompassing the worldwide distribution of this aquatic fern. Twenty-two primers yielded a total of 486 useful RAPD markers. Bandsharing data were used to generate a dendrogram with UPGMA, which reflects previous classification based on other criteria. A similar grouping of accessions were obtained with PCA analysis. Valuable results on genetic variation in *Azolla* have also been obtained with the DAF method.[184]

4. General Considerations

DNA fingerprinting methods have so far been exploited only to a lesser extent in systematic investigations of wild species (examples on cultivated species are given below in Section III.A). However, these methods are apparently capable of producing useful data for species delineation in apomicts like blackberries and dandelions. DNA fingerprinting may become a useful tool for illustration of interspecific relationships also among sexual species as demonstrated in some other examples. Nevertheless, great care should be taken in the interpretation of these data, see Section II.A.5. Whatever fingerprint technique is used, evaluating a high number of characters (i.e., band positions) by using numerous probes/primers is highly recommended for obtaining valid estimates of genetic relatedness.

An important point to consider is the difference between phenetic and phylogenetic relationship (see Chapter 4, Section V.B.4). Several studies have shown that DNA fingerprint data can be used successfully to analyze phenetic relationships, e.g., by subjecting bandsharing values to PCA or PCO analyses. Phylogeny reconstruction, on the other hand, relies on specific characteristics in the molecular markers to be used, as discussed by Doyle[162] and also Soltis et al.[648] For this purpose, RFLPs and sequencing data obtained from homologous parts of the investigated genomes have proven very useful.[31,536,650,674,675,795] However, to what extent DNA fingerprint data can be subjected to meaningful phylogenetic interpretations remains to be investigated. In the meantime, we might do well to heed the cautionary views expressed by Williams and St. Clair[781] on the interpretation of RAPD data: "The dendrogram is a representation of a complex set of interrelationships rather than a phylogeny that implies unidirectional descent."

III. CULTIVATED PLANT SPECIES

A. CULTIVAR IDENTIFICATION AND ESTIMATION OF GENETIC RELATEDNESS

Unequivocal identification of plant cultivars is important for practical breeding purposes as well as for related areas like plant proprietary rights protection. Moreover, assessment of genetic diversity among cultivars and their wild relatives has recently attracted increased attention in efforts to cope with the commonly encountered reduction of diversity resulting from the breeding process. Cultivar identification and estimation of genetic relatedness has now been achieved with DNA fingerprinting in a large number of crops. Only a few representative examples are given here, for additional information see Appendices 1 and 2.

1. Vegetatively Propagated Cultivars

Apple — A variety of different fingerprinting studies have been performed on cultivar identification in the apple. DNA samples isolated from two or three trees each of the apple cultivars Golden Delicious, Jonathan, Red Delicious, and Rome Beauty *(Malus × domestica)* were hybridized by Nybom et al.[518] to the M13 probe in the very first attempt to apply DNA fingerprinting to this crop. Each one of five restriction enzymes provided cultivar-specific fragment patterns. Within-cultivar variation was only found in Rome Beauty and was most likely due to misidentification of the plant material. Later on, three *in vitro* cultures of apple rootstocks could be distinguished by hybridization to the M13 probe as well as to a poly(GT) probe.[519]

RAPDs have also been applied to the identification of apple cultivars. In one study by Koller et al.,[367] 11 different cultivars were distinguished using only two primers. Minor within-cultivar variation was sometimes noted for the five trees examined of each cultivar, but only when the PCR-reactions were run separately. In another study, two primers were used to distinguish eight apple cultivars, whereas different accessions of the same cultivar yielded identical fingerprints.[482] Using longer primers and polyacrylamide gel electrophoresis, most of the apple varieties of the *M. domestica* type could be told apart.[269] Not surprisingly, variation was found to be more prominent among some ornamental apple cultivars which represent different species. Cultivars as well as several representatives of wild apple species were also successfully differentiated with RAPDs, and the resulting bandsharing differences were used for the construction of dendrograms.[168]

A study of ornamental apple trees and their offspring showed that finger-print data were closely associated with known levels of genetic relatedness.[509] Fruits from five different genotypes had been collected after open pollination. From each of these maternal parents, 6 to 12 seedlings were analyzed by hybridization to the M13 probe. Comparisons between a maternal genotype and its offspring yielded an overall mean of 71% bandsharing, whereas comparisons among siblings and/or half sibs yielded 63% and unrelated individuals 41% bandsharing.

Blackberries and raspberries — Several *Rubus* cultivars have been analyzed with DNA fingerprinting. In an early study by Nybom et al.,[517] DNA samples from eight different blackberry and two raspberry cultivars were hybridized to the M13 probe. All cultivars had unique fragment patterns, and no within-cultivar variation was encountered. The average bandsharing between cultivars was 55%. Pairwise comparisons of bandsharing values yielded values of 63 to 90% among sibling cultivars, 60 to 89% between parents and offspring, and 7 to 44% among unrelated cultivars. The lowest values were obtained in comparisons between the blackberry cultivars on the one hand and the red raspberry cv. Titan as well as the purple raspberry cv. Lowden Sweet on the other hand. Similar findings were reported in a later DNA fingerprint study of blackberry and raspberry cultivars.[513]

In yet another *Rubus* study, DNA samples were hybridized by Parent and Pagé[541] to a 22-mer probe corresponding to the core sequence of the human myoglobin minisatellite. With this probe, all 13 red raspberry and 2 purple raspberry cultivars investigated were successfully distinguished. Bandsharing values ranged from 24 to 70%, with a mean of 47%, and were generally in accordance with the degree of interrelatedness as deduced from pedigrees. As expected, no variation was encountered within cultivars when analyzing the possible effects of age of the raspberry plantation, developmental stage during the growing season, or position of the sampled leaf on the stem. The same cultivars were subsequently analyzed with RAPDs.[542] All 15 cultivars could be distinguished using three primers. Again, no influence of environmental factors on band patterns was observed.

Carnation — Comparatively few DNA fingerprint studies have been carried out so far on ornamental crops. In carnation, some cultivars are propagated by seeds and others are propagated vegetatively. Nine cultivars and breeding lines of carnation, *Dianthus caryophyllus* and *Dianthus* hybrids, were successfully identified by Tzuri et al.[703] through hybridization of DNA samples to minisatellite and oligonucleotide probes. The human minisatellite probe 33.15 produced the best resolution of scoreable bands. In a subsequent study, several different carnation genotypes were analyzed with the human minisatellite probes 33.6 and 33.15 to investigate their genetic relatedness.[711] Generally, carnation cultivars are grouped into three main categories based on phenotypic characteristics: "standard", "spray", and "dwarf". Bandsharing values were generally high when individuals within the "standard" and "spray" categories were compared. Both of these are derived from *D. caryophyllus*. Considerably lower bandsharing values were obtained among cultivars in the "dwarf" category, possibly due to its hybrid origin. To assess the level of variation between these categories, bulk samples containg pooled cultivars from each category were analyzed. The highest bandsharing values were obtained between "standard" and "spray". The lowest values were obtained instead when a plant from a wild Alaskan species was compared to any of the cultivated plants. Comparisons between standard and spray categories on the one hand and dwarf on the other hand yielded intermediate bandsharing values. Taken together, bandsharing results obtained by DNA fingerprinting reflected the inferred relationships between and within categories of carnations.

Chrysanthemum — Ornamental chrysanthemums derived from the polyploid species *Dendranthema grandiflora* are usually propagated by terminal, vegetative cuttings. A set of 18 different cultivars were analyzed by Wolff and Peters-van Rijn[794] with RAPDs using eight primers. Levels of genetic variation were quite high, and one to three primers proved sufficient to distinguish all the investigated cultivars. Genetic diversity among species was studied using 6 primers and 15 individuals, representing 13 different species of the genera *Dendranthema* and *Chrysanthemum* (see Figure 18, Chapter 4). The mean bandsharing value for comparisons among species was 49%, which is significantly lower than the 66% encountered among cultivars of *D. grandiflora*. No relationship was observed between the ploidy level and the number of amplification products, which may be explained by the competitive behavior of RAPD primers (see Chapter 2, Section III.C).

Banana — In a comparative study by Kaemmer et al.,[340] 15 banana species and cultivars were analyzed by hybridization-based fingerprinting with a set of oligonucleotide probes as well as by PCR-based fingerprinting with the RAPD and the DAF methods. The samples included diploids, triploids, and tetraploids, comprising genomes derived from two different species; *Musa acuminata* (A genome) and *M. balbisiana* (B genome). Considerable polymorphism was demonstrated between genotypes with the three hybridization probes (GATA)$_4$, (GTG)$_5$, and (CA)$_8$, whereas no variation was found within cultivars, suggesting that the DNA fingerprints are somatically stable (Figure 24). Polymorphic patterns from these 15 genotypes were also obtained using single or pairwise combinations of short oligonucleotide primers in RAPD and DAF experiments, followed by separation of the amplification products on agarose or polyacrylamide gels and visualization with ethidium bromide or silver staining, respectively. Remarkably, a combination of two primers often yielded more information than using single primers. Both oligonucleotide hybridization and PCR amplification were useful for the discrimination of genotypes and also allowed the detection of bands characteristic for the A and B genome, respectively. A 0/1 matrix was derived from the banding patterns, based on 51 bands for the hybridizations and on 55 bands for the PCR amplifications, respectively, and analyzed by the parsimony programs MIX and DOLLO of the PHYLIP (Felsenstein) computer package (see Chapter 4, Section V.B.4). Though branching pattern and grouping of genotypes varied slightly between

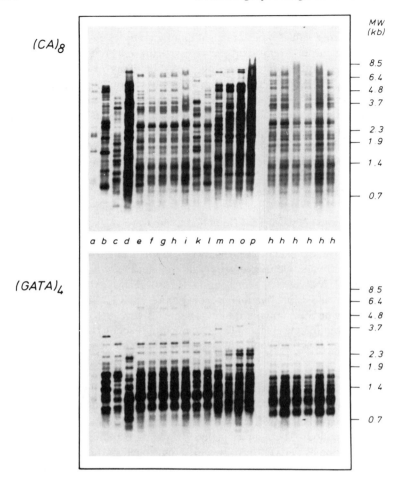

FIGURE 24. Oligonucleotide fingerprinting in banana: banding patterns are polymorphic between and stable within cultivars. Genomic DNA was isolated from individual plants of 15 *Musa* cultivars representing different genotypes, digested with *Hin*fI, separated on an agarose gel, and in gel-hybridized to the indicated ^{32}P-labeled oligonucleotide probes (left panel of each autorad, lanes a to p). Fingerprints of six individual plants derived from the same cultivar ("Grand Nain"; AAA genotype, corresponding to lane h) are shown in the right panel of each autorad. The same gel was used for both hybridization probes. Size markers (kb) are indicated.

different most-parsimonious trees, the genetic relatedness suggested by the majority of the dendrograms was in good agreement with the evolutionary history of the banana species and cultivars included in this study.

2. Seed-Propagated Cultivars

Cocoa — As in many other perennial crops, proper cultivar identification in cocoa, *Theobroma cacao*, traditionally relies on morphological characters that cannot be assessed until tree maturity. Using RAPD analysis with ten primers, a set of 13 genotypes could be successfully distinguished at the seedling stage by Wilde et al.[778] Wild cocoa species proved to be relatively dissimilar to the cultivated genotypes, whereas the latter exhibited various levels of relatedness. Bandsharing values among the investigated plants ranged from 51 to 86%.

Commercial cultivation of cocoa is based on a very restricted gene pool. To assess the amount and distribution of genetic variation in the species, Russell et al.[598] analyzed a total of 25 accessions from three populations from Peru and Ecuador with RAPDs using nine primers. PCO analysis clearly discriminated between the geographical origins of the three cocoa populations. Shannon's index of phenotypic diversity was used to quantify the level of polymorphism detected and to partition it into between- and within-population components. On average, diversity was higher within rather than between populations.

Papaya — Papaya cultivars, *Carica papaya,* and some related species were investigated by Sharon et al.[624] through hybridization of DNA samples with several different probes: R18.1 [a bovine probe containing a poly(GT) sequence]; the minisatellites M13, 33.15, and 33.6; and the simple sequence oligonucleotide (GTG)$_5$. Bandsharing values varied between 63 and 92% for the *Dra*I digested samples, depending on the probe. A study by Stiles et al.[663] instead made use of RAPDs to analyze ten papaya cultivars. Minimum similarity was as high as 70%, which was taken as evidence for a narrow basis of germplasm. All the examined cultivars could be distinguished with the RAPD analysis. However, within-cultivar variability was not examined, except in one case where a male as well as a female plant could be distinguished. The bandsharing data were used to generate dendrograms by UPGMA cluster analysis, as well as a parsimony analysis with the PAUP (Swofford) computer program. Using simple matching or the Jaccard's coefficient for evaluating the bandsharing data did not make any difference. Interestingly, the result obtained by UPGMA was more consistent with the known pedigrees than the parsimony analysis. However, possible occurrence of within-cultivar variation, as suggested by differences between a male and a female plant from the same cultivar, renders the deductions somewhat uncertain.

Brassica — Many economically important crop species are found in the genus *Brassica,* e.g., cabbage, turnip, and rapeseed. One approach to identify cultivars and estimate relatedness in *Brassica* species made use of hybridization to simple sequence probes. In the course of these studies, high levels of genetic variation were found between and within different cultivars of rapeseed, *B. napus,* with the (GATA)$_4$ probe, whereas five other probes proved to be rather uninformative for this cross-pollinating species (Figure 25).[555,756,757] A study by Poulsen et al.[557] of the differential abundance of simple repetitive sequences in species of *Brassica* and related genera showed that the amphidiploid genomes of *B. juncea, B. carinata,* and *B. napus* each harbor intermediate amounts of (GATA)$_4$- and (GACA)$_4$-detected repeats as compared to their diploid progenitors. Moreover, these results suggested that *B. nigra* is more closely related to *Sinapis arvensis* than to other *Brassica* species, thus supporting recent chloroplast, mitochondrial, and nuclear RFLP analyses.

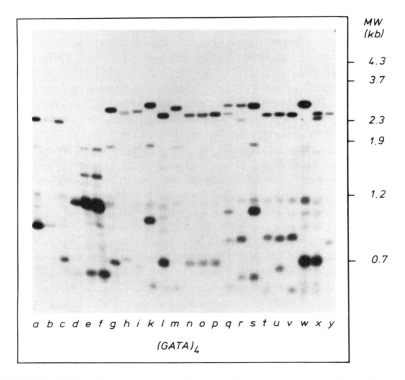

FIGURE 25. Within-cultivar variation of oilseed rape (*Brassica napus* var. *oleifera*). Genomic DNA was isolated from three individual plants of each cultivar: lanes a to c Ceres, d–f Jet neuf, g–i Tower, m–o Rally, p–r Diplom, q–s Hanna, t–v Line, and w–y Topas. DNA samples were digested with *Taq*I, separated on an agarose gel, and in gel-hybridized to a ^{32}P-labeled (GATA)$_4$ probe. Size markers (kb) are indicated.

In a RAPD study by Hu and Quiros[300] of *B. oleracea*, 4 primers were used to discriminate among 14 broccoli and 12 cauliflower cultivars. Three to five plants were pooled for each cultivar studied. The average number of polymorphic bands was 14.5 for comparisons between broccoli and cauliflower cultivars, 5.8 among broccoli cultivars, and 7.9 among cauliflower cultivars. Larger differences for each crop were found when cultivars from different seed companies, rather than from the same company, were compared.

In another study using RAPD methodology, 22 *B. oleracea* genotypes, representing 10 of the 14 recognized botanical varieties, were investigated by Kresovich et al.[373] with 25 primers. Two individuals of *B. rapa* were included as a reference and for comparisons among species. More than 140 reproducible, polymorphic bands were generated providing a unique fingerprint for each genotype. Levels of genetic variation between different cultivars of *B. oleracea* were, on the whole, in good accordance with known relationships. Thus, cultivars from the same botanical variety usually grouped together. The only exception was *B. oleracea* var. *costata* (Portuguese cabbage and kale), which is characterized by landraces that appear to be rather diverse in comparison to the other varieties. Comparisons between *B. oleracea* cultivars yielded bandsharing values of 65 to 94%, whereas comparisons between these cultivars on the one hand and cultivars of *B. napus* on the other hand yielded bandsharing values of only 18 to 47%.

In a third study using RAPDs in *Brassica*, cultivars from several different species were analyzed by Demeke et al.[149] together with white mustard, *Sinapis alba*, and radish, *Raphanus sativus*. DNA from four to six plants from each of 19 cultivars were pooled. With 25 primers preselected for being highly efficient at separating genotypes, 284 polymorphic bands were obtained. These bands were scored not only for absence vs. presence, but also for levels of intensity. PCO analysis performed on eight *Brassica* taxa, as well as *R. sativus* and *S. alba*, revealed the well-known relationships between diploid and amphidiploid *Brassica* taxa, whereas *R. sativus* and *S. alba* behaved quite distinctly. Five cabbage and four cauliflower cultivars were similarly analyzed with 13 primers. The two groups were readily separated, with the cabbage cultivars appearing to be genetically more diverse than the cauliflower cultivars. Finally, variation among six plants of *B. carinata* cv. Dodola was analyzed with 13 different primers. Six out of a total of 69 bands proved to be polymorphic among these six individuals, suggesting that pooling of DNA samples may unintentionally disguise significant within-cultivar variation.

A very interesting comparison of RFLP and RAPD markers for germplasm evaluation was carried out by Thormann and Osborn[693] on *Brassica* material comprising five accessions each of *B. oleracea, B. rapa,* and *B. napus;* one each of *B. nigra, B. carinata,* and *B. juncea;* and one of *Raphanus sativus*. Using RFLPs obtained by cDNA, genomic DNA clones, and RAPDs, a total of 432, 482, and 364 bands were scored, respectively. Similarity index matrices were calculated and used for UPGMA cluster analyses. The three different data sets provided similar information on within-species genetic relationships. However, cluster arrangement between species based on RAPDs differed from those based on the two RFLP data sets. A Mantel matrix correspondence test revealed a correlation between the two RFLP matrices of $r = 0.97$, whereas the RAPD matrix yielded $r = 0.87$ and $r = 0.88$ when compared to the other two. The authors suspected that the discrepancies were caused by the erroneous assumption that identical-size RAPD products from accessions of different species represent amplification products from homologous loci. This was tested by hybridizing Southern blots of RAPD gels with ^{32}P-labeled excised RAPD products. No less than 30% of the comigrating bands proved to be nonhomologous when the probe and the Southern blot were from different species.

Sugar beet — Different breeding lines of sugar beet, *Beta vulgaris* var. *altissima*, were hybridized to various simple sequence oligonucleotide probes with $(GATA)_4$ showing the most clear-cut results in terms of banding pattern and discrimination power.[48,618,757] $(GATA)_4$ allowed Schmidt et al.[618] to clearly differentiate four double-haploid breeding lines of var. *altissima* as well as of some other cultivars and one accession of the wild relative *B. vulgaris* subsp.

maritima. While no variation was observed within double-haploid breeding lines, variation was demonstrated to occur within cultivars. Therefore, breeding line- but not cultivar-specific fingerprints can be obtained.

A RAPD study of sugar beet also reported high levels of variation.[502] However, as was the case with oligonucleotide fingerprinting, the variation occurred both between and within cultivars, suggesting that varietal identification in sugar beet will be difficult with these approaches.

Legumes — Besides cereals, legumes are probably the best-characterized plant family in terms of DNA fingerprint approaches. A few examples will be given here to highlight important results. Hybridization with simple sequence probes has been used to analyze different accessions of the self-pollinating legume crop chickpea. High levels of variation were observed by Weising and co-workers with some, but not all probes (see Figure 7, Chapter 2).[623,755–757,759–761] Not only accessions, but also individual plants could be distinguished with (GATA)$_4$ and (CA)$_8$. Some probes, on the other hand, yielded fragment patterns that were identical between all cultivars (Figure 7, Chapter 2).[623,761] Such probes were instead highly useful for distinguishing between a set of eight different *Cicer* species.

High levels of polymorphism were observed in a study of 54 bean, *Phaseolus vulgaris*, accessions by Gepts et al.[217] Levels of diversity revealed by M13 hybridization (one enzyme-probe combination) were comparable to those detected by RFLPs (36 enzyme-probe combinations) and higher than those detected by isozymes (10 polymorphic enzymes). By comparing wild bean populations to cultivated races, it could also be demonstrated that the gene pool of the cultivated bean is characterized by a reduced genetic diversity. Differences among some commercial cultivars became only detectable after digestion with several enzymes and probing with different minisatellites.

PCR-based methods have also proved useful in fingerprinting beans. However, in a study by Skroch et al.[635] of 10 snap bean genotypes, more than half of the 400 RAPD primers used were unable to detect any polymorphism. RAPDs were also used to analyze a set of 285 different bean genotypes, yielding 88 informative markers.[635] PCA analysis revealed a pattern that is consistent with the separation of the common bean germplasm into two genetically distinct centers of diversity — Andean and Mesoamerican — with the snap bean genotypes forming a bridge between these two groups.

Halward et al.[256,257] used ten primers in a RAPD study on two cultivars of peanut, *Arachis hypogaea,* and 25 unadapted germplasm lines. Surprisingly, no variation was encountered in this material in spite of the fact that the collections were obtained from a geographically very large area and represented all four known botanical varieties. Moreover, identical DNA amplification product profiles were found also in *A. monticola,* the wild allotetraploid progenitor of cultivated peanut. These results suggest that the genetic variability in cultivated peanut is very low. In contrast, *A. glabrata,* a tetraploid perennial species from another section, exhibited clearly deviating DNA amplification product patterns. Moreover, a set of wild and cultivated *Arachis* species, as well as a synthetic amphidiploid, yielded high levels of variation in another study, especially for comparisons involving different genome types.[389]

Twenty cultivars and accessions representing four species of the Australian forage tree genus *Stylosanthes* were analyzed by Kazan et al.[346] with RAPDs using 22 arbitrary primers. The total of about 200 amplification products exhibited low levels of polymorphisms within species (0 to 16%), but higher levels among species (28 to 46%). Very few polymorphisms (0 to 2%) were detected among the five plants investigated from each cultivar or accession. These results were further elucidated by a cluster analysis, where the four main clusters each comprised one species. The estimates of genetic relatedness obtained with the RAPDs were in good agreement with other studies from morphology, cytology, and enzyme electrophoresis. Moreover, it was noted that amplification products that were polymorphic within species were usually weakly staining, whereas those polymorphic between species were usually quite intense. An in-depth study of genetic variation in the *S. guianensis* complex was performed with 45 accessions.[347] Again, very low levels of genetic variation were recorded within accessions, whereas the higher levels of variation encountered between accessions were in good accordance with previous taxonomical classifications.

Alfalfa, *Medicago sativa*, is an outcrossing species, and, therefore, cultivars are genetically heterogeneous. In addition to single plant analyses, bulked genomic samples with equal amounts of DNA from five or seven plants per population were amplified with RAPD primers by Yu and Pauls.[827] The profiles resulting from bulked DNA consisted in most cases of a combination of all bands exhibited by the individual plants. The order of the genetic distance values in pairwise comparisons of three cultivars was the same regardless of whether based on bulked samples or on individual plants.

Celery — Various cultivars of *Apium graveolens,* representing three different botanical varieties: i.e., 21 celery (var. *dulce*), one celeriac (var. *rapaceum*), and one annual smallage (var. *secalinum*), were analyzed with RAPDs by Yang and Quiros.[818] In a first set of experiments, within-cultivar variation was studied on three celery cultivars using 14 primers and two different approaches: (1) by bulking DNA from 5 to 50 plants from each cultivar and (2) by analyzing 12 individual plants from each cultivar. Two established cultivars turned out to be homogeneous, whereas the third, a variable breeding line, exhibited substantial levels of genetic variation with both approaches. Since bulking proved to yield satisfactorial results, bulks of 5 to 20 individuals were used for the main study involving 28 arbitrary primers and 23 cultivars. There were 309 bands detected, of which 96 were variable. However, only 29 of these were reproducible and therefore selected as useful polymorphic markers. All cultivars could be distinguished, with the celeriac and the annual smallage being more dissimilar to each other and to the celery cultivars than the latter were among themselves. The PAUP program was used to construct a dendrogram based on parsimony, according to which the celery cultivars were divided into three groups, while four cultivars remained unassigned outside these clusters.

This grouping was in good agreement with previous results based on six stem protein markers.

Tomato — Genetic diversity in cultivated tomato as assessed by RFLPs and allozymes is generally reported to be rather low. However, Vosman et al.[732] were able to distinguish all 15 investigated tomato cultivars by hybridization of DNA samples with the (GATA)$_4$ probe and also, though with a lower number of polymorphic bands, with the (GACA)$_4$ probe. Within-cultivar variation was not encountered, neither in the true-breeding cultivar Moneymaker nor in the hybrid cultivar Calypso.

In another study, three species of wild tomato as well as several cultivars were investigated by hybridization with different simple sequence oligonucleotide probes[341a] (Figure 26). (GGAT)$_4$ proved to be the most efficient probe for cultivar identification, showing high levels of between- and low levels of within-cultivar variation. Bandsharing data were analyzed with different computer programs, e.g., the distance matrix program CLORD and the parsimony program MIX (Figure 26). While branching patterns and the grouping of genotypes varied slightly between different dendrograms, the suggested relationships are in good agreement with data previously reported from investigations of chloroplast and nuclear RFLPs, as well as the proposed origin of cultivated tomato.

Williams and St. Clair[781] explored the diversity within and between various groups of tomato in a RAPD study on 46 accessions chosen to encompass old and new cultivars as well as South American regional cultivars and wild material. Two accessions of *Lycopersicon cheesmani* from the Galapagos Islands were also included to provide an outgroup for the phylogenetic analysis. A total of 215 RAPD markers were scored, 80 of which were polymorphic. Levels of genetic variation, as deduced form the RAPD data, were clearly associated with the plant breeding history of tomatoes. Similar groupings were obtained from dendrograms constructed with, respectively, UPGMA clustering and the PAUP parsimony program.

In an attempt to compare the efficiency of generating genetic markers with allozymes, RFLPs, and RAPDs, Foolad et al.[196] studied three tomato cultivars and one accession of the wild species *L. pennelli*, with the latter methodology yielding numerous polymorphic bands. It was concluded that RAPD analysis appears to be the most promising method for future construction of intraspecific molecular genetic maps in tomato.

Allium — Seven cultivars of onion, *A. cepa*, and one each of Japanese bunching onion, *A. fistulosum*, chive, *A. schoenoprasum,* and leek, *A. ampeloprasum,* as well as one plant of the wild species *A. roylei,* were analyzed with RAPDs by Wilkie et al.[779] Seven out of 20 primers revealed polymorphisms between cultivars of *A. cepa*. Between the different species, wide variation in DNA fragment patterns was observed with nearly every primer tested. Genetic distances based on bandsharing data were calculated, and cluster analysis was applied to generate a dendrogram. The genetic relationships indicated were in broad agreement with previous classifications based on morphology and cpDNA analyses. Homology between sequences representing

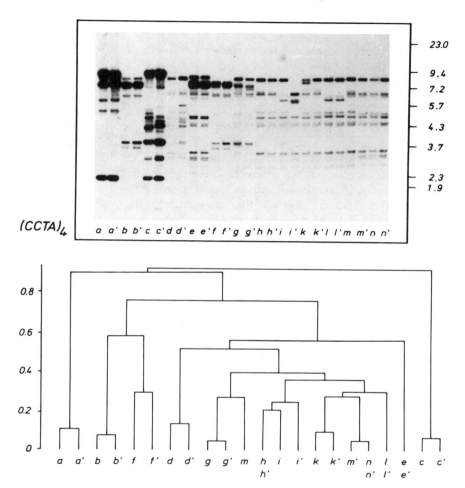

FIGURE 26. Estimation of relatedness between different genotypes of wild and cultivated tomato by oligonucleotide fingerprinting. Upper panel: wild species, varieties and modern cultivars of tomato: a, a' *Lycopersicon hirsutum*; b, b' *L. pimpinellifolium*; c, c' *L. peruvianum*; and d, d' to n, n' different cultivars of *L. esculentum*. Genomic DNA from individual plants (two per cultivar) was digested with *Hin*fI, separated on agarose gels, and in gel-hybridized to different ^{32}P-labeled oligonucleotide probes. Only the results obtained with the (CCTA)$_4$ probe are shown here. Size markers (kb) are indicated. Lower panel: unrooted dendrogram obtained by cluster analysis. Data were compiled as a (0/1) matrix from 168 scored fingerprint bands revealed by the probes (GATA)$_4$, (GTG)$_5$, (GGAT)$_4$, (CCTA)$_4$, (GAAGTGGG)$_2$, and (CAG)$_5$. The relative genetic distance is given above the dendrogram. The grouping of tomato species and varieties according to fingerprint data is consistent with current views of tomato phylogeny and breeding history.

common DNA amplification products from different cultivars was confirmed by Southern blot analysis of RAPD gels using DNA from isolated amplification products as probes. For some of these probes, a decreased signal intensity was noted when they were hybridized to DNA from other species, suggesting copy number polymorphisms or, alternatively, differences in overall amplification efficiency. However, a lower sequence homology between probe and sample DNA would also yield fainter bands.

Cereals — A substantial number of DNA fingerprint studies have been carried out in the economically important cereal crops. Some of these have been confined to discrimination of different cultivars and/or breeding lines, whereas others have also aimed at exploring levels of genetic relatedness in the investigated material. In one of the first reports on plant DNA fingerprinting, Dallas[137] was able to distinguish rice cultivars using the 33.6 and 33.15 probes. Intracultivar variation was not observed. Also simple sequence oligonucleotide probes like (GATA)$_4$ have since proven very efficient for fingerprinting rice.[566]

Separation of rice cultivars was also demonstrated in the first paper on AP-PCR.[766] In a later study on rice using the AP-PCR approach, the universal M13 sequence primer as well as a 17-mer primer were shown to result in unusually large numbers of polymorphisms between cultivars, especially when the amplified products were run on a polyacrylamide gel and silver stained.[537] In a RAPD study, Fukuoka et al.[203] analyzed 16 rice accessions with 36 10-mer primers; 28 yielded a total of 116 polymorphic amplification products. All accessions, representing the three ecospecies generally recognized in *Oryza sativa*; i.e., *Indica, Japonica,* and *Javanica*, were uniquely characterized with at least one of these RAPD markers. Bandsharing data were subjected to a cluster analysis. The two *Indica* accessions formed one relatively isolated cluster, whereas the two Javanese accessions grouped between the *Indica* and the *Japonica* clusters, but closer to the latter. The large cluster of 14 Japanese accessions could be further subdivided into two main groups, which also reflected known relationships.

Pedigree relationships in spring barley were studied by Tinker et al.[696] using RAPDs. A set of 27 inbred lines with varying degrees of common ancestry were analyzed with 33 primers. Bandsharing data were used to compute genetic distances, which were then compared to kinship coefficients as deduced from the known pedigrees. A linear relationship between the resulting matrices was evident, but low kinship coefficients gave only poor predictions of genetic distance. Cluster analysis showed that groups of inbred lines based on the pedigree data were generally similar to those based on the RAPD data, with some notable exceptions.

Disappointingly low levels of between-cultivar variation were encountered in a RAPD study of 12 bread wheat cultivars, *Triticum aestivum,* analyzed by Devos and Gale.[153] The authors concluded that in wheat polymorphisms were visualized at similarly low levels with RAPDs as with traditional RFLPs. A different approach was taken by He et al.[277] who analyzed 13 cultivars and breeding lines of wheat. Here, pairs of arbitrary primers were used for the amplification, and resolving the products with DGGE resulted in high levels of polymorphisms. Submitting the same samples to agarose gel electrophoresis resulted in little or no reproducible polymorphisms.

A set of 14 cultivars of soft red winter wheat and 2 cultivars of hard red winter wheat were distinguished by Dweikat et al.[172] using five single primers and three combinations of two primers each and subsequent DGGE of the resulting amplification products. Again, using conventional agarose gel electrophoresis instead of DGGE resulted in little or no polymorphism between cultivars. The 32% polymorphic DNA bands obtained with DGGE pointed to an 85% average similarity between cultivars. A UPGMA-generated dendrogram indicated two main groups, i.e., A (soft winter wheats) and B (hard winter wheats) with two subgroups within A. In the same study, interesting results with RAPDs were reported also for barley and for oats, *Avena sativa*. In a study on diploid wheat, Vierling and Nguyen[730] analyzed the extent of intra- and interspecific genetic variation using several genotypes of the primitive wheat cultivar *T. monococcum* and the wild species *T. urartu*. Data from a RAPD analysis with 60 primers were used to construct dendrograms with UPGMA to illustrate the relationship between the various wheat genotypes. Quite expectedly, levels of genetic variation were shown to be lower in cultivated *T. monococcum* with 41 to 81% bandsharing than in the wild species *T. urartu* with 32 to 68% bandsharing. Clusters obtained within the *T. urartu* dendrogram were in good accordance with the geographic and climatic origin of the genotypes investigated. Similarly, a set of 20 wild emmer wheat accessions, *T. turgidum* subsp. *dicoccoides*, proved to be considerably more variable than a set of 10 durum wheat cultivars, *T. turgidum* subsp. *durum*, when analyzed with 40 RAPD primers.[337]

Maize is normally outbreeding, but inbred lines can be produced and have been very useful for development of hybrid cultivars. Six inbred maize lines were analyzed by Welsh et al.[769] with AP-PCR using the radioactively labeled universal M13 sequence primer. With this approach, one to four diagnostic bands were produced for each line. Moreover, these polymorphisms were combined in offspring resulting from pairwise crossing of the inbred lines.

Very low levels of overall genetic variation were encountered by Tao et al.[683] in a RAPD study on grain sorghum, *Sorghum bicolor*. The mean frequency of polymorphic bands was 0.117, which is similar to results obtained in an RFLP analysis on the same material. A cluster analysis was performed on the RAPD data using UPGMA. Some distinct clusters consistent with established classification could be discerned. However, striking discrepancies were also noted, especially with regard to geographic origin.

3. Vegetative Sports and Other Mutants

Somatic mutations affecting plant structure and productivity give rise to so-called "sports". Some of these constitute an improvement of the original genotype and have thus been registered and patented as varieties of their own. However, most sports deviate from the original variety only in minor characteristics and may thus be very difficult to distinguish genetically. Hybridization to minisatellite probes has proved successful in detecting somatic mutations, e.g., in human cancer research.[507] A similar approach was attempted in a study of 15 different sports of Red Delicious apples.[510] No qualitative differences in fragment patterns were found. However, some minor differences in relative fragment intensity were noted, which might be explained by the chimaeric origin of many Red Delicious mutations. Similarly, no variation was found among sports of several other apple varieties using the RAPD methodology.[482] This is really not surprising since it was estimated that only about 0.002% of the entire genome was screened, making the detection of one or a few mutations quite unlikely. Mulcahy et al.[482] thus recommended that attempts to obtain markers for point mutations should employ sibling populations segregating for the mutation of interest. These populations can then be analyzed with the "bulked segregant analysis" as described by Michelmore et al.[461]

In the vegetatively propagated chrysanthemum, sports with, e.g., different flower colors are much sought after, but sometimes difficult to distinguish unambiguously from one another. Thirteen different clones derived from the same original genotype were investigated with RAPDs using 27 primers.[794] Some marginal variation for the fainter bands was noted for one of these primers but otherwise they yielded completely identical fragment patterns.

There are also some examples where fingerprint techniques have been successfully used for mutant discrimination. A mutated banana plant was thus distinguished by Kaemmer et al.[340] from its parental line with a RAPD primer, either by itself or in combination with another primer. Similarly, a mutated cherry plant, *Prunus cerasus × canescens*, was differentiated from its parental line by 1 out of the 40 RAPD primers tested.[817] Two ethylmethanesulfonate (EMS)-induced supernodulating near-isogenic soybean lines and the cultivar Bragg, from which they originated, yielded identical DAF fingerprints when analyzed by the usual approach.[91] Interestingly, however, restriction enzyme digestion of the DNA samples prior to amplification instead revealed numerous polymorphisms, closely linked to the supernodulation locus as deduced from analysis of F_2 progeny.

4. *In Vitro*-Propagated Plant Material and Somaclonal Variation

Tissue-culture techniques are commonly used in many plant species, and it is well-known that a proportion of regenerants often differs from the parental type, a phenomenon called "somaclonal variation". This variation is thought to originate either from the release of genetic diversity preexisting in the explant or from *de novo*-acquired variability originating during cell-line dedifferentiation or callus maintenance *in vitro*. However, as the following examples show, such variation is rarely visualized with the DNA fingerprint techniques.

The first effort to monitor the effects of somatic cell manipulation with minisatellite DNA hybridization was made by Dallas[137] who compared control plants of a rice cultivar with progeny regenerated from protoplasts. One very faint fragment was found in all regenerated plants, but not in the control. All the other fragments were identical in all plants investigated, suggesting that the regeneration process had no major effect on the hybridization pattern. Hybridization to M13 and simple sequence probes of DNA samples from protoplast-derived apple plants[512a] and strawberry plants, *Fragaria × ananassa,*[522] showed no variation.

(GATA)₄

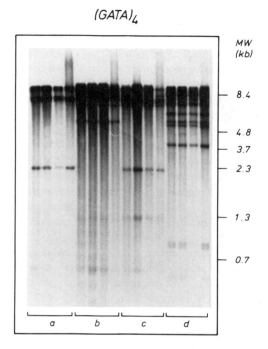

FIGURE 27. Stability of clone-specific oligonucleotide fingerprints during micropropagation of pharmaceutically important yarrow plants (*Achillea millefolium* species complex). Genomic DNA was isolated from cultured plant material in intervals of 3 months. During each interval, microshoots were passaged twice onto fresh medium. DNA samples were digested with *Taq*I, separated on an agarose gel, and in gel-hybridized to a ^{32}P-labeled (GATA)$_4$ probe. Lanes a to c are different clones of *A. asplenifolia*, and lane d is *A. roseoalba*. The four lanes shown for each of the clones represent different time points of harvest of cultured material (i.e., 0, 3, 6, and 9 months, from left to right). Size markers (kb) are indicated.

Several micropropagated clones of the medical plant *Achillea millefolium* were screened for polymorphisms by hybridization-based fingerprinting with simple sequence probes.[735a] While, e.g., (GATA)$_4$ fingerprints allowed the clear discrimination of clones, no intraclonal variability was observed after several rounds of subculturing (Figure 27). Similar results have been reported also for cultivated beet[618] and red raspberry.[292,541]

Plants of basket willow, *Salix viminalis*, originating from the same callus culture, exhibited identical DNA fingerprints upon hybridization with the M13 probe, whereas different clones could be told apart.[709] Material of tomato, consisting of 95 callus lines of cultivar Moneymaker and of 45 second-generation plants obtained from first-generation regenerants of tissue-cultured plants of Moneymaker, was analyzed by hybridization to the (GATA)$_4$ probe.[732] All samples yielded identical DNA fingerprints, even though several of the second-generation plants showed morphological variation like dwarf growth and tetraploidy as well as anthocyanin and chlorophyll aberrations.

A successful attempt to document somaclonal variation in *Brassica napus* was reported by Poulsen et al.[555] who compared three protoplast regenerants derived from the same plant of cultivar Topas. Two of these regenerants were morphologically similar, whereas the third was more compact. DNA samples were digested with various enzymes and hybridized to the (GATA)$_4$ probe. The morphologically deviating plant exhibited a size change of a *Taq*I fragment and a loss of a *Hpa*II fragment when compared to the other two plants.

Another successful application of DNA fingerprinting for the detection of somaclonal variation was reported by Nelke et al.[496] for red clover, *Trifolium pratense*. Hybridization of the 33.6 probe to genomic DNA, digested with various enzymes, could thus demonstrate several band differences between the original genotype and a regenerative somaclonal variant. However, none of the discriminating bands were linked to the regenerative trait.

Genetic stability of *in vitro*-propagated plant material has been monitored also with RAPD technology. Whereas these markers allowed the detection of preexisting polymorphisms among different fescue species as well as among donor genotypes of meadow fescue, *Festuca pratensis*, there was no newly generated variation detected in protoplast-derived plants as compared to control plants regenerated from suspension culture.[714] RAPDs were also used to assess the genetic stability of somatic embryogenesis-derived populations of black spruce.[309] No genetic instability in 30 somatic embryos of each of three cell lines could be detected with these markers.

A RAPD study by Brown et al.[74] made use of tissue culture-derived plants and seed-grown controls in wheat. While most of the investigated plants appeared to be phenotypically normal, some differences nevertheless did occur between regenerants and controls, especially in diploids. To clarify to what extent this phenotypical variation was related to the tissue culture method, RAPD analyses were performed with genomic DNA from a wheat line maintained as either callus or as a suspension culture. Regardless of culture method, noticeable differences in RAPD patterns were obtained relative to the controls as well as between the different methods of culture. Individual tobacco protoplasts derived from a single, stably transformed plant were also analyzed in the same study. In a first series of experiments, each protoplast turned out to have an apparently unique DNA amplification product pattern. However, including additional purification steps to remove proteins prior to PCR amplification resulted in considerably more homogeneous RAPD patterns. The authors suspected that the remaining variation was due to loss of DNA during isolation rather than to inherent differences between the protoplasts.

5. General Considerations

The many examples outlined above illustrate how cultivar identification can be achieved from DNA fingerprint data, especially in material characterized by high levels of genetic variation between cultivars and little or no variation within. Such examples are typically found in vegetatively propagated cultivars, but also in seed-propagated cultivars derived through selfing (e.g., most cereals, many legumes, and tomato). However, a different situation is encountered in cross-pollinating cultivars, exemplified by the genus *Brassica*. Here, considerable levels of intracultivar variability may occur which interfere with cultivar identification.[555] This problem has been addressed in various ways; in some studies (representing the easiest way to avoid this problem), only one specimen is examined from each cultivar. Others pool the DNA samples of a number of specimens and analyze them as an entity,[827] whereas still others analyze several individual plants and estimate the levels of variability within as well as between cultivars.[346,347] For in-depth studies, we strongly recommend the latter approach.

As yet, spontaneous somatic mutations, which often give rise to commercially important sports, have not been documented with DNA fingerprint methods.[482,510,794] On the contrary, artificially induced mutations have been detected with DNA fingerprinting in several cases,[91,340,817] which may be due to larger genomic rearrangements.

Tissue culture techniques generally do not give rise to variation in DNA fingerprint patterns. Accordingly, these molecular methods would be very suitable for cell line authentication at laboratories where large numbers of *in vitro*-cultured samples are handled. This approach has already been taken by the European Collection of Animal Cell Cultures.[657]

In the future, molecular markers will most likely play an increasingly important role in developing, acquiring, and enforcing plant proprietary rights.[335] However, several important criteria must be met before any descriptor can be considered as practically acceptable.[640] Among these are (1) high power of discrimination, (2) minimal interaction with environment (including laboratory procedures), (3) capability to generate data of equivalent meaning across laboratories, (4) capability to estimate genetic distances between cultivars, (5) information on genetic location and control, and (6) public availability of the methodology. Whereas the requirement for discriminatory power is easily met with DNA fingerprinting, there are still problems with most of the other requirements. Thus, reproducibility of, e.g., RAPD patterns has proved difficult between laboratories.[549] We strongly recommend that the potential of the various fingerprint methods described in this book, as well as of some more recent approaches mentioned among future prospects in Chapter 7, are carefully investigated before any decisions are made as to what kind of molecular markers should be used in legal matters.

Estimation of relatedness between cultivars with DNA fingerprinting is more complicated than mere identification, as already pointed out in Section II.A.5. However, the interesting results obtained with minisatellite and simple sequence hybridization, as well as with PCR-based methods, suggest that DNA fingerprinting may be a helpful tool in studies of genetic relatedness, as long as prudent caution is exerted in the evaluation of results.

B. PATERNITY TESTING AND THE IDENTIFICATION OF HYBRIDS

1. Seed-Propagated Plants and Sexual Hybrids

Whereas paternity determination has been one of the major applications of DNA fingerprinting in humans and many animal species, comparatively few studies have been presented in plants. The first report by Nybom and Schaal[515] deals with apple seedlings derived from open pollination of cvs. Golden Delicious and Jonathan in an orchard where these two were grown together with cv. Red Delicious. DNA fingerprints of the seedlings were generated by hybridization with the M13 probe (Figure 28). Of the 30 Golden Delicious seedlings analyzed, 5 appeared to have been sired by Jonathan and 24 by Red Delicious. One seedling showed some fragments not encountered in any of the putative parents and thus might have arisen by pollination from an outside source or by mutation. Of the 34 Jonathan seedlings, 9 appeared to have been sired by Golden Delicious, 16 by Red Delicious, 5 by either Golden Delicious or Red Delicious, and 2 by selfing, whereas another 2 exhibited fragments not encountered elsewhere. The analysis was greatly facilitated by the identification of a multiallelic locus with at least five alleles.

F_1 hybrids, obtained by crossing inbred parental lines, have become valuable cultivars in many outcrossing plant species where they produce homogeneous crops of high quality. However, self-incompatibility of the maternal parent is often not sufficiently strict to completely exclude selfing in some crops, e.g., cauliflower. In order to estimate the extent of selfing vs. outcrossing, a PCR protocol with arbitrary primers of 10 to 16 basepairs (bp) was developed by Boury et al.[61] Parental lines of cauliflower were distinguished with about half of the primers tested, and thus the resulting offspring could be classified into hybrids or inbreds. The PCR results were in good accordance with those obtained from later morphological evaluation of adult offspring plants, but could be obtained at a much earlier stage.

Though apomictic, *Rubus* species may also set some seeds sexually, especially when crosspollinated with a different species.[508] Therefore, it has been suggested that a large part of the vast number of existing blackberry species have been derived through such sexual hybridization events. Experimental cross-pollinations involving various blackberry species have been performed to study the relationship between apomixis and hybridization.[508] However, it is quite difficult to determine unambiguously whether the offspring are in fact derived from cross-pollination or from apomictic seed set unless the parental species involved deviate morphologically to a considerable extent. Recently, a number of blackberry seedlings, derived through experimental interspecific crosses within subspecies *Erecti,* were successfully fingerprinted by hybridizing DNA samples to the M13 probe as well as to a poly(GT) probe.[512a] The two probes yielded fully consistent results — i.e., samples could either be differentiated with each of the probes or with neither. Of the 59 offspring plants obtained from pollination of a nonhybridogenous plant (i.e., derived through apomixis), 54 exhibited a DNA fragment pattern identical to that of the seed

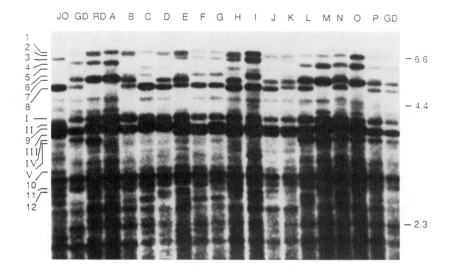

FIGURE 28. Paternity analysis in apple. DNA samples were digested with *Rsa*I, separated on an agarose gel, Southern blotted, and hybridized with a ^{32}P-labeled M13 probe. GD = Golden Delicious; JO = Jonathan; RD = Red Delicious; A, B, and L to P are seven seedlings that have GD as seed parent; and C to K are nine seedlings that have JO as seed parent. Pollen parents were presumedly GD, JO, or RD. Fragments utilized in this study are indicated to the left, I to V apparently belong to the same locus. Size markers (kb) are indicated to the right. (From Nybom, H. and Schaal, B.A., *Theor. Appl. Genet.*, 79: 763–768, 1990. With Permission.)

parent and were thus themselves also generated by apomixis, whereas the remaining 5 plants contained additional bands derived from the pollen parent, suggesting an average level of 8% sexuality in the present material. On the other hand, in plants resulting from pollination of hybridogenous plants (i.e., derived through sexual seed set after interspecific cross-pollination), 6 out of 10 were themselves also sexually derived. DNA fingerprinting may thus prove a fast and efficient technique for evaluation of progeny plants that would otherwise have to be grown for at least a year before characterization with morphological traits is possible.

Kentucky bluegrass, *Poa pratensis*, is another facultatively apomictic and polyploid (2n = 24 to 124) species, which sometimes produces aberrant progeny. This may have at least nine different genetic origins, depending on gametic ploidy level and whether fertilization was self, cross, or absent. By application of silver-stained RAPD markers, progeny resulting from cross-fertilization could be accurately identified in a study by Huff and Bara[303] and the inheritance of parental genomes quantified.

A promising application of DNA fingerprinting is the detection of gene introgression in cultivated species. Dihaploid genotypes have been obtained in, e.g., potato by pollination of tetraploid *Solanum tuberosum* cultivars with diploid species like *S. phureja*. However, the resulting plants may not always be parthenogenetic in origin as previously suggested. A RAPD analysis carried out by Waugh et al.[740] showed that some of the investigated dihaploid potato plants had also inherited at least some amplification products from *S. phureja*. Therefore, double fertilization must have taken place, most likely followed by preferential elimination of *S. phureja* chromosomes. Germplasm introgression was successfully monitored by RAPDs also in trispecies hybrids involving alfalfa and some wild *Medicago* species,[440] in chromosome addition lines resulting from interspecific hybridization in *Brassica*,[668] and in studies of wheat/alien chromosome substitution lines.[153,357]

2. Somatic Hybrids

Somatic hybrids can be obtained by fusion of protoplasts from different cultivars or even species. Simple sequence oligonucleotide DNA hybridization was used by Poulsen et al.[556] to analyze "resynthesized" *Brassica napus* derived through somatic hybridization of protoplasts from *B. oleracea* and *B. campestris*. While the DNA fragment patterns of different hybrid plants were identical, they did not simply represent a combination of the parental patterns. Instead, the absence of parental bands as well as the presence of new bands suggested that elimination and/or rearrangements occurred during or after the fusion of the two genomes. Some F_1 progeny plants of selfed hybrids were also analyzed, showing patterns more or less identical to the patterns of their parents. Therefore, the resynthesized amphidiploid *B. napus* genome, once formed, appears to be stable.

RAPDs were used by Baird et al.[30] to characterize both inter- and intraspecific somatic hybrids of potato. One set of plants had been obtained by hybridization of *Solanum tuberosum* and *S. brevidens*, whereas the other hybrid was formed with two different dihaploid *S. tuberosum* clones. Since RAPD markers are dominant, a useful single primer must generate an amplification product typical for each donor genotype so as to confirm the presence of both genomes in putative hybrids. Alternatively, two primers can be used sequentially if each of them detects products specific for one parent only. The hybridity of the somatic cell fusion lines was demonstrated with both of these models. A few primers, however, generated products in the parents which were not present in the hybrids. It was suggested that the combination of two different templates may lead to aberrations due to competition in the PCR reaction. Thus, one template might be amplified more efficiently than the other. However, genomic rearrangements, as suggested for the *B. napus* study outlined above, may also be involved.

In a similar RAPD analysis, five putative somatic hybrids between *S. tuberosum* and *S. brevidens* were investigated together with their parental genotypes and two protoplast regenerants of *S. tuberosum*.[813] Four out of five primers produced significantly different DNA amplification product profiles for the parental genotypes. The hybrids exhibited a combination of the two parental profiles, except for one primer which, for unknown reasons, instead revealed the same profile in the hybrids as in one of the parents.

3. General Considerations

For analysis of paternity in sexual hybrids, hybridization fingerprinting as well as RAPDs appear to be quite useful. However, all studies presented so far have dealt with situations where the putative fathers are rather few. Studies of mating systems in natural populations would in most cases involve a much larger number of putative fathers. Lewis and Snow[401] have suggested that we shall need to score more than 50 RAPD loci for each offspring for a paternity exclusion analysis. Though sounding quite laborious, such a number is actually realistic since more than one locus can usually be scored from one primer, suitable primers can be identified in preliminary surveys, and a considerable degree of automation is possible. Another problem is that RAPD patterns are, to some extent, affected by genetic background — i.e., parallel bands may or may not appear in a somatic hybrid or F_1, whereas even nonparental bands may occur in the offspring.[577] Therefore, hybridization-based fingerprinting may prove to be more reliable for paternity and hybrid testing than the PCR-based methods, at least when high levels of accuracy are demanded.

Paternity testing of somatic hybrids appears to be relatively complicated, with the absence of parental bands as well as the presence of new bands having been reported.[30,556,813] These results were obtained with hybridization fingerprinting as well as with RAPDs, suggesting that genomic rearrangements may be responsible.

IV. INHERITANCE OF DNA FINGERPRINTS, LINKAGE ANALYSIS, AND GENOME MAPPING IN PLANTS

The analysis of DNA fragment segregation to the progeny is extremely valuable for studies of genetic relatedness, regardless of which particular DNA fingerprint method is used. For example, in studies of birds, segregation analysis is often regarded as a prerequisite, and only those fragments showing the expected Mendelian inheritance are included in calculations of relatedness.[266] High levels of linkage between bands may distort the results, and such loci should be avoided or treated as a single locus. In one extreme case, 80% of the detected bands in a bird fingerprint derived from minisatellite hybridization were found to belong to the same locus.[266] Corresponding studies in plants have detected much lower levels of linkage so far, but appropriate caution is recommended.

The major aim of evaluating genetic linkage data, however, is to produce genetic maps. Such maps, comprising closely spaced DNA (and other kinds of) markers, are very useful for genome analysis (see review in Young[824]). A comprehensive description of the procedures used in a recent mapping project in *Arabidopsis* is given by Reiter et al.[573]

The first step to generate a linkage map usually involves screening for the most dissimilar (but still interfertile) genotypes of the species of interest. These genotypes are then crossed, and suitable progeny populations are analyzed for segregation and linkage of polymorphic markers. Linkage data are usually evaluated with the help of computer programs (see Chapter 4, Section V.B.5). DNA markers that are linked to a trait of interest can be used both as a starting point for gene cloning and to monitor trait introgression in plant breeding programs, i.e., marker-assisted selection. Until recently, the molecular markers used for genetic mapping were mainly RFLPs derived from low copy number genomic or cDNA clones. However, RFLP analysis is a time-consuming and also expensive technique. Lately, RAPDs have been introduced to genome mapping with notable success. One major problem is that RAPD markers are inherited in a dominant manner, thus limiting their value in F_2 populations.[784,785] There are various ways of circumventing this problem in some crops, such as mapping backcross populations or populations derived from inbred homozygous parents,[96] analyzing haploid progeny,[702] or using recombinant inbred lines.[572,573] For targetting specific regions of the genome, NILs (near-isogenic lines),[432] bulked DNA samples,[461] or populations already mapped with RFLPs,[222] have been made use of. Another possibility is to use monoploids when available. In this way, potato plants, anther derived from a single heterozygous diploid clone, could be investigated instead of seedling populations derived from crosses.[631]

A. MINISATELLITE AND SIMPLE SEQUENCE HYBRIDIZATION

Whereas both alleles at a locus can be scored with RFLPs, the complex DNA fragment patterns obtained with multilocus fingerprint probes usually do not allow us to identify allelic fragments. Therefore, we can rarely determine whether a given fragment occurs in a homozygous or a heterozygous state. By analysis of suitable populations such as an F_2 generation derived from two inbred lines, informative fragment segregation data can nevertheless be obtained. The DNA fragments present in the offspring are then usually found in one or sometimes both of the parental genotypes. Most studies carried out so far on the inheritance of fingerprint patterns unfortunately rely on very low numbers of offspring. Nevertheless, Mendelian inheritance has been reported in the vast majority of cases, whereas linkage between fragments was only rarely observed.[137,391,435,515,624,703,732] Genetic maps based on multilocus hybridization fingerprints have been performed in humans and animals,[339,765] as well as in fungi,[593] but not yet in plants.

B. PCR-BASED METHODS

PCR-based fingerprinting has quickly become a much-used tool in genome mapping in spite of some problems: (1) RAPDs are usually dominantly inherited; codominance and allelic relationships between pairs of amplification products have been reported in a few cases only, (2) efforts to relate staining intensity of the amplification products with homozygosity vs. heterozygosity have mostly been unsuccessful, and (3) non-Mendelian segregation ratios have frequently been reported for RAPD fragments. Deviations from Mendelian segregation, however, have been reported also for other markers such as RFLPs and allozymes, especially in connection with interspecific crosses, backcrosses, or self-incompatibility genes. Most importantly, PCR-based methods often enable us to analyze large numbers of offspring in a relatively short time, generating polymorphic markers and genetic maps much faster than other methods.[572,573]

Gymnosperms — Since genetic analysis and screening of forest trees is severely limited by their large size and long generation time, methods to improve the efficiency of early selection for, e.g., wood properties would be highly desirable. Lately, the RAPD technology has been introduced to genetic mapping in gymnosperms. In a study by Carlson et al.[96] of Douglas-fir, *Pseudotzuga menziesii,* and white spruce, theoretical considerations for the use of dominant molecular markers in genome mapping were addressed. Since levels of heterozygosity are generally high in conifers, segregating progeny often represent a backcross situation between a heterozygote and a homozygous recessive and thus lend themselves very well to RAPD analysis. Three crosses of Douglas-fir were investigated with six primers, yielding ten segregating RAPD markers. Of these, seven fitted a 1:1 segregation ratio expected for a backcross situation, whereas the other three did not. In the white spruce progeny, four primers detected three polymorphic markers, but their distribution was not statistically tested due to the low numbers of available offspring.

Tulsieram et al.[702] took advantage of the fact that conifer seeds have a relatively large haploid megagametophyte, which is completely maternal in origin. Therefore, megagametophyte DNA was analyzed in a batch of seeds from a single white spruce tree. Initial screening of 300 primers revealed 69 primers yielding polymorphic patterns for five DNA samples. However, only 42 of these revealed bands that segregated in a 1:1 ratio when applied to all 43 seeds analyzed in the following step. Segregation ratios of some bands were found to be highly skewed. Of the remaining 61 loci, 47 could be mapped to 12 linkage groups. These groups covered roughly a third of the genome and possibly represented all 12 chromosomes in the haploid set. Distribution of RAPD markers were studied also by Bucci and Menozzi[79] in 34 megagametophytes from a single spruce tree, *Picea abies.* Out of 165 polymorphic bands, 54 fitted a 1:1 segregation ratio and were considered Mendelian traits. Another 37 bands apparently also fitted a 1:1 segregation, but were faintly stained and not unambiguously scoreable. Though levels of segregation distortion for RAPD markers in spruce appear to be rather high, this phenomenon has been reported also in allozyme and RFLP studies.[79,96]

Mapping projects on gymnosperms and other woody plant species are now in progress at several laboratories around the world.[97,152,236–238,242,324,475,491,493,497,526,634,715]

Arabidopsis — Lately, the herbaceous weedy species *A. thaliana* has become a favorite organism in many plant molecular biology laboratories, mainly because of its small genome size and short generation time. In addition, it reproduces by selfing, thereby producing homozygous lines that are very suitable for genetic mapping studies. Several RFLP maps of this species have already been presented. Reiter and co-workers[572,573] applied RAPD methodology to develop a high-density genetic linkage map containing 252 RAPD markers with only four months' labor. A recombinant inbred population, which constituted a permanent set of fixed genotypes, was used for this study. All 46 plants analyzed were highly homozygous, but each for different alleles. Initially, about 1200 primers were tested, most of them 10-mer oligonucleotides. Of these, 245 detected polymorphisms between the two parental lines. A total of 392 polymorphic markers were obtained, 225 of which segregated in a Mendelian fashion. The remainder either segregated in a non-Mendelian ratio or were not reproducible. To determine the genomic abundance of RAPD products, 18 of these were labeled and hybridized to genomic DNA blots. Nine of the RAPD products hybridized to sequences present at three copies or less in the genome, whereas the other nine hybridized to sequences present at three to ten copies. Thus, RAPD markers may be converted into low copy RFLP probes, targetting a specific genomic region.

The RAPD markers were integrated into two already existing RFLP maps by remapping some RFLP clones with the same 46 lines as used in the RAPD work. Five linkage groups were identified and assigned to the five *Arabidopsis* chromosomes. Finally, in an attempt to increase the marker density in a specific region of interest, two DNA pools were created. According to the strategy of Giovanonni et al.,[222] these pools had been designed from the recombinant inbred lines in such a way that polymorphisms between them could be expected to map to chromosome 1. In fact, testing 384 additional primers yielded 32 more polymorphisms, of which 23 mapped to chromosome 1. Taken together, this study clearly demonstrates that RAPD markers can be used for the fast creation of high-density genetic maps.

Sugar beet — A mitochondrial gene, which confers cytoplasmic male sterility, has become a useful tool in plant breeding of sugar beet. Recently, Lorenz et al.[413] showed that two NILs, one male sterile and one male fertile, could be discriminated with PCR-based fingerprinting. Amplification of mitochondrial DNA from these lines, with the M13 repeat sequence as primer, thus revealed additional bands of about 1600 and 3000 bp in the male-sterile plant. Also (GATA)$_4$, as well as one out of four arbitrarily selected 10-mer primers, yielded fragment differences between the male-sterile and male-fertile lines.[413]

Silene — Up till now, few markers for genes of known function have been available for plant population studies. Pooling strategies allow us to target molecular markers to specific traits in the absence of a complete linkage map. A recent example concerns the evaluation of sex-specific markers in a dioecious plant species. To that end, Mulcahy et al.[481] used RAPD methodology to generate markers for the Y-chromosome of the dioecious herb *S. latifolia*. Two leaf tissue pools were created, containing leaf material from 18 female siblings and from 18 male siblings, respectively. DNA was extracted from each pool and subjected to RAPD analysis. Four out of 60 primers gave rise to male-specific amplification products, which were subsequently shown to provide reliable markers for the Y-chromosome. Thus, sexing of *S. latifolia* specimens can now be undertaken already at the seedling stage, which should be a major stimulus for molecular and population studies of other dioecious plants. A general technique for isolation of sex-specific markers was outlined by Griffiths and Tiwari.[241]

Rosaceae — A recently published genome map of apple by Weeden and co-workers[283,748] contains 360 markers, the majority of which have been obtained by RAPD analysis using a total of 64 primers or primer combinations. Several other genome mapping projects involving RAPDs are now well under way in apple (Figure 29)[210,356] and also in pear, *Pyrus communis*,[748] and peach, *Prunus persica*.[1,108,155,237,238,747] Attempts are also being made to find RAPD markers for specific loci, such as the scab resistance gene *Vf* in apple.[219,816]

Legumes — Numerous RAPD-based mapping experiments have been performed in the legume family. In their pioneering paper on the RAPD technique, Williams et al.[782] were able to generate markers that could be placed on an existing RFLP map for soybean. They used single 10-mer primers and 66 F_2 individuals segregating from a cross between the *Glycine max* cv. Bonus and an accession from the wild species *G. soja*. Each polymorphism was scored as a dominant marker and correlated with segregation data for previously mapped RFLP markers. Subsequent RFLP analysis using excised RAPD products as probes verified map positions for most of the RAPD markers. Some RAPD polymorphisms were found to hybridize to mid- or highly repetitive DNA. This indicates that PCR polymorphisms can be useful for generating markers in genomic regions which are not easily accessible to RFLP analysis due to the presence of repetitive DNA sequences. Another study in soybean was aimed at targeting a specific gene. The DAF methodology thus allowed the identification of markers that cosegregated with the *nts* locus (which controls nodule formation) in F_2 families derived from crosses between mutant soybean lines and *G. soja*.[91]

Alfalfa is one of the most widely cultivated forage crops, with diploid as well as tetraploid (more common) cultivars. Segregation of RAPD markers in

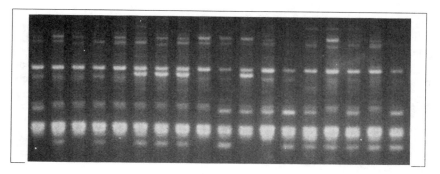

FIGURE 29. Genome mapping in apple using RAPDs. Amplification products were generated from apple seedlings of a cross between cvs. Scijoy and Redfree with the primer OPJ10 (Operon Technologies, Inc.), separated on an agarose gel, and stained with ethidium bromide. The fragments scored for mapping purposes are indicated by arrows, the upper fragment (0.65 kb) segregating 1:1 and the lower fragment (0.5 kb) 3:1. (Courtesy of Dr. Susan E. Gardiner.)

diploid, cultivated alfalfa was monitored by Echt et al.[173] in a backcross population. Thirteen primers were used to analyze the two parental genotypes and 78 backcross progeny. Because of the heterozygous nature of the parents, two types of segregation ratios were expected in the offspring, 3:1 and 1:1, depending on whether both or only one of the parents carried the marker in question. Of the RAPD markers, 28 of 37 segregated as dominant Mendelian traits according to chi-square analysis. The proportion of RAPD markers that did not segregate in the expected manner was comparable to the proportion of allozyme and RFLP markers previously reported to segregate abnormally in alfalfa. Recently, tetraploid alfalfa has also been used for mapping studies. Yu and Pauls[828] showed that segragation ratios of RAPD markers in an F_1 population were generally consistent with random chromosome and random chromated segregation in meiosis. The authors proposed some general strategies for molecular mapping in tetraploids.

A third example in alfalfa will possibly enable marker-assisted selection in future plant breeding programs. A RAPD marker was identified by Yu and Pauls[826] for a gene that induces somatic embryogenesis. An F_1 population consisting of 83 plants derived from a cross between one embryogenic and one non-embryogenic line was evaluated for the ability to produce somatic embryos and subsequently was screened with 100 RAPD primers. Of these, 49 produced DNA amplification product patterns that were polymorphic for the two parental genotypes. The embryogenic parent generated a total of 35 polymorphic bands, of which 19 showed reproducible segregation. Linkage analysis suggested that one of these bands is linked to the gene for somatic embryogenesis.

Segregation of RAPD markers in *Stylosanthes* was recorded by Kazan et al.[348] in an F_2 family derived from an interspecific cross between *S. hamata* and *S. scabra*. A total of 73 RAPD markers were selected, since they were polymorphic between the parental genotypes and easily scored. Of these, 55 markers fitted the expected 3:1 segregation ratio. Eight markers deviated significantly from the 3:1 ratio, and another ten markers failed to segregate in the offspring. Linkage analysis revealed ten linkage groups containing a total of 44 RAPD loci.

A bulked segregant analysis approach[461] was used by Miklas et al.[462] in an effort to identify a RAPD marker for the *Up2* allele, which confers resistance to bean rust in the common bean. Two DNA bulks were created, one consisting of three individuals with the resistance allele and the other consisting of three individuals lacking this allele. Only 1 out of 931 products amplified by the 167 10-mer primers was polymorphic between the two bulks. This primer was then used to amplify DNA from 84 individuals in a cross segregating for the *Up2* allele. No recombination was observed between this product and the allele in question, suggesting that they are tightly linked. Moreover, it was shown that the same amplification product occurs in all investigated lines emanating from the Andean gene pool, regardless of whether these also carry the *Up2* allele. On the contrary, in the Mesoamerican lines, the product was found only in those to which *Up2* had been purposely transferred, suggesting that the identified marker can be used to monitor the introgression of the *Up2* allele into bean germplasm of Mesoamerican origin. Recently, two additional RAPD markers for the bean rust resistance gene block were identified.[255] Analysis of a collection of resistant and susceptible cultivars and experimental lines revealed that some recombination between these markers and the gene block has occurred.

Tomato — Several different approaches have been described toward genomic mapping of tomato with RAPDs. In one study, three chromosome 6-specific RAPD markers were identified by Klein-Lankhorst et al.[361] The material analyzed consisted of plants of *Lycopersicon esculentum, L. pennellii,* and an *L. esculentum* substitution line harboring chromosome 6 from *L. pennellii*. The obtained RAPD markers were subsequently checked by excising the PCR products, labeling them, and then using them as probes to produce RFLPs. The markers were found to be linked to morphological characters, and one of them to the nematode resistance gene *Mi*.

Martin et al.[432] made use of NILs which differ in only a small part of the genome. NILs are usually initiated by crossing a cultivar with a wild species possessing a desirable character like a disease resistance gene. By repeated backcrossing to the cultivar, a new line is created which is almost identical to the original cultivar, but which has introgressed the disease resistance phenotype. Two NILs were analyzed with 144 primers, and seven amplified products were identified that were present in one line, but not the other. Since the two lines differed for the presence of a *Pseudomonas* resistance gene (*Pto*), four of the RAPD markers were then investigated for linkage to that particular gene. Three markers appeared to be tightly linked to the *Pto* gene in question. The polymorphic PCR products were excised, radioactively labeled, and also used as probes to detect RFLPs between the NILs. Since this initial study, several successful efforts have been made to further saturate the *Pto* region with RAPD markers.[95,433]

Giovanonni et al.[222] made use of populations for which a high-resolution RFLP map had already been established. Two pools of DNA from individuals homozygous for different alleles from a targetted chromosome interval, defined by two or more linked RFLP markers, were constructed from members of the mapping population. This strategy made it possible to use the same population to map virtually any area of interest. For the specific project presented, two DNA pools containing 7 to 14 F_2 individuals were chosen for each of two different intervals to be mapped. The DNA pools were screened with 200 primers, resulting in three polymorphic markers. These markers were excised, reamplified, radioactively labeled, and used as a probe to detect RFLPs between the DNA pools. Two of the markers were shown to be tightly linked to one each of the selected intervals, whereas the third marker mapped to the correct chromosome, but at a considerable distance from the target interval.

Lettuce — NILs as well as pooling strategies have been exploited for mapping purposes in cultivated lettuce, *Lactuca sativa*. Paran et al.[540] applied

212 primers to a set of three NILs. Two of these pairs differed in a region containing two linked loci, *Dm1* and *Dm3*, which are involved in the resistance reaction toward downy mildew. The third pair instead differed in a region containing another mildew resistance locus, *Dm11*. Linkage was determined by scoring the differences of RAPD markers between the NILs and confirmed by screening various segregating populations. Four RAPD markers were found to be linked to the *Dm1* + *Dm3* region. Another six markers were linked to the *Dm11* region. A seventh marker, which differed in the *Dm11* region, turned out to be a false positive since it did not segregate as expected. When used as an RFLP probe, this marker was shown to hybridize to repetitive DNA. Recently, the linked RAPD markers have been transformed into SCARs (sequence characterized amplified regions; see Chapter 7, Section II.C) to enhance their utility and reliability in mapping.[351,539]

The successful application of pooling strategies in RAPD-based mapping, i.e., the so-called "bulked segregant analysis", was first demonstrated in lettuce by Michelmore et al.[461] Two bulked DNA samples were generated from a segregating population derived from a single cross. Each pool, or bulk, contained individuals that were identical for a particular trait, but arbitrary for all unlinked regions. Altogether 66 F_2 individuals were analyzed with 14 to 20 plants in each bulk. These bulks were analyzed by RAPD, and three amplification products were generated that proved to be linked to a gene for resistance to downy mildew, *Dm5/8*.

Cereals — A large number of genome mapping studies have also been carried out in cereals, e.g., an in-depth study by Heun and Helentjaris[284] on the inheritance of RAPDs in maize. Whereas most other studies used rather homogeneous materials, this investigation was set up to include five inbred parental lines of widely different genetic origins. A total of 140 amplification products, derived from 21 primers, were scored into different categories depending on presence/absence as well as intensity of fragments, which allowed the distinction between unambiguous and quantitative polymorphisms. Analysis of the 5 parental lines together with 16 F_1 progenies (6 of which were reciprocals) resulted in the classification of 37.1% of the amplification products as unambiguously polymorphic, 42.1% as quantitatively polymorphic, and the remainder as monomorphic. No amplification products were found in the F_1 plants, which were not also encountered in one of the parental lines. Of the polymorphic products, more than 90% appeared to be inherited to the F_1 generation in the expected manner. RAPD fragments of the greatest intensity showed the most predictable behavior. Of the unambiguously polymorphic situations, 95% could be interpreted genetically by assuming complete dominance of the presence of the parental product, while 3.2% of the F_1 plants exhibited a RAPD that was intermediate between the two parents. In at least part of the remaining 1.8%, products appear to have outperformed others in the amplification process. For quantitative polymorphisms, values of 88.1% for complete dominance and 5.0% for partial dominance were obtained. The remaining cases, i.e., behaving neither completely nor partially dominant, at least cannot be explained by amplification of cytoplasmic DNA, since analysis of the six reciprocal crosses ruled out uniparental inheritance.

Some quantitative trait loci (QTLs) were recently mapped by Pé et al.[545] in maize by combining RFLP and RAPD analyses of 150 F_2 individuals obtained from crossing sensitive and resistant inbred parental lines. Five genomic regions were found, each containing one or several QTLs involved in the resistance to the pathogen *Gibberella zeae*, which causes stalk and ear rot. Analysis of resistance was performed on F_3 families, resulting from selfing of the F_2 plants. The amount of phenotypic variability explained by these five QTLs is 20% according to multiple regression analysis.

Whereas most publications report of successful results with the fingerprinting techniques, some adverse reactions have also been expressed. A composite map of the barley genome, including molecular, allozyme, and morphological markers, was published recently by Kleinhofs et al.[360] Of the 295 mapped loci, 14 were RAPD markers. These had come at a great expense of time and effort. The RAPD markers often showed distorted segregation ratios and were difficult to reproduce. In spite of speed and high levels of polymorphism, the authors do not recommend RAPDs for mapping the barley genome.

On the contrary, Tinker et al.[696] report almost no segregation distortion in a RAPD study in barley using 20 double-haploid spring barley lines from the F_1 of a biparental cross. The inheritance of 19 RAPD markers was studied, including 1 which appeared to be codominant. Chi-square tests were performed independently for each of the polymorphisms, only one of which yielded a significant deviation from the expected 1:1 ratio. To test the genetic independence of segregating markers, linkage analysis was performed. Three groups with tight linkage were detected among eight of these polymorphisms. Two groups consisted of pairs of loci which had a 5% recombination ratio. The third group consisted of four loci, three of which did not recombine at all and a fourth with a recombination ratio of 5%. The completely linked loci were products of separate primers.

A combination of NILs and bulked segregant analysis was used to identify RAPD markers linked to a barley gene, which confers resistance to the fungus *Rhynchosporium secalis*.[35] Cosegregation of the linked markers with chromosome arm-specific RFLPs confirmed the location of the *Rh* locus to the long arm of barley chromosome 3.

Ronald et al.[594] used NILs of *Oryza sativa* to obtain markers for the rice bacterial blight disease resistance locus *Xa21*. From the 985 primers screened, two RAPD markers were obtained that differentiated between the two lines. These markers were also shown to cosegregate with the resistance locus in an F_2 population of 386 progenies. The amplification products were cloned and hybridized to Southern blots prepared from the two NILs. Both probes hybridized to the two NILs, but at different positions, indicating that these RAPD products had been duplicated specifically at the *Xa21* locus.

Until now, genetic mapping of polyploid species has been difficult. However, successful results were achieved by Al-Janabi et al.[9] using AP-PCR and single dose markers. A progeny was analyzed from a cross between accession "SES 208" of *Saccharum spontaneum*, a polyploid (2n = 64) relative of cultivated sugarcane, and a diploidized haploid derived through anther culture of SES 208. The selected 127 primers revealed 2160 markers, of which 279 were present in the male parent and absent in the female. Of these, 208 markers segregated 1:1 and were used for mapping, resulting in 42 linkage groups with at least 2 markers and an additional 32 unlinked markers. An autopolyploid origin of SES 208 was indicated, since (1) the ratio of single dose markers to higher dose markers fitted the assumption of auto-octaploidy and (2) no repulsion-phase linkages were detected.

C. GENERAL CONSIDERATIONS

While few inheritance and mapping studies have been performed yet with hybridization-based multilocus fingerprints, PCR-based methods are already well established as tools for mapping purposes in a wide variety of plant species. Thus, marker-assisted selection in breeding programs as well as map-based gene cloning on the basis of RAPD markers will become feasible in near future. Some drawbacks of genome mapping with RAPDs should also be noted. (1) Crossing programs and progeny populations have to be carefully designed. Thus, backcross populations, recombinant inbreds, anther-derived haploids, or megagametophytes are preferred over F_2 populations; (2) the information content of each RAPD marker is relatively low as compared to RFLP markers, due to their dominant nature and low allele number; (3) RAPD maps are not really species specific, instead one map is derived for the father and one for the mother of a specific cross; and (4) comparable maps in related species may not be generated from a common set of RAPDs, as demonstrated by Kesseli et al.[351] who found no shared bands between cultivated lettuce and the wild, partially compatible relative, *Lactuca saligna*. However, this lack of homology between genomes offers an obvious advantage of RAPDs when constructing maps in autopolyploid species. Instead of the complex banding patterns with alternate alleles commonly encountered with RFLPs, RAPDs in most cases will probably segregate in simple 3:1 ratios.[351]

An explanation of non-Mendelian segregation, put forward specifically for RAPDs, is that fragments actually amplified by the PCR probably represent the most successful products among many more competing candidates for amplification. The overall genetic background thus may determine which candidates are amplified, resulting in the occurrence of "epistatic effects" as suggested, but not actually encountered, by Heun and Helentjaris.[284] This phenomenon would seriously decrease the relevance of RAPD genetic maps, since markers obtained in one population might behave very differently in another.

Different strategies have been presented that facilitate the detection of linkage between PCR-derived fingerprint markers and a particular trait in the absence of a genetic map. Among these, NILs and various pooling strategies have a great potential for the fast characterization of a trait of interest.

V. FUNGI

Fungi are becoming increasingly important for a variety of industrial purposes, and many species are serious pathogens of plants, domestic animals, and humans. In several areas of research, the precise and unequivocal identification, discrimination, and characterization of fungal species, races, isolates, populations, and pathotypes is of prime importance. However, this is a difficult or even impossible task if the characterization solely relies on growth characteristics, sexual compatibility, and morphological and biochemical criteria. Molecular markers have therefore been looked for, and a wide variety of molecular techniques is now available to study genetic variation within fungi. These techniques include, e.g., the study of allozymes, RFLPs, and electrophoretic karyotypes (which involves the separation of whole chromosomes by pulsed-field gel electrophoresis). In the late 1980s, hybridization-based DNA fingerprinting was introduced for the analysis of fungal genomes.[64,611] Since then, DNA fingerprinting has become one of the favorite methods for diagnostic and epidemiological studies of human, plant, and insect pathogenic fungi; for the characterization of nonpathogenic and industrially important fungi; and for the evaluation of taxonomic and phylogenetic problems.

To date, mainly three kinds of DNA probes have been used for fingerprinting studies in fungi: (1) anonymous repetitive DNA probes derived from the fungal genome under investigation; (2) minisatellite probes, mostly derived from the human or wild-type M13 phage genome; and (3) synthetic oligonucleotide probes complementary to simple repetitive sequences.

The majority of DNA fingerprinting studies in fungi have relied on cloned genomic probes.[72,263,359,611] Generally applicable and commercially available probes such as minisatellites or simple repetitive oligonucleotides, in comparison to the situation in plant genome research, have been used to a lesser extent. Nevertheless, quite promising results were obtained with this kind of probes, when, e.g., the M13 probe and the synthetic oligonucleotides $(CA)_8$, $(CT)_8$, $(CAC)_5$, $(GTG)_5$, $(GACA)_4$, and $(GATA)_4$ were hybridized by Meyer et al.[452a,453,455,459] to restriction-digested DNA from a wide range of fungal taxa. More than 70 species of zygomycetes, ascomycetes, and deuteromycetes (fungi imperfecti) representing 18 genera of filamentous fungi (*Absidia, Aspergillus, Chaetomium, Emericella, Eupenicillium, Eurotium, Fusarium, Geosmithia, Gibberella, Leptosphaeria, Mucor, Parasitella, Penicillium, Stachybotrys, Trichoderma, Trichothecium, Ulocladium,* and *Verticillium*) and 5 genera of yeast-like fungi (*Arxula, Candida, Cryptococcus, Kluyveromyces,* and *Saccharomyces*) were included in these survey studies. Polymorphic target sequences of the tandem repeat probes were detected in all species studied. As was observed in plants,[757,759,761] the informativeness of fingerprint patterns depended strongly on the optimal combination of species, hybridization probe, and restriction enzyme.

With the advent of PCR, a new repertoire of typing techniques was added for fungal genome analysis.[782,783] Similar to the situation in plants, arbitrary primers which amplify anonymous, polymorphic DNA fragments have received the most attention, and a multitude of studies have been performed using the RAPD technology. More recently, it was shown that minisatellite core sequences and simple repetitive oligonucleotides may not serve only as informative probes for traditional DNA fingerprinting, but also as PCR primers. Polymorphic DNA fragment patterns have thus been obtained with these kinds of primers from more than 40 species representing eight fungal genera (*Candida, Cryptococcus, Fusarium, Leptosphaeria, Mortierella, Penicillium, Saccharomyces,* and *Trichoderma*).[459,460,619] In many cases, the resulting profiles varied even among individual strains.

Both hybridization- and PCR-derived fingerprint patterns were found to be highly reproducible. The occurrence of multilocus profiles in a large variety of species suggested that tandemly repeated target sequences detected by the oligonucleotides used as hybridization probes or PCR primers, respectively, are ubiquitously represented among fungi.

In the following sections, we present a variety of examples for the application of fingerprint techniques in different areas of fungal research. For a more comprehensive overview, the reader should refer to Appendices 3 and 4.

A. HUMAN PATHOGENIC FUNGI: DIAGNOSTICS AND EPIDEMIOLOGY

Several species of filamentous fungi and yeasts are part of the normal body flora, but they can also become serious human pathogens. Under normal conditions, fungal growth is suppressed, and disease only occurs when the ability of the fungus to cause an infection exceeds the host-specific mechanisms to prevent invasion or to eradicate the infecting fungus. The spectrum of diseases caused by human fungal pathogens ranges from superficial skin infections and single organ mycoses to severe systemic infections. Pathogenic fungi also continue to be a frequent source of hospital-acquired (=nosocomial) infections. Opportunistic mycoses are potentially life-threatening infections which are becoming increasingly dangerous in immunocompromised patients, such as those with AIDS, transplants, and hematologic malignancies.

Considering the importance of mycotic infections, there is a surprising lack of reliable identification methods. Commonly used typing methods of medically important fungi are based on the analysis of morphologic and physiologic parameters which is often difficult and time consuming. Accurate, simple, fast, and reproducible methods are therefore of prime importance for (1) the identification of fungal pathogens; (2) the determination of the origin of infections, the routes of acquisition, and transmission of strains in an increasing number of nosocomial infections; and (3) the tracing of antifungal drug resistance. If one compares different typing methods, DNA fingerprinting techniques have clearly emerged as one of the most powerful of all epidemiological tools, provided that an appropriate probe is available.

Candidiasis and cryptococcosis are among the most important fungal diseases of humans. The use of DNA fingerprinting for diagnostic and epidemiological studies in both diseases is outlined in the following sections.

1. *Candida*

The spectrum of diseases caused by *C. albicans* ranges from superficial skin infections to severe systemic diseases. Since this fungus is also a common source of hospital-acquired infections, strain-typing methods are especially important for studies on transmission. The earliest approach to fingerprint different strains of *C. albicans* relied on mid-repetitive sequences. Several clones were isolated from a genomic library of *C. albicans* (e.g., Ca3, Ca5, and Ca7), and used to type *C. albicans* and *C. tropicalis* strains.[644] Some probes proved to be species specific (e.g., Ca3 for *C. albicans* and Ct13-8 for *C. tropicalis*).

Ca3 and Ct13-8 were henceforth used in numerous fingerprint studies. In an epidemiological investigation on vaginal isolates from different patients, each patient was shown to carry a different strain.[644] In a later study, Soll et al.[645] demonstrated that a systemic infection can be caused by multiple *Candida* strains in a single immunocompromised patient. To investigate the mitotic stability of the Ca3 hybridization patterns, strains were passaged over 400 generations. Ca3 fingerprints were analyzed each 100 generations by Schmid et al.[615] and remained absolutely stable. On the other hand, 46 unrelated tester strains yielded different Ca3 patterns. These results (stability within and polymorphism between strains) suggested that Ca3 is a very effective probe for strain discrimination. The same authors also evaluated the relatedness between strains on the basis of similarity coefficients using the newly developed computer program DENDRON.[615]

Colonization of several body sites was investigated by Soll et al.,[647] analyzing carriage and strain relatedness in 52 healthy women at 17 anatomical locations. This study demonstrated that healthy persons could be carriers of a pathogenic fungal strain without being infected. Half of the test persons carried the organism simultaneously in more than one of the surveyed locations. In a comparative study on AIDS and non-AIDS patients, Schmid et al.[616] showed that the genetic diversity among strains was significantly lower in AIDS patients than in non-AIDS patients. Apparently, the same strain persisted through reinfection in patients with AIDS. Finally, it was shown that female and male partners carried the same *Candida* strains, suggesting that the infecting strains were transmitted from the infected women to their partners.[617]

Scherer and Stevens[611] isolated another repetitive DNA sequence from the *Candida* genome. This so-called "27A probe" was found to be species specific for *C. albicans* and proved very useful for DNA fingerprinting. Stability of fingerprint patterns was demonstrated by passaging four different strains over 300 generations. In an epidemiological study, *Candida* isolates were taken from six different body sites of several patients. One of the patients was infected by one specific strain, four carried a set of closely related strains, and only one patient appeared to be infected by two different strains. A great variety of strains was identified when several infected patients from a single hospital were examined.[611]

Using the same probe, an epidemiological study by Fox et al.[197] of 110 isolates from 63 immunocompromised hosts showed different fingerprint patterns for 60 out of 63 patients. No intrapatient variation was observed, regardless of the anatomical isolation site. The 27A probe was also applied to investigate an outbreak of a *Candida* infection involving five infants receiving total parental nutrition in a neonatal intensive care unit.[627] The infection was found to be caused by three strains of *C. albicans* and one strain each of *C. tropicalis* and *C. parapsilosis*. By comparing the isolates derived from the patients with strains isolated from equipment used by the nurses, it was found that the infecting strains were most likely transmitted by contaminated syringes.

More recently, DNA fingerprinting of *C. albicans* was also performed using the repetitive DNA sequence poly(GT). Ten vaginal *C. albicans* isolates were clearly discriminated by Wilkinson et al.[780] A radioanalytical imaging computer system was used to evaluate the fingerprint patterns, allowing an easy application of the technique to routine epidemiological studies of candidiasis.

In connection with the increasing number of immunocompromised patients, uncommon *Candida* species are becoming more and more important. However, the usefulness of the species-specific Ca3 and 27A probes is limited for these more exotic species. A main advantage of simple sequence-complementary probes such as poly(GT) is that their application is not restricted to a certain species. Poly(GT) sequences are ubiquitous in eukaryotes, and a $(GT)_n$ probe has previously shown to be informative for strain typing in yeasts.[736] Synthetic oligonucleotides composed of simple repetitive motifs provide an even more versatile tool for DNA fingerprinting of uncommon *Candida* species. Using the probes $(GACA)_4$, $(GATA)_4$, $(GGAT)_4$, $(GTG)_5$, and $(GT)_8$ in combination with the restriction enzyme *Eco*RI, Sullivan et al.[670] analyzed a variety of isolates belonging to the species *C. tropicalis, Torulopsis (Candida) glabrata, C. krusei,* and *C. parapsilosis*. Except for the three isolates from *C. parapsilosis*, all isolates could be clearly discriminated. The most informative probe in this study was $(GT)_8$. Rehybridization with the 27A probe (see above) showed that the simple sequence probes are applicable to a wider range of *Candida* species and also have a higher discriminatory potential than the 27A probe for strain typing of *C. albicans*.

Finally, PCR-based methods have also been introduced to the characterization of *Candida* isolates. Lehmann et al.[396] used single 10-mer RAPD primers to determine the genetic relatedness among medically important *Candida* species. Distinctive and reproducible PCR products were observed for isolates of *C. albicans, C. lusitaniae, C. tropicalis,* and *T. (C.) glabrata*. Different isolates of the same species showed intraspecific DNA polymorphisms.

Several oligonucleotides specific for short tandem repeats were tested by Niesters et al.[504] as single PCR primers for the differentiation of isolates of the four major human pathogenic *Candida* species [*C. albicans, T. (C.) glabrata, C. tropicalis,* and *C. krusei*] as well as of the less pathogenic species *C. guilliermondii, C. stellatoidea, C. parapsilosis,* and *C. pseudotropicalis*. The oligonucleotides $(AC)_{10}$, $(CAC)_7$, and $(GACA)_5$ produced informative banding patterns, while a smear was generated with a poly(GT) primer.

In another attempt of PCR fingerprinting, two different oligonucleotides, i.e., the core sequence of the M13 minisatellite repeat and the simple repeated motif $(GACA)_4$, were successfully used by Schönian et al.[619] as single primers to amplify polymorphic DNA sequences from the genomes of several *C. albicans* isolates. It was possible to differentiate *C. albicans* isolates obtained from healthy dental clients and from patients in different intensive care units (Figure 30). PCR fingerprints of *C. albicans* isolates from immunocompromised patients were more similar to each other than those derived from dental clients. In some cases, multiple isolates from different body sites of the same patient yielded different patterns, suggesting that candidiasis could be caused by several infecting strains. The high reproducibility and speed of the technique makes PCR fingerprinting with M13 and simple sequence primers a useful tool for surveying large numbers of pathogens in epidemiological studies.

2. *Cryptococcus*

The basidiomycetous, encapsulated yeast *Cryptococcus neoformans* is among the most prevalent life-threatening mycotic agents. It is found in the environment worldwide. Two genetically distinct varieties of *C. neoformans* are recognized: *C. neoformans* var. *neoformans* (serotypes A and D) and *C. neoformans* var. *gattii* (serotypes B and C).

A dispersed repetitive DNA element cloned from *C. neoformans* (CNRE-1) was used as a genomic DNA probe to fingerprint clinical isolates from all four serotypes[655] and was shown to recognize multiple polymorphisms among the isolates. The varieties *neoformans* and *gattii* were clearly distinguishable. Of all isolates from a single hospital, 50% were closely related.

A species-specific (CND 1.7) and a variety-specific (CND 1.4) mid-repetitive genomic DNA probe from *C. neoformans* were applied to epidemiological studies on AIDS and non-AIDS patients.[554] CND 1.7 did not cross-hybridize with other yeast-like fungal pathogens, and CND 1.4 hybridized specifically with isolates of the var. *neoformans*. A combination of both probes made it possible to distinguish the serotypes A and D from B and C and simultaneously provided a strain-specific pattern. These probes made it possible to identify an individual *C. neoformans* strain as well as to distinguish reinfections from new infections.

Varma and Kwon-Chung[725] used a URA5 gene sequence associated with telomeres as a genomic DNA probe (UT-4p) for fingerprint studies of *C. neoformans*. DNA fingerprints specific for each serotype were found. To determine the species specificity of UT-4p, the probe was also hybridized to DNA of several other *Cryptococcus* species as well as isolates from *Ustilago maydis* and *Candida albicans*. None of these species yielded hybridization patterns similar to those of *Cryptococcus neoformans*. DNA fingerprints were stable and reproducible during continuous subculturing and after drug application.

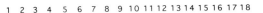

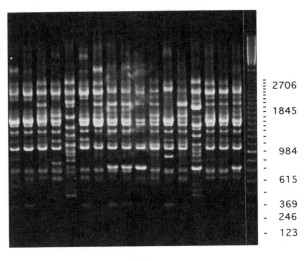

M13

FIGURE 30. PCR-based DNA fingerprints used to investigate a nosocomial outbreak of candidiasis. The M13 core sequence (GAGGGTGGXGGXTCT) was used as a primer to generate fragment patterns of 17 *Candida* isolates from patients in different intensive care units (ICU) at the Charité hospital in Berlin. Amplification products were separated on an agarose gel and stained with ethidium bromide. Lanes 5 and 14 are *C. glabrata*; the remainder are *C. albicans*. Lanes 1 to 4 are isolates PS, HJ, WS, and GZ1 obtained from patients on ICU 1; lane 5 is the isolate GZ2 obtained from patient GZ who was transferred from ICU 1 to ICU N; lanes 6 to 10 are isolates HC, SA, WW, BH, and GB obtained from patients on ICU IVb; and lanes 11 to 17 are isolates PE3, SH1, MOH, KH, KB, HW, and LH obtained from patients on ICU V. Size markers (123-bp ladder, GIBCO-BRL) are indicated in lane 18.

More than 100 clinical and environmental isolates of *C. neoformans*, including isolates from patients with and without AIDS, were studied by Meyer et al.[452a,459,460] using PCR-based fingerprinting. Oligonucleotides corresponding to the core sequence of the M13 probe, $(GACA)_4$ and $(GTG)_5$, were used as single PCR primers. Specific PCR fingerprints were generated for all isolates. Highly reproducible patterns were obtained, and the fingerprint profile was the same whether crude or cesium chloride (CsCl) purified DNA was used as a template for the PCR. Besides discriminating individual strains, specific banding patterns were observed for three of the four serotypes (Figure 31). Only the serotypes B and C showed very similar patterns, in accordance with results obtained with other techniques. The banding pattern of most isolates was very similar to that of the serotype A reference strain, consistent with other surveys which indicate that 80% of all *Cryptococcus* infections in the U.S. are caused by strains of this serotype. The stability of the PCR fingerprint patterns was tested by weekly passages of the strains for 6 months *in vitro*, reextracting DNA several times during this period, and repeating the PCR. No changes in banding patterns were observed (see also Chapter 4, Section IV.B., Figure 19). The possibility to generate a strain-specific profile and concomitantly determine the serotype of clinical isolates within a single experiment will greatly facilitate future epidemiological studies of these fungal pathogens.

3. *Sporothrix*

Sporotrichosis is a chronic cutaneous fungal infection. The disease is caused by the imperfect fungus *Sporothrix schenkii*, which enters the body through the skin after traumatic inoculation. Cooper et al.[124] used a genomic hybridization probe for DNA fingerprinting in an epidemiological study of clinical and environmental isolates derived from a sporotrichosis epidemic in 1988. All infections of the outbreak were found to be associated with the handling of seedlings packaged in *Sphagnum* moss. It could be shown that one of the six morphologically and physiologically distinct *Sporothrix* groups obtained from the environment (*S. schenkii*) was responsible for the epidemic in several states of the U.S.

4. *Pneumocystis*

P. carinii is an opportunistic pathogen that is a major cause of morbidity and mortality among AIDS patients. Using a repeated DNA sequence (Rp3-1) from *P. carinii* as a hybridization probe, a high number of DNA polymorphisms were found between different *P. carinii* isolates.[666] Rp3-1 target sequences were present on each of the *P. carinii* chromosomes, indicating that this probe may be useful for further DNA fingerprinting investigations of this medically important species.

5. *Aspergillus*

The imperfect fungus *A. fumigatus* is the most common air-borne fungal pathogen in humans, causing aspergillosis. A number of research groups have studied DNA typing techniques which would allow epidemiological investigations of this important pathogen.

In one study, PCR with single RAPD primers was successfully used to distinguish clinical isolates of *A. fumigatus*.[27] Among the 44 10-mer primers tested, 22 generated distinct, reproducible RAPD patterns, and 15 primers were able to differentiate between isolates. Loudon et al.[414] used five different single 8-mer RAPD primers to type 19 clinical isolates of *A. fumigatus*. Two of these primers discriminated the isolates satisfactorily. Different patterns derived from isolates of the same patient were observed in some cases, suggesting that aspergillosis could be caused by multiple infections with different strains of *A. fumigatus*.

Several repetitive DNA sequences cloned from *A. fumigatus* were tested by Girardin et al.[223] for their potential as hybridization probes for DNA fingerprinting. Through this approach it was possible to discriminate between all clinical and environmental *A. fumigatus* isolates. The probes were found to be

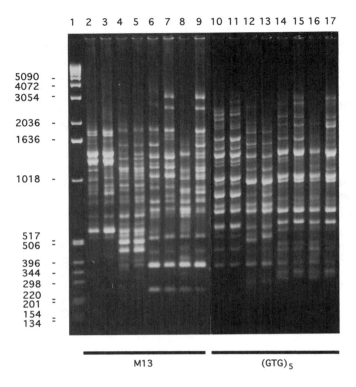

FIGURE 31. PCR-based DNA fingerprints used to identify strains and serotypes of *Cryptococcus neoformans*. The M13 core sequence (GAGGGTGGXGGXTCT) or the simple sequence (GTG)$_5$ were used as single primers to generate fragment patterns from genomic DNA of various *C. neoformans* isolates. Amplification products were separated on an agarose gel and stained with ethidium bromide. Lanes 2/10 and 3/11 *C. neoformans* var. *neoformans* (serotype A), strains 110 and 15; lanes 4/12 and 5/13 *C. neoformans* var. *neoformans* (serotype D), strains 3501 and 3502, lanes 6/14 and 7/15 *C. neoformans* var. *gattii* (serotype B), strains n32 and UCLA 373-B; lanes 8/16 and 9/17 *C. neoformans* var. *gattii* (serotype C), strains n33 and UCLA 380-C. Size markers (1 kb ladder, GIBCO-BRL) in lane 1.

species specific and did not hybridize to DNA of *A. nidulans, A. flavus, A. ochraceus,* and *A. restrictus.*

6. *Histoplasma*

H. capsulatum is a respiratory and systemic pathogen in humans and animals. While mainly found in tropical regions, the increasing number of immunocompromised patients also raised the importance of histoplasmosis in temperate areas. The diversity among clinical isolates of *H. capsulatum* was investigated by Kersulyte et al.[350] with PCR and a variety of single arbitrary primers. Using long (more than 17-mer) as well as short (10-mer) primers, they found that *H. capsulatum* is genetically more divergent than expected from previous studies on mitochondrial RFLPs. From one particular RFLP class, 29 clinical isolates could be clearly distinguished by RAPD analysis. Laboratory-derived variants yielded fingerprints which were identical to those of their wild-type parents, despite years of separate propagation. This shows that RAPD profiles are highly stable during *in vitro* subculturing.

B. PLANT PATHOGENIC FUNGI: POPULATION GENETICS, EPIDEMIOLOGY, AND PATHOTYPE IDENTIFICATION

Many fungal species are plant pathogens, and some cause highly persistent and devastating crop diseases. Commercial distribution of plant material has facilitated the spreading of the associated pathogen even between several continents. Traditionally, plant pathogenic fungi are identified by morphological, biochemical, and allozyme characteristics. However, these methods have proved to be insufficient. In recent years, the introduction of hybridization- and PCR-based DNA typing techniques has greatly facilitated the exact characterization, identification, and early diagnosis of pathogens, as well as studies on their epidemiology. In this section, some examples for the application of DNA fingerprinting to the characterization of plant pathogenic fungi are given in alphabetical order of the genera. For a more comprehensive overview, refer to Appendices 3 and 4.

1. Ascochyta

Synthetic oligonucleotides specific for simple repetitive sequences were used as hybridization probes to study the chickpea pathogen *A. rabiei*.[49,758] The probes $(GATA)_4$, $(GTG)_5$, $(CA)_8$, and $(TCC)_5$ made it possible to distinguish five of six isolates collected in Syria. The DNA fingerprinting data supported a previous classification according to the aggressiveness of the isolates, based on their virulence on a set of host cultivars. The fingerprint patterns were somatically stable over 1 year of *in vitro* culture. In contrast to the situation encountered with other fungi,[453,455] *Ascochyta* fingerprint patterns obtained by four-bp cutters proved to be more informative and easier to screen than those obtained by the six-bp cutter *Eco*RI. PCR with four different arbitrary primers (either singly or in pairwise combination) was carried out on the same isolates.[341] A comparison of the PCR-based method with oligonucleotide fingerprinting showed that the discriminating capacity of the latter method was more powerful.

In a more recent study, 50 *A. rabiei* isolates were hierarchically sampled from a single chickpea field in Tunisia and fingerprinted by hybridization with simple repetitive oligonucleotides.[477a] Twelve different fingerprint haplotypes were observed among 50 isolates, with an irregular distribution of haplotypes across the field. In some cases, different haplotypes were identified within a single lesion of the host plant (Figure 32).

2. Colletotrichum

Species of the genus *Colletotrichum* infect a wide range of crops causing a variety of diseases. Serious yield losses are predominantly recorded from tropical and subtropical regions. The high genetic variability of the pathogen renders resistance breeding difficult. Fast and reliable methods to characterize genotypes within pathogen populations would be very helpful.

The first successful attempt to fingerprint *C. gloeosporioides* was reported by Braithwaite and Manners.[64] The authors were able to distinguish two different pathogen types by hybridization-based fingerprinting with the human minisatellite probes 33.6 and 33.15. Rodriguez and Yoder[583] isolated and characterized an interspersed repeat element from the genome of *C. lindemuthianum*. This element (GcpR) consisted of tandemly arranged CAX triplets in which X was commonly G or A. Using GcpR as a probe, unique hybridization patterns were obtained for each of eight species of *Colletotrichum* as well as for several representatives of a wide range of phylogenetically divergent fungal taxa. This extensive cross-hybridization may be caused by the ubiquitous appearance of simple repetitive sequences [here exemplified by the $(CAX)_n$ motif] in plants and fungi.

Later on, PCR-based techniques were mainly used for genome analysis of *Colletotrichum* species. Thus, the molecular variation in a worldwide collection of *C. gloeosporioides* isolates derived from different hosts was studied by Mills et al.[469] using RAPD analysis. A total of 40 different single 10-mer primers were tested. The complexity of the banding patterns varied between the primers; some produced multiple bands, and others produced only a single or a small number of bands. Variation was found between as well as within isolates from a particular host species. With the exception of the fungal isolates from mango which gave similar RAPD patterns irrespective of their geographic origin, the banding patterns allowed grouping of the isolates according to host species and geographical location.

Guthrie et al.[248] used a similar approach to study the genetic variability between isolates of *C. graminicola* from a variety of geographical locations all over the world. High levels of heterogeneity were found for all investigated locations except for Puerto Rico. The homogeneity of Puerto Rican isolates may reflect an isolated gene pool. The authors used image processing to exclude low-intensity, hard-to-score bands from the analysis. Thus, bands were only scored if their intensity was above a certain threshold.

Fungal isolates that cause disease in members of a particular plant family are referred to as forma specialis (f. sp.). Single RAPD (10-mer) primers were used by Vaillancourt and Hanau[710] to distinguish between *C. graminicola* strains of two different formae speciales which are associated either with maize or sorghum. The resulting RAPD patterns were highly reproducible and closely related to host specificity, even among isolates originating from different hemispheres. Thus, RAPD banding patterns could be used to reliably and unambiguously distinguish the two groups. Bandsharing similarity coefficients amounted to 45% between the isolates of two formae speciales and to 85% within each forma specialis. These results indicate that maize and sorghum isolates of *C. graminicola* represent two distinct and separate genetic lineages that are reproductively isolated.

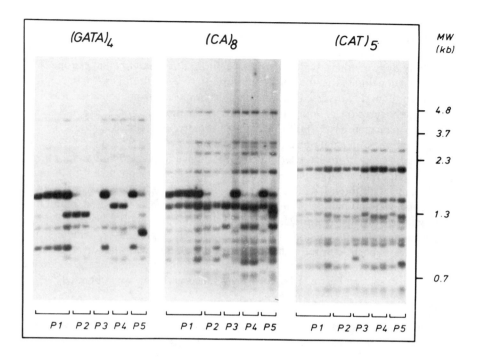

FIGURE 32. Oligonucleotide fingerprinting for the identification of fungal haplotypes. *Ascochyta rabiei* isolates were taken from five chickpea plants (P1 to P5) in a field in the Beja region in Tunisia. Genomic DNA was purified from single-spored isolates, digested with *Hin*fI, separated on an agarose gel, and in gel-hybridized with the indicated ^{32}P-labeled probes. For the individual host plants P2 to P5, one isolate from each of two lesions was examined per plant. For plant P1, two isolates (in neighboring lanes) were examined from each of the two lesions. In some cases, different fungal haplotypes were isolated from the same lesion. Size markers (kb) are indicated.

3. *Erysiphe*

Another important plant pathogen, *E. graminis*, causes the powdery mildew disease of barley. In 1989, a genomic DNA probe (E9) was isolated from *E. graminis* f. sp. *hordei* which detected about 30 polymorphic bands when hybridized to *Hin*fI-digested *E. graminis* DNA.[72,524] Since the E9 probe did not cross-hybridize to barley DNA, it was possible to isolate *E. graminis* DNA along with plant material. The E9 probe was used for a DNA fingerprint analysis of 97 isolates of *E. graminis* f. sp. *hordei* sampled from barley seedlings.[72] It was shown that a large group of 39 isolates with very similar virulence phenotypes shared identical fingerprints and thus probably represented a single clone. Several sets of other clones were also identified and grouped by cluster analysis. No association was found between an observed hybridization pattern and any other marker such as virulence or resistance to antifungal agents.

The E9 probe was subsequently used by Brown et al.[73] to investigate a powdery mildew epidemic on the British Isles, infecting barley cultivars carrying the new resistance allele Mla13. The epidemic was found to be initiated mainly by two clones, which possibly originated from continental Europe. The authors concluded that the genetic uniformity of agricultural crops, in contrast to natural, highly diverse plant populations, imposes strong selection for virulent pathogen clones. Populations of airborne pathogens like *E. graminis* f. sp. *hordei* can then easily build up, spread rapidly, and finally cause a severe epidemic from a single or a few clones.

Wolfe et al.[790] investigated about 600 isolates of *E. graminis* f. sp. *hordei* collected from a large part of Europe and detected high levels of genetic variation with three RAPD primers. A wide range of haplotypes spread over Europe were found, as well as substantial variation also within individual fields. The distribution of DNA polymorphisms indicated that, on the average, approximately 80% of the total allelic variation is present on a local scale. However, substructuring of the pathogen population was also demonstrated, since some DNA loci showed significant geographical heterogeneity. Evidence of an association between DNA and virulence variation was obtained.

4. *Fusarium*

Fusarium is a heterogeneous genus that includes many plant pathogenic species all around the world. Many *Fusarium* strains are characterized by a high degree of host specificity. Using traditional methods, the discrimination of races, species, and formae speciales is usually difficult.

DNA fingerprint techniques were introduced to genome analysis of *Fusarium* isolates in 1991. Twenty different RAPD primers (10-mer to 57-mer) were used by Crowhurst et al.[131] to assess genetic variability between 21 New Zealand isolates of *F. solani* f. sp. *cucurbitae*. On the basis of amplification products, the 21 isolates could be differentiated into two distinct groups which corresponded to the mating types MPI (11 isolates) and MPV (4 isolates). Several isolates, which could not be assigned to either mating type by traditional means due to their failure to cross with known tester isolates, could be placed in one of the two mating types by means of their RAPD patterns. All primers gave essentially the same groupings, but the patterns obtained with the 10-mer primers were usually superior to those obtained with longer primers since they contained a lower number of faint bands. These results suggest that any arbitrary primer of any length may prove useful in detecting polymorphisms between any fungal isolates.

In an attempt to characterize different races of *F. oxysporum* f. sp. *pisi*, Grajal-Martin et al.[235] performed a RAPD study using 14 10-mer primers. Fifty-two bands were produced which revealed high levels of genetic variability. Grouping of the isolates according to the Wagner parsimony option of the PHYLIP software package showed that only race 2 was genetically homogeneous, while races 1, 5, and 6 showed much greater variability.

In another RAPD study, 17 *F. graminearum* strains were analyzed with a variety of primers.[532] Surprisingly low levels of variability were observed. The majority of primers yielded one of two common patterns. However, all strains could be identified by the combined RAPD profiles. Recently, hybridization of samples from four different *Fusarium* species to the M13 probe was shown to reveal considerable levels of intraspecific polymorphism.[502a]

5. Leptosphaeria

L. maculans, the sexual stage of *Phoma lingam*, is a pathogen for many crucifers and causes systemic infections of the oilseed rape *Brassica napus* var. *oleifera*. The pathogen may affect plant growth until devastation of the crop. The simple sequence-complementary oligonucleotides $(CT)_8$, $(GTG)_5$, and $(GACA)_4$ as well as the M13 probe were used to investigate 16 different isolates of the pathogen by hybridization-based fingerprinting.[405,455,457] High levels of intraspecific DNA polymorphism were detected with all the probes tested. In addition to distinguishing all isolates clearly and unequivocally, the banding patterns also allowed their classification into two groups comprising aggressive and nonaggressive pathotypes, respectively (Figure 33). In the course of this study, a previously unassigned strain could be grouped with the aggressive isolates (Figure 33A).[457] Identical groupings were obtained by using $(GACA)_4$ as a single PCR primer (Figure 33B).[404] The two major aggressivity types of *L. maculans* could also be distinguished by RAPD analysis using single arbitrary 9- and 10-mer primers (Figure 33C).[608]

In similar experiments, Goodwin and Annis[227] were able to divide Canadian isolates of *L. maculans* into three distinct groups by using 10-mer RAPD primers. The first group contained all isolates of the aggressive type. Isolates of the nonaggressive type from western Canada were clustered in the second group. The third group contained all avirulent isolates from Ontario. These results supported the hypothesis that the aggressive pathotype was only recently introduced into Canada and has diverged relatively little, whereas nonaggressive strains have been present in Canada for a longer time and have diverged through geographic isolation.

According to inoculation tests on a differential host set, five pathotypes of *L. maculans* (four aggressive and one nonaggressive) can be defined. While hybridization- and PCR-based DNA fingerprinting and RAPD patterns allow the discrimination of aggressive vs. nonaggressive types, no correlation between banding patterns and pathotype grouping has been observed within the aggressive types.[457,608]

Both PCR-based methods described above appear to be very powerful tools for the early diagnosis of the pathogen because they are easy, fast, and inexpensive to carry out. However, the results also showed that PCR-based techniques will not always individualize or identify strains by one PCR fragment pattern alone. In comparison to the PCR methods, DNA fingerprinting based on hybridization showed a slightly higher degree of reproducibility and a higher differentiation potential for very closely related isolates.

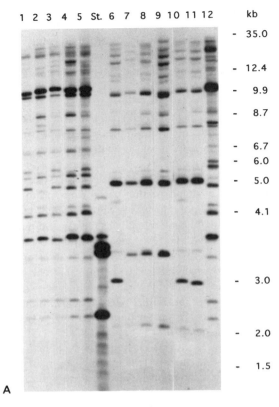

A

(GACA)₄ - Probe

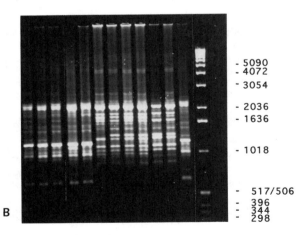

B

(GACA)₄ - Primer

FIGURE 33.

1 2 3 4 5 6 7 8 9 10 11 12 13 14

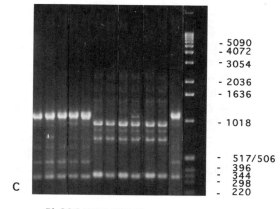

5'-GGCATCGGCC-3' - Primer

FIGURE 33. Comparison of three DNA fingerprint techniques for differentiating aggressive and nonaggressive strains of *Leptosphaeria maculans*. (A) Hybridization-based fingerprinting: genomic DNA was digested with *Bam*HI, separated on an agarose gel, Southern blotted, and hybridized with a ^{32}P-labeled (GACA)$_4$ probe. (B) PCR-based fingerprinting: genomic DNA was amplified using (GACA)$_4$ as a primer, separated on an agarose gel, and stained with ethidium bromide. (C) PCR-based fingerprinting (RAPD): genomic DNA was amplified instead using the arbitrarily chosen sequence 5'-GGCATCGGCC-3' as a primer. Lanes 1 to 5 and 12 are *L. maculans* isolates belonging to the aggressive pathotype group [IIa1, V2, MIX7, IV2, IX4, and BSA#63698 (W4E87)]; and lanes 6 to 11 are *L. maculans* isolates belonging to the nonaggressive pathotype group (V1, VII3, NV6, NXI10, SIII2, SV1). Size markers are indicated in lane St (human DNA digested with *Hin*fI) and lane 13 (1 kb ladder, GIBCO-BRL). Control without template DNA is in lane 14. (Figure 33C courtesy of C. Schäfer.)

6. *Magnaporthe*

The ascomycete *M. grisea* causes one of the most devastating diseases in rice, the rice blast. This disease occurs in almost all rice-growing areas around the world. Since rice is one of the major food sources in developing countries, development of control mechanisms for this disease has great significance for world agriculture. In 1989, a family of dispersed repeats (so-called "MGR sequences") was cloned from the *M. grisea* genome by Hamer et al.[263] Hybridization probes derived from these MGR sequences proved to be highly informative for DNA fingerprinting and have been widely exploited for genome analysis of the pathogen. Most importantly, these probes proved to be specific for rice-infecting isolates of *M. grisea*.[59,263] Results from DNA fingerprinting suggested that rice-pathogenic isolates are monophyletic and have been distributed throughout the world along with their host. The use of MGR probes not only made it possible to distinguish isolates and identify pathotypes, but also to assess phylogenetic relationships within and between various pathotype groups.[400]

Another application of MGR probes was to screen segregating populations from a cross between pathogens and nonpathogens of rice. Linkage analysis of MGR target fragments allowed the mapping of pathogenicity genes, which is an important step toward the isolation of these genes by chromosome walking techniques (see Section V.G).[262,593,712]

In a recent study, microgeographic variation of the pathogen was examined by hierarchical sampling of *M. grisea* isolates from two fields in Arkansas.[812] Seven distinct fingerprint groups were identified. Genetic variation was found to be distributed mostly between sample locations within a field rather than between fields. Since some fingerprint groups were composed of isolates that were heterogeneous with respect to virulence, the authors suspected a complex relationship between pathotypes and fingerprint groups in U.S.

7. *Ophiostoma*

O. ulmi causes the Dutch elm disease on several elm species, e.g., *Ulmus americana*. Hintz et al.[289] used the human minisatellite probe 33.6 to investigate ten isolates of *O. ulmi* from several geographical locations around the world, all belonging to the aggressive subgroup. DNA fingerprints were obtained by hybridization of the 33.6 probe to *Hae*III, *Hin*fI, or *Pst*I digested genomic DNA of *O. ulmi*. The patterns were polymorphic, but to an unexpectedly low degree. Though the majority of the fragments were conserved between *O. ulmi* isolates, there was sufficient variation to divide the isolates into three populations. All three enzyme-probe combinations gave consistent results. A high similarity of the banding patterns between isolates from different geographic locations was found, suggesting a worldwide clonal propagation of a small number of extremely successful genotypes.

8. *Phytophthora*

Species of the genus *Phytophthora* are responsible for soilborne diseases in a wide variety of host plants. One of the agronomically most important members of the genus is *P. infestans*, which causes potato and tomato blight. This pathogen is considered the most damaging potato pathogen worldwide. Several studies have been undertaken to apply DNA fingerprinting to *Phytophthora* species.

Repetitive DNA sequences were isolated from *P. citrophthora* and subsequently used as genomic DNA probes for *P. citrophthora* and *P. parasitica*.[228,229] The two DNA probes distinguished all tested isolates of *P. citrophthora* from a variety of hosts, except cacao. Hybridization signals were neither obtained with isolates of other *Phytophthora* species causing similar diseases of citrus nor with DNA isolated from their plant host.

In a later study, Goodwin et al.[230] cloned genomic fragments from *P. infestans*. The probes RG7 and RG57 revealed polymorphic fingerprints when hybridized to restriction-digested DNA from seven different Mexican isolates of *P. infestans* as well as from five other species (*P. colocasiae, P. phaseoli, P. mirabilis, P. hibernalis,* and *P. ilicis*). The banding patterns produced by both probes were somatically stable. Linkage analysis in test-cross populations showed that all except two bands revealed by the probe RG57 were unlinked to each other, while the RG7-detected bands showed higher levels of cosegregation and/or allelism.

Most recently, the RG57 probe was used by Drenth et al.[166] to estimate genetic diversity among isolates of *P. infestans* collected from potato and tomato fields. Fourteen fields were analyzed which were distributed over six regions in the Netherlands. As was also found with *Magnaporthe grisea*[812] and *Septoria tritici*,[443] the DNA fingerprints revealed a high level of genetic diversity (52%), within fields; another 8% was due to differences among fields within regions, and 40% between regions. Canonical variate analysis was used to group the genotypes according to their fingerprint patterns. Regardless of the geographic distance between fields, isolates from commercial potato were grouped together. Some genotypes were found on both tomato and potato.

Whisson et al.[775] isolated random clones from a genomic library of *P. megasperma* f. sp. *glycinea* and used them as hybridization probes to evaluate the genetic relatedness among Australian and North American isolates of this pathogen. Two probes were identified which provided highly diagnostic multilocus fingerprints. Cluster analysis of fingerprint data derived from several probes generated a dendrogram that separated one particular race of the U. S. from nine other isolates that grouped according to their geographic origin (i.e., Australia vs. U.S.). The high diversity of the American isolates on one hand and the close relationships between these geographically separated groups on the other hand, led the authors to suggest that the Australian isolates may have been introduced from the U.S.

9. *Sclerotinia*

A dispersed repeated fragment (pLK44.20) was cloned from the important rapeseed pathogen *S. sclerotiorum* and used for population and epidemiological studies.[364,365] Similar to other genomic clones used as fingerprint probes for fungi,[230,593] this repetitive DNA probe shows some properties of a retrotransposon. Initially, 63 *S. sclerotiorum* isolates were investigated by Kohn et al.[365] from transects of two fields of oilseed rape (*Brassica napus*) in Ontario. The fingerprint data from this study demonstrated that the field populations of *S. sclerotiorum* were composed of a heterogeneous mixture of distinct genotypes, each of which was relatively conserved. The authors suggested that conserved genotypes were either maintained by clonal propagation or by homothallic sexual reproduction. These results were supported by a larger epidemiological survey of *S. sclerotiorum* over several rapeseed fields in western Canada (isolates sampled over a distance of up to 3200 km). In this study, some clones were found distributed over long geographic distances, while field populations were generally composed of a clonal mosaic.[364]

C. INSECT-PATHOGENIC FUNGI

Fungi are not only pathogens of plants and humans, they can also infect insects. However, very few fungal pathogens of insects have been characterized yet by molecular techniques. In a recent investigation, RAPD primers were used by Strongman and MacKay[667] to characterize isolates of the fungal entomopathogen *Hirsutella longicolla* and to distinguish the two varieties *H. longicolla* var. *longicolla* and var. *cornuta* from each other. The results obtained with 83 primers clearly showed that these varieties are genetically distinct. Two different RAPD patterns corresponded to the two varieties. Four isolates, the identity of which was previously unknown because of the absence of distinctive morphological characters, could be assigned by their RAPD patterns to either *H. longicolla* var. *cornuta* or *H. longicolla* var. *longicolla*. However, discrimination among strains within a variety was not possible.

Another insect-pathogenic fungus is *Beauveria bassiana*, which attacks several species of grasshoppers. A series of four DNA probes, which specifically hybridize to this fungus only, have been developed from moderately repetitive genomic DNA.[280] By using lower stringency hybridization conditions, fragment patterns were obtained also with other species in the genus.

D. NONPATHOGENIC FUNGI

While most of the fingerprint applications in fungi are dedicated to human and plant pathogenic fungi, some papers also report on the use of DNA fingerprinting techniques to characterize nonpathogenic fungi. Some examples are given in this section (see also Section V.H).

One report described the use of PCR with arbitrary primers and other molecular markers to estimate the age and the size of a fungal organism. Genomic RFLP probes as well as 10-mer RAPD primers were used by Smith et al.[639] to analyze field isolates of the basidiomycete *Armillaria bulbosa* from a 15-ha plot in northern Michigan. RFLP as well as RAPD patterns were invariant in all diploid and vegetative isolates of a large sampling area, indicating that all isolates represent a single clone and are probably derived from a single individual. The authors hypothesized that this clone reached its present size through vegetative growth. By measuring the yearly rhizomorph growth rate, the age of the clone was estimated to be 1500 years, which makes it one of the oldest and largest organisms on earth.

RAPD primers were also used by Jacobson et al.[311] to investigate nine field isolates of the ectomycorrhizal fungus *Suillus granulatus* from a 300-m² study site in Virginia. The five primers indicated that this population was composed of at least eight genotypes, exhibiting various degrees of presumptive genetic relationship. Somatic incompatibility tests were also applied, but were considerably less efficient at telling the genotypes apart. There was no correlation between the results obtained by these two methods. Instead some of the compatible pairs of isolates proved to have very divergent RAPD patterns.

Endophytic fungi reside completely within plant tissues, without causing obvious damage. In many cases, endophytes produce toxins which are deleterious to insect larvae. It was hypothesized that mutualistic association with fungal endophytes could serve to defend long-lived plants against herbivores. *Rhabdocline parkeri* is a fungal endophyte which causes latent infections in the needles of Douglas-fir. The fungus has been found to inhabit nearly every tree in western Oregon and Washington. McCutcheon and Carroll[441] used a set of 10-mer RAPD primers to estimate genotypic diversity between different isolates of this endophyte. High levels of polymorphisms were observed between isolates from old trees (several genotypes occurred within a single needle) as compared to young, isolated trees. Less genotype diversity in the endophyte population was found to correlate with high susceptibility to herbivorous insects, thereby confirming the hypothesis that genotypically different fungi produce mixtures of toxins which act synergistically against herbivores. The RAPD patterns of a recently collected strain were identical to those of a strain collected from the same tree 5 years earlier, indicating that *R. parkeri* strains persist in the field over long periods.

E. COMMERCIALLY IMPORTANT FUNGI

Fungi and especially yeasts have been used for industrial purposes throughout centuries. Some fungal species are exploited for the production of bread, cheese, wine, and beer. Others are important enzyme or antibiotic producers. With the increasing industrialization followed by the competition between producers, it has become more and more important to protect an industrially used fungal strain and to control its identity during the production process. In the past few years, patent applications for biological materials constantly increased, but reliable markers are rather few. DNA fingerprinting techniques can provide a unique strain profile, allowing the identification of any industrial strain at any time. Thus, a specific DNA fingerprint can be part of patent applications in future.[612a]

In addition to the identification of industrially used fungal stains, molecular markers can provide genetic information for fungal breeders to improve strains that are commercially used for food production. Although DNA fingerprinting techniques are still mainly used to investigate medically or phytopathologically important fungal species, some papers already report on the application of DNA fingerprinting to characterize commercially used fungi.

1. Yeasts

In one of the very first papers on DNA fingerprinting in fungi, Walmsley et al.[736] suggested the application of this technology for the characterization of yeast. A poly(GT) hybridization probe was used to distinguish between several yeast species and strains, including brewer's yeast (*Saccharomyces cerevisiae*). Polymorphic hybridization patterns were found for all investigated yeasts. The patterns could be distinguished by the number, distribution, and intensity of the bands. The authors found that sequences flanking poly(GT) repeats in *S. cerevisiae* can also be used to distinguish subgroups of yeasts. Because of the comparatively simple nature of the generated patterns, such probes are particularly suited for routine differentiation of only one or two different organisms, e.g., if a mixed culture is analyzed or contaminations with other organisms should be monitored.

A total of 34 different yeast strains used in biotechnology, derived from 10 species were investigated by Lieckfeldt et al.[405] using the M13 minisatellite and the synthetic oligonucleotides $(GTG)_5$, $(GACA)_4$, and $(CT)_8$ for hybridization fingerprinting. DNA polymorphisms were generated from all species studied, and strain-specific bands could be identified for all isolates tested (Figure 34). Such strain-specific DNA fingerprints were suggested to provide an important identity criterion for patent application. Characteristic bands were also found for each of the investigated species. The 21 brewery strains of *S. cerevisiae* showed DNA profiles which unambiguously distinguished them from strains of other species derived from the same genus or from other genera (i.e., *Arxula, Candida,* and *Kluyveromyces*).

These results were confirmed and extended by PCR-based fingerprinting, where the oligonucleotides $(GTG)_5$, $(GACA)_4$, and an oligonucleotide corresponding to the core sequence of the M13 minisatellite were used as single primers.[405] Twenty-three biotechnologically important *S. cerevisiae* strains, including the strains from the previous study, could be discriminated from each other and also from *S. pastorianus, S. bayanus,* and *S. willianus.* However, with respect to the individualization of closely related strains, the hybridization-based fingerprinting approach appeared to have a higher discriminatory potential than PCR using the same oligonucleotides.

In a comparative study, Kunze et al.[378] used three different methods (pulsed-field gel electrophoresis, hybridization-based DNA fingerprinting, and protein electrophoresis) to classify brewer's yeast strains. Both pulsed-field gel electrophoresis and hybridization fingerprinting were found to be appropriate to distinguish between strains, whereas protein electrophoresis was less effective. Upon hybridization to the oligonucleotides $(GT)_{10}$ and $(GACA)_4$, polymorphisms were observed between all investigated isolates. The banding patterns differentiated between all 12 *Saccharomyces* strains, including 9 brewery strains of *S. cerevisiae*, the type culture as well as another strain of *S. cerevisiae* var. *diastaticus*, and one strain of *S. cerevisiae* var. *uvarum*.

2. Mushrooms

Very recently, DNA fingerprinting technology has been introduced to the genomic characterization of commercially produced basidiomycetes, i.e., edible mushrooms. One of the most important species in this context, the button mushroom *Agaricus bisporus*, suffers from the absence of outcrossing between strains. Because of its unusual sexual behavior, i.e., secondary homothallism, the basidiospores receive two instead of one meiotic nucleus and germinate into heterokaryotic, self-fertile mycelia that reconstruct the parental genotype. A consequence of this lack of recombination is that commercially used strains are extremely limited in genetic resources. Single 10-mer RAPD primers were used by Khush et al.[353] to differentiate between strains of *A. bisporus*, to

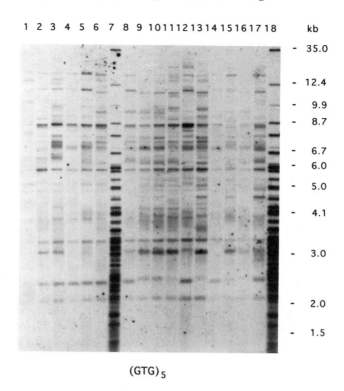

(GTG)$_5$

FIGURE 34. Genetic variation among different strains of brewer's yeasts demonstrated with hybridization-based fingerprinting. Genomic DNA of *Saccharomyces cerevisiae* was digested with *Bam*HI, separated on an agarose gel, Southern blotted, and hybridized with a ^{32}P-labeled (GTG)$_5$ probe. Lanes 1 to 6 are wine yeasts (CW3, CW8, CW16, CW21, CW26, and CW23), and lanes 8 to 17 are distilling yeasts (CBr2, CBr8, CBr9, CBr17, CBr21, CBr24, CBr22, CBr33, CBr40, and CBr46). Size markers (kb), human DNA digested with *Hin*fI, are indicated in lanes 7 and 18.

identify the two haploid nuclear components of a heterokaryotic strain, to verify artificially synthesized heterokaryons, and to monitor the transmission of genetic loci to the progeny. The RAPD markers have a high potential for strain characterization and improvement because of the large number of loci targetted by each primer and the simplicity with which they are generated. The application of this technique to a wider scale of isolates should make it possible to obtain a chromosome map of *A. bisporus* and to design a breeding strategy that considers the genetic peculiarities of this species.

In another recent study, different strains of a second commercially produced mushroom, the shiitake fungus *Lentinula edodes*, were typed by arbitrarily primed PCR using the M13 sequencing primer.[381] Almost every tested strain had a unique DNA profile, however, with some bands shared between all strains.

F. TAXONOMY AND PHYLOGENY

Knowledge of the genetic relationships between different fungal taxa and the identity of a fungal species is of basic significance for the efficient investigation of fungal ecology and phylogeny. However, the identity of a biological species in fungi is not as evident as it is in other eukaryotic organisms. The main reason for this is that many fungi have complex, but poorly understood systems of genetic recombination or no known sexual reproduction at all. Therefore, fungal species are often defined by morphological and biochemical parameters, which are expected to reflect natural biological species divisions. Since the biological species concept is only properly based on genetics,[65] unequivocal classification of fungi (and particularly imperfect fungi) by such criteria is often very difficult. DNA polymorphisms in general and the advent of DNA fingerprinting based on hybridization and PCR in particular suggested a new and powerful means to investigate taxonomic and phylogenetic problems. A few recent studies focusing on this topic are exemplified below. However, taxonomical aspects are also treated by some of the studies outlined in the previous sections (e.g., *Hirsutella*;[667] Section C).

The genus *Trichoderma* represents a fungal group which is particularly difficult to classify. Taxonomical classification based on morphological characters led to the grouping of different aggregates.[578] Some *Trichoderma* species are economically very important because of their ability to produce large quantities of cellulases and other hydrolytic enzymes, whereas other species are used as biocontrol agents. Most of the cellulase work is currently based on a particular species from the South Pacific, *T. reesei* QM6a. From this strain, hundreds of mutants were generated which have become the almost exclusive source for cellulase production during the past 30 years. Nevertheless, the taxonomic position of *T. reesei* is still unclear. Rifai[578] and more recently Bissett[52] considered *T. reesei* to belong to *T. longibrachiatum*. This is, however, both a morphologically and nutritionally distinct species.

Meyer et al.[455,456,458] hybridized the synthetic oligonucleotides $(CT)_8$, $(GTG)_5$, and $(GACA)_4$ and the M13 probe to restriction-digested genomic DNA to analyze the phylogenetic relationship among nine species of the *Trichoderma* aggregate, including three strains of *T. reesei*. On the basis of the fingerprint patterns obtained, the *Trichoderma* aggregate could be reclassified. Clear evidence was obtained that *T. reesei* is distinct from *T. longibrachiatum*. The three mutants of *T. reesei* proved to be indistinguishable from each other. Besides these two groups, *T. saturnisporum*, *T. virgatum*, and *T. harzianum* could also be clearly distinguished (Figure 35B). These observations were subsequently confirmed by PCR-based fingerprinting using oligonucleotides consisting of simple repeated sequences as single primers (Figure 35A).[456,459]

Another study dedicated to fungal taxonomy made use of genomic clones derived from the plant pathogen *Fusarium oxysporum* to generate distinctive, polymorphic hybridization patterns from different *F. oxysporum* isolates.[359] *F. oxysporum* has one of the broadest host ranges of any plant pathogenic fungus

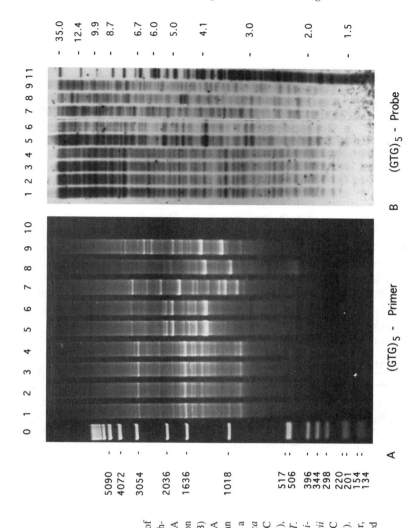

FIGURE 35. Genetic variation among isolates of *Trichoderma* using two different DNA fingerprint techniques. (A) PCR-based fingerprinting: genomic DNA was amplified using (GTG)$_5$ as a primer, separated on an agarose gel and stained with ethidium bromide. (B) Hybridization-based fingerprinting: genomic DNA samples were digested with *Bam*HI, separated on an agarose gel, Southern blotted, and hybridized with a ^{32}P-labeled (GTG)$_5$ probe. Lane 1 *T. reesei* (*T. todica* ATCC #36396), lane 2 *T. reesei* QM6a (ATCC #13631), lane 3 *T. reesei* QM9414 (ATCC #26921), lane 4 *T. reesei* QM9123 (ATCC #24449), lane 5 *T. longibrachiatum* (CBS #816.68), lane 6 *T. longibrachiatum* (TU-Vienna), lane 7 *T. pseudokoningii* (ATCC #24961), lane 8 *T. saturnisporum* (ATCC #18903), and lane 9 *T. harzianum* (ATCC #36042). Size markers are indicated in lane 0 (1 kb ladder, GIBCO-BRL) and lane 11 (human DNA digested with *Hin*fI). Control without DNA is in lane 10.

and is subdivided into numerous formae speciales (see Section B.4). DNA fingerprints were used to evaluate the genetic relationships among 30 different crucifer-infecting strains using the PAUP parsimony software package. The results indicated that members of a particular forma specialis are closely related to each other, suggesting a common ancestry.

In a recent study, RAPD analysis was applied to the classification of different populations of the ascomycete *Hypoxylon truncatum* sensu Miller. Yoon and Glawe[822] used 20 arbitrarily chosen primers. These resulted in 99 informative characters that were used to assess genetic variation between 54 isolates by UPGMA cluster analysis. The resulting phenogram clearly divided the isolates into two distinct groups which corresponded to different stromatal types. The RAPD results suggested that the two forms of stromata in *H. truncatum* actually reflect distinct species.

G. INHERITANCE OF DNA FINGERPRINTS, LINKAGE ANALYSIS, AND GENETIC MAPPING

Hybridization- and PCR-based DNA fingerprint patterns are generally inherited in a Mendelian fashion. In principle, markers generated by these techniques can therefore be exploited for linkage analysis and genetic mapping purposes. In practice, however, the dominant behavior of RAPD markers and the insufficient allelic information provided by multilocus fingerprint patterns cause some limitations for mapping applications in diploid organisms (see Chapter 2, Section III.C; Chapter 4, Section IV). These limitations are less of a problem in fungi, where a haploid lifestyle prevails. Several studies already have demonstrated the applicability of hybridization- and PCR-based fingerprinting for linkage analysis and genetic mapping in fungi.

One RAPD study was dedicated to the construction of a genetic map of *Neurospora crassa*.[783] DNA samples from two isolates of *N. crassa* used as parental strains were amplified with single RAPD primers. The segregation of polymorphic RAPD markers was studied in 38 haploid progenies of a cross between the two parental strains. It was shown that the RAPD markers mapped to dispersed, generally unlinked loci. A genetic map of the *N. crassa* genome based on RAPD and RFLP markers is under construction.

Genetic mapping with the help of hybridization-based DNA fingerprinting was, e.g., demonstrated for the MGR sequences of the rice blast fungus *Magnaporthe grisea* (see also Section V.B.6).[262,593] A rice-pathogenic strain of this fungus was crossed with a nonpathogenic strain, and the progeny was utilized as a mapping population. The two parents differed strongly in their content of MGR-complementary sequences; while the MGR586 probe produced 57 scorable hybridization bands from the pathogenic strain, only one band was present in the nonpathogenic strain. The majority of the MGR bands segregated in 1:1 ratios (presence vs. absence) in the progeny. Linkage analysis indicated a random distribution of MGR sequences across the genome and allowed the construction of a genetic map. Eight linkage groups were detected, and several genes could be mapped in relation to the MGR markers. For example, the nucleolar organizer (*Rdn*1) mapped to linkage group A, the mating type locus mapped to linkage group H, and the putative pathogenicity locus *Smo*1 mapped to linkage group C. Linkage maps derived from MGR sequences may prove very useful for future studies on genome organization in *M. grisea*, especially concerning the map-based cloning of pathogenicity genes.

Segregation and linkage analyses of fingerprint bands were also performed in the chestnut blight fungus, *Cryphonectria parasitica*.[464] A genomic repetitive sequence used as probe hybridized to 6 to 17 restriction fragments per fungal isolate. In laboratory as well as field crosses, the expected 1:1 segregation of fingerprints was observed. Tight linkage between bands was detected in a few cases.

H. GENERAL CONSIDERATIONS

As demonstrated by the large number of publications in the past few years, hybridization- and PCR-based fingerprinting techniques have a wide range of applications in the fungal kingdom. To summarize the results outlined in the previous sections, both methods have proven their high potential to:

1. Analyze the identity, occurrence, and geographic distribution of, e.g., strains, formae speciales, pathotypes, and clinical isolates.
2. Distinguish sexual from asexual reproduction, and thereby identify clonal lineages.
3. Evaluate genetic diversity and relatedness within and between populations.
4. Monitor the epidemiology of human and phytopathogenic fungi.
5. Clarify taxonomic and phylogenetic problems.
6. Provide identification criteria for patenting industrial or commercially important strains.
7. Provide polymorphic markers for genetic linkage maps.

In view of so many application areas, DNA fingerprinting of fungi should preferably be as easy-to-perform as possible. This would attract many mycologists with only limited molecular biological experience to make use of these techniques. The majority of fingerprint studies on fungi has been carried out with genomic hybridization probes so far. However, the use of this kind of probes is limited by their high specificity for certain species and the requirement for molecular cloning. In recent years, alternative approaches have been introduced which make use of versatile, ubiquitously informative, and commercially available probes (i.e., minisatellites or simple tandem repeat probes). To our experience, this kind of probe is applicable to a wide variety of fungal species, and we strongly recommend to test one of these probes before involving laborious cloning work.

The introduction of PCR with arbitrary and tandem repeat primers has further contributed to the development of easier and more widely applicable DNA typing techniques for fungi. An important aspect of reliable DNA typing is the reproducibility of results. In this respect, the use of tandem repeats originally designed as probes for DNA hybridization [e.g., the M13 minisatellite core sequence or the oligonucleotides $(CT)_8$, $(GTG)_5$, and $(GACA)_4$] as single PCR primers appears to be especially promising. The higher specificity of target recognition by these primers might reduce the influence of competition between primers (as observed in RAPDs; see Chapter 2, Section III.C), and thus result in a higher reproducibility of fragment patterns.

VI. CONCLUSIONS

In the past few years, DNA fingerprinting investigations in plants and fungi have been taken up at hundreds of laboratories around the world. An ever-increasing number of reports are being published each year, describing results and pointing toward new directions for this promising technique. From the data thus far, some generalizations may be suggested. These methods have been used for three major topics of investigation: (1) genetic identity; (2) genetic relatedness, including paternity studies; and (3) genetic mapping.

Genotype identification by DNA fingerprinting is often quite efficient, regardless of the method chosen. This is especially true if the investigated genotypes have arisen by recombination of relatively heterozygous parental genotypes. Plants and fungi derived through several generations of inbreeding or through somatic mutations are, on the other hand, usually not easily distinguished from each other.

If hybridization to minisatellite or simple sequence oligonucleotide probes is the method of choice, some efforts should be directed toward finding a restriction enzyme that will produce a suitable number of bands. However, in most cases, genotype identification per se is not dependent on which enzyme is chosen. As for choice of probes, it appears that minisatellite DNA probes are often very efficient at detecting variation among recombined genotypes. Different simple sequence probes, on the other hand, usually detect varying levels of genetic variation. Choice of probe should therefore be directed by the aim of the investigation, and has to be done empirically.

For the PCR-based methods, efficiency of discriminating different genotypes varies considerably among primers. This is especially true for the RAPD method using mostly 10-mer primers in combination with agarose gel electrophoresis. Usually, it is advisable to screen a considerable number of primers on a few samples in order to find an optimal set, which can then be used for the entire material under study. Additional polymorphisms can be detected by a variety of means, e.g., by separating the amplification products on polyacrylamide gels followed by silver staining, or exploiting alternative separation principles such as in DGGE. Pre- or post-PCR restriction of DNA is another means of increasing levels of detected polymorphism.

Longer primers, such as the M13 sequencing primer used in AP-PCR or primers identical to sequences commonly used for hybridization-based fingerprinting [e.g., the M13 repeat and $(GACA)_4$], appear to produce levels of polymorphisms that are quite comparable to those obtained by hybridization to the M13 (or other minisatellite) probe. Also, the DAF method, which involves polyacrylamide gel electrophoresis and silver staining, usually results in large numbers of bands. Thus, both of these latter methods generate very complex banding patterns, which are sometimes difficult to score unambiguously. Nevertheless, for identification of samples run on the same gel, these complex patterns may be quite useful.

With all PCR-based methods, great care must be taken to make sure that reaction conditions such as concentrations of DNA, enzyme, and primer are kept constant throughout the study, since both number and intensity of amplification products are strongly affected by a variety of parameters. Otherwise, quantification of genetic diversity may give erroneously high values as a result of variation introduced by the RAPD technology itself. This is especially true when faint amplification products are included in the analysis. Performing duplicate analyses may help to eliminate artifactual polymorphisms.

Most reports are not confined to the differentiation of genotypes. Frequently, differences and similarities between DNA fragment or amplification product profiles are quantified and used to estimate relatedness between genotypes. This is, however, a difficult task. Most assumptions, on which the statistical treatments of fingerprint data are based, are not met or at least are incompletely investigated. Nevertheless, such quantifications are much sought after in, e.g., studies of genetic relatedness in wild as well as in cultivated plants. Apart from optimizing the experimental conditions in the laboratory, great care should also be taken to include sufficient numbers of markers as well as of plants on each level of variation, i.e., plants within a population and populations within a species for wild plants. Correspondingly, a suitable number of plants should be investigated for estimation of cultivar homogeneity and for assessments of relatedness between cultivars. Pooling of DNA from several plants tends to obliterate heterogeneity within, e.g., populations or cultivars, and should be avoided under most circumstances, unless special reasons apply.[827] Comparison of genetic relatedness data arrived at with different methods like, e.g., RFLPs with genomic DNA and cDNA clones and RAPDs, appear to be reasonably well correlated, but definitely not identical.[693] This should warn us that the data obtained with these methods are, at best, only approximations and must be regarded as such. For more information on comparison of methods, see Chapter 6.

Genome mapping with various molecular markers is becoming increasingly popular. In only a few years, RAPDs have developed into a much-used method in laboratories around the world. Whereas some drawbacks certainly exist (see Section IV.C), this approach has nevertheless proved worthwhile in a large number of studies.

Chapter 6

COMPARISON OF METHODS FOR DETECTING GENETIC VARIATION

In this chapter, we will compare some techniques described in Chapters 2, 4 , 5, and 7 for their discriminatory power, cost, and ease of use. For additional reviews on the choice of molecular techniques in population biology and systematics of plants and fungi, see Bachmann,[28] Bostock et al.,[60] Doyle,[162] Hillis and Moritz,[286] Hoelzel,[290] McDonald and McDermott,[445] and Soltis et al.[648]

I. MORPHOLOGICAL CHARACTERS AND ALLOZYMES VS. DNA MARKERS

Morphological characters have long been used to identify species, families, and genera. Moreover, morphological as well as life history traits have been the subject of numerous studies in population genetics and agriculture, where fitness and yield are the biologically and evolutionary important factors.[188] Levels of variability can be estimated for morphological characters, their response to selection and their genetic background can be determined, and genetic correlations and selection forces in the past can be inferred.[791] Contrary to molecular markers, morphological characters are, however, often strongly influenced by the environment, and, consequently, special breeding programs and experimental designs are needed to distinguish genotypic from phenotypic variation. Moreover, for small nonflowering plant species (e.g., mosses and algae) as well as fungi, it is frequently difficult to find a sufficient number of morphological characters for a comprehensive systematic study. Molecular methods, on the other hand, provide almost unlimited numbers of potential markers.

Another much-used method for detecting and estimating genetic variation is allozyme electrophoresis (for a short description, see Chapter 2, Section I.B). There are three general advantages of allozyme studies as compared to many kinds of DNA markers: (1) the low cost for chemicals and labor; (2) the user friendliness — many individuals can be scored for several allozyme loci within a short timespan; and (3) the fact that allozyme markers are codominant — both alleles in a diploid organism are usually clearly identifiable, and heterozygotes can be discriminated from homozygotes, which is a prerequisite for estimation of allele frequencies in population genetic studies.

However, there are also a number of limitations to allozyme studies. Thus, a new allele will only be detected as a polymorphism if a nucleotide substitution has resulted in an amino acid substitution, which in its turn affects the electrophoretic mobility of the studied molecule. Because of the redundancy of the genetic code and the fact that not every amino acid replacement leads to a charge difference, only 30% of all nucleotide substitutions result in polymorphic fragment patterns. Therefore, allozyme analysis underestimates the genetic variability. Using allozymes also restricts the study to those parts of the DNA which code for stainable enzymes. This is not necessarily a random sample of the genome. Another problem is that many plant species are polyploid, and the analysis of allozyme patterns of polyploids can be extremely difficult. In a few cases it has also been shown that allozymes differ in one or more physiological respects[71] and, therefore, may not be evolutionary neutral. A more practical aspect is that the plant tissue intended for allozyme studies has to be processed shortly after harvest since proteins usually are quite unstable. In contrast, DNA-based methods allow a longer time span between harvest and processing (see Chapter 4, Section II.A.1), implying that these methods may be preferred if the plant material is collected in remote areas.

In conclusion, generating molecular markers at the DNA level as opposed to using morphological characters or allozymes has several advantages: (1) since the genotype of an organism is examined directly, environmental and developmental influences on the phenotype are not a problem; (2) since different regions of DNA evolve at different rates, appropriate regions may be chosen for a given study (e.g., highly variable regions for paternity studies or cultivar identification and regions of limited variability for phylogenetic investigations); (3) an almost unlimited number of detectable polymorphisms exist; and (4) a variety of techniques have been developed, each of which has the potential to provide suitable markers for a particular problem.

II. DIFFERENT KINDS OF MOLECULAR MARKERS

A considerable number of molecular methods, including allozymes as well as DNA markers, are now available for the detection of genetic variation (see Chapter 2, Section I, for a short survey). However, the level of variation which is detected differs considerably between these methods. Other factors such as user friendliness and costs are also important for the choice of method.

A. DISCRIMINATORY POWER AND CONSISTENCY

Consistent results have been reported in most cases where two or more different marker techniques were applied to investigate the same material. Thus, low levels of genetic variation detected with one technique have often been confirmed by applying additional methods. For example, in the rare endemic *Lactoris fernandeziana*, allozymes, rDNA, and random amplified polymorphic DNAs (RAPDs) showed similarly low levels of polymorphism,[66] and few polymorphisms were observed with either RAPDs or allozymes in a study on *Pinus resinosa*.[478] Comparable levels of variation were also found for RAPDs vs. restriction fragment length polymorphisms (RFLPs) in barley[696] and for RAPDs vs. allozymes in 2n and 4n breeding lines of potato.[560]

In other studies, various methods provided different levels of sensitivity, while the results were still consistent. This is, e.g., illustrated by an investigation on three *Plantago* species with different breeding systems, where genetic variation was analyzed by morphological criteria, allozymes, chloroplast (cp)DNA RFLPs, and hybridization fingerprinting with the M13 probe.[792,793,795,798] In these studies by Wolff and co-workers, the mating system and the history of the investigated populations (e.g., the occurrence of bottlenecks) proved to be major determinants of the variability of all kinds of characters. While the morphological variability appeared to be additionally influenced by natural selection, the three types of molecular markers gave consistent estimates of the distribution of genetic variability. Thus, the inbreeding *P. major* exhibited low levels of genetic variation within, but extensive variation between populations, whereas the outcrossing *P. lanceolata* showed high within- and low between-population variability (see Figure 21, Chapter 5). The discriminatory potential of the molecular markers was considerably different; M13 fingerprinting showed the highest levels of variation and chloroplast RFLPs the lowest, with allozymes in between.

Another example of differential marker sensitivity was provided by Ishii et al.[310] who compared genetic distances between rice species using mitochondrial (mt) DNA, cpDNA, and nuclear genome RFLPs. It appeared that the cpDNA sequences were only half as variable as the other two marker types. Also, cpDNA and mtDNA were more conserved within the taxon than the nuclear genome.

In many cases, DNA fingerprinting has proved to be more sensitive at detecting genetic variation than other methods. This is best illustrated by several studies in tomato. Tomato cultivars are known for their low levels of genetic variation, which is caused by a bottleneck in their breeding history. The low amount of genetic diversity is reflected by the rare occurrence of allozyme polymorphisms and RFLPs (see, e.g., Foolad et al.[196] for a discussion) which require the use of interspecific crosses to generate an RFLP-based genetic map.[681] More recently, DNA fingerprinting was successfully applied to the tomato genome, and RAPD analysis[196,781] as well as hybridization to minisatellite[69] and simple sequence probes[341a,732] (see Figure 26, Chapter 5) resulted in high levels of polymorphism. An interesting variation of the fingerprinting approach was also applied by Broun et al.[70] to tomato; telomeric probes were hybridized to restriction fragments separated by pulsed-field electrophoresis and proved to be successful in distinguishing tomato cultivars.

Sometimes different methods can complement each other to give a more comprehensive understanding of the biological process. Thus, the formation of a hybrid species in the genus *Iris* was reconstructed by Arnold[20] using RAPD data which could differentiate the original species and cpDNA data from which the direction of gene flow could be deduced.

A general conclusion from the examples given above as well as from numerous other studies is that whatever molecular markers of varying discriminatory power are used in a particular species, they usually appear to evolve in a correlated way. Thus, populations with a high level of, e.g., allozyme variability usually exhibit high levels of polymorphism also with the DNA finger-

TABLE 5
Levels of Discrimination for Several Molecular Techniques

Technique	>Genus	Genus/species	Species
Sequencing conserved genes	+	+/–	–
cpDNA (RFLP)	+	++	– (+)
rDNA (RFLP)	+/–	+	+
Sequencing variable sequences	+/–	+	+/–
Nuclear DNA (RFLP)	–	+	+
Allozymes	–	+	+
PCR-based fingerprinting	–	+/–	++
Hybridization-based fingerprinting	–	+/–	++

Note: ++, highly useful; +, useful; +/–, useful in some cases; and – generally not useful.

printing techniques. There are, however, a few exceptions to this rule (see, e.g., Wayne et al.[742]). One important exception concerns hybridization-based fingerprinting with simple sequence probes. As outlined in Chapter 2, Section II.D, short tandem repeats may undergo a particular kind of mutation process (i.e., dynamic mutation)[575] which can result in high and unpredictable mutation rates and inconsistencies with genetic variation levels derived from other kinds of markers.

It may be concluded that molecular methods show a considerable range in discriminatory power, and, therefore, the subject of a particular investigation should be decisive in the choice of methodology. For species delineation, cpDNA studies may be a good choice. Identification of individual genotypes, on the other side of the spectrum, probably requires one of the different DNA fingerprinting techniques. A rough overview of the levels of discrimination usually obtained with the molecular techniques mentioned in this chapter is given in Table 5. To our experience, hybridization-based DNA fingerprinting will often give the highest level of discrimination (although this to some extent depends on the probe used), followed by polymerase chain reaction (PCR)-based DNA fingerprinting (RAPDs) and RFLPs. A very high level of variability is also obtained with microsatellite PCR, see Chapter 7. Allozymes generally detect lower levels of variability and are restricted to a maximum of 20 to 40 stainable enzymes.

B. COSTS
The approximate costs for chemicals and disposables are shown in Table 6 for several methods. We have tried to be realistic, e.g., we assumed that multiple digestions (three) or multiple RAPD reactions (ten) will be set up from each extracted sample and that Southern blots will be used six times. Costs are in U.S. dollars, although they are based on prices in the Netherlands. According to our estimations, RFLPs, RAPDs, and PCR cost approximately the same (in chemicals) per sample (but see also Ragot and Hoisington[564]). Hybridization-

TABLE 6
Approximate Cost of Some Molecular Techniques Per Sample and Per Experiment in U.S. $

Procedure	RFLPs	Hybridization-based fingerprinting	PCR-based fingerprinting
DNA extraction	0.32	0.32	0.06
Digestion, per sample	0.25–2.75	0.25–2.75	—
PCR reaction, per sample	—	—	1.10
Electrophoresis + blotting	0.10	0.25	0.20
Hybridization	0.70	1.40–3.50	—
Detection	0.06	0.12	0.07
Total per sample, per experiment	1.40–4.00	2.30–7.00	1.45

based fingerprinting is somewhat more expensive, mainly because more agarose and larger membranes are needed (unless in-gel hybridization is performed). We should keep in mind, however, that with DNA fingerprinting many bands can be scored simultaneously. Sequencing is the most expensive technique, but again many nucleotides (up to 300) can be determined at a time.

Our calculations do not include the cost for labor, laboratory equipment, and space. Hybridization-based techniques (RFLPs, hybridization-based DNA fingerprinting) are generally much more time- and labor-consuming than PCR-based techniques. For RFLP analysis, a genomic or cDNA library has to be constructed and probes have to be tested. We did not take these costs into account. Also, techniques which do not require separate isotope facilities, e.g., PCR-based techniques and hybridization with nonradioactively labeled probes, may in some cases be cheaper. Sharing RAPD primers, simple sequence primers, and hybridization probes with other laboratories will also lower the costs.

C. CONCLUSIONS

Allozyme analysis is presently the easiest and cheapest of the methods compared in this chapter, provided that sufficient polymorphism is encountered. In many kinds of studies, however, the use of DNA fingerprinting, either hybridization or PCR based, is necessary to obtain a sufficient number of polymorphic markers. As PCR- and hybridization-based DNA fingerprinting uses probes and primers that can be applied in a wide variety of species without prior sequence information, they appear to be the most useful complement to the allozyme technique.

Some species are comparatively difficult to analyze, e.g., polyploids or species with a large genome size. In such cases, DNA fingerprinting techniques may be more useful than allozymes and RFLPs, as demonstrated on, e.g., onion[779] and conifers,[702] which have large genomes, and on polyploid species and breeding lines of *Rubus*[370] and alfalfa,[828] respectively. For genetic analyses in polyploid species, the approach of single dose restriction fragment analysis may be chosen.[808] Such fragments originate from one parent and are only singly present in this parent, resulting in a transmission to 50% of the offspring.

Concluding the previous discussions, the following points should be taken into account when deciding on the most suitable technique.

1. Which markers will result in the most appropriate levels of discrimination?
2. Do results need to be transferred across laboratories? If so, reliable and reproducible methods such as allozymes, hybridization-based fingerprinting, and RFLPs are the techniques of choice.
3. How much money is available for the project? Allozyme analysis is usually the cheapest alternative, provided that sufficiently high levels of polymorphism are detected.
4. How much time is available for the project? Allozymes and PCR-based fingerprinting will usually give the fastest results. Hybridization-based fingerprinting takes longer, while techniques requiring molecular cloning are the most time consuming.
5. Is sufficient expertise available? If not, techniques involving cloning and/or sequencing should perhaps be avoided.
6. What are the specific problems inherent to the organism under study? Availability of only little or partially degraded DNA may necessitate the use of PCR-based methods.
7. Is the mode of marker inheritance relevant for the aim of the study? If homo- and heterozygotes have to be discerned, codominant markers such as single-locus RFLPs, allozymes, and PCR-amplified microsatellites (see Chapter 7) will probably be the markers of choice.

Chapter 7

FUTURE PROSPECTS

The preceding chapters have described in some detail how to use hybridization- and PCR-based DNA fingerprint techniques and have also provided an overview of the fields of research these techniques are presently applied to. Plant biologists and mycologists interested in using molecular techniques should benefit from the progress made since DNA fingerprinting was first used in plants and fungi. The introduction of DNA fingerprint techniques to previously nonmolecular biology laboratories has been facilitated by at least three factors. (1) A multitude of highly informative simple sequence- and minisatellite-derived multilocus probes is now available for plant and fungal DNA fingerprinting. Many of these, such as the "Jeffreys" probes 33.15 and 33.6, the M13 repeat probe, and an assortment of different simple sequence oligonucleotides, are commercially available. This obviates the need for establishing molecular cloning facilities in the laboratory. (2) A variety of nonradioactive labeling and detection techniques have been developed, allowing us to perform hybridization- and PCR-based DNA fingerprint analyses in the absence of isotope facilities. (3) A probably unlimited reservoir of informative RAPD primers is available. Primers can be purchased commercially and are easily exchanged between laboratories.

The present chapter on future developments of DNA fingerprinting in plants and fungi will mainly focus on two fields: (1) the technical improvement of existing methods and (2) the introduction of locus-specific typing strategies (which should be referred to as "DNA profiling" rather than DNA fingerprinting; see Jeffreys et al.[323]), which overcome the limitations inherent to multilocus fingerprinting. Some of these locus-specific approaches have already been established for human and animal systems and, recently, also for plants, whereas others [e.g., the analysis of sequence characterized amplified regions (SCARs) and selectively amplified restriction fragments] are fairly new.

I. IMPROVEMENT OF EXISTING METHODS

One important approach to improve current DNA fingerprinting protocols involves the **separation** of fragments. Most protocols use agarose gel electrophoresis to separate DNA fragments according to their size. While this method is easy to perform, a main disadvantage is its limited resolving power, especially in the low molecular weight range. In many cases, a higher resolution of banding patterns is desirable, e.g., for distinguishing closely related genotypes.

An obvious way to increase the resolution of PCR- or hybridization-generated DNA fingerprint patterns is to use polyacrylamide instead of agarose gels (see, e.g., Caetano-Anollés et al.[86,87] and numerous other reports; Chapter 4, Section IV.C). Alternatively, one may also try one of the newly developed separation media (e.g., Metaphor™ agarose, FMC Bioproducts; Visigel™, Stratagene). Other recent approaches are more sophisticated and rely on the observation that the mobility of a **single-stranded** DNA fragment on a poly-acrylamide gel is not only dependent on its size, but also on its base composition. Similar-sized DNA fragments that differ in sequence are therefore separated from each other after denaturation, resulting in so-called single-strand conformation polymorphisms (**SSCPs**).[529,530] This conformation dependence of DNA mobility on gels is also exploited by a technique called dena-turing gradient gel electrophoresis (**DGGE**).[487] In this method, polyacrylamide gels are prepared which contain a gradient of a denaturant (usually formamide/urea). Double-stranded DNA molecules that migrate through the gradient gel become single stranded at a position which depends on their melting point (which is sequence dependent). When this position is reached, denaturation results in a sharp reduction of mobility. The fragment virtually "stops".

The different separation principles (i. e. sequence vs. size dependence of mobility) make it possible to combine DGGE and ordinary polyacrylamide gel electrophoresis to a highly resolving two-dimensional fingerprint tech-nique.[483,704–706] Such two-dimensional techniques have not yet been applied to plant and fungal genome analysis. However, since the DGGE technique alone has already been shown to distinguish RAPD patterns from very closely related strains in several cereal species[172] (see Chapter 5, Section III.A.3), two-dimen-sional techniques may provide a valuable alternative when low levels of polymorphism are encountered with other techniques. One should nevertheless keep in mind that DGGE and two-dimensional approaches are more sophisti-cated and require additional equipment as compared to ordinary agarose elec-trophoresis. Moreover, computer-based image analysis facilities are required for the reliable evaluation of complex, two-dimensional spot patterns.

Another sophisticated technique for separating DNA fragments is known as **capillary electrophoresis**.[232] DNA molecules are electrophoresed at high volt-ages in ultrathin capillars (20 to 100 µm internal diameter) and detected by ultraviolet (UV) absorbance. The potential of capillary electrophoresis to sepa-rate different alleles from PCR-amplified human minisatellites has been dem-onstrated.[546] Though capillary electrophoresis exhibits several advantages such as high speed (less than 30 min) and high resolution and requires only minute amounts of DNA, the high costs of the instrument will probably prevent its general application in the near future.

Improvements are not only to be expected concerning the separation, but also the **detection** of fragments. For hybridization-based DNA fingerprinting, ^{32}P and autoradiography are still the routine tools for signal detection in the majority of laboratories. However, the fast development of reliable nonradio-active detection methods in the past few years (see Chapter 4, Section III.H.3) suggests that this situation will change in the near future and that chemilumi-nescence-based assays will be generally applicable and accepted. In cases where radioactive techniques are retained, the use of phosphorimaging systems may become a useful alternative to classical autoradiography (see Chapter 4, Section III.H.2).[333]

For the detection of PCR products, it seems likely that ethidium bromide staining and radioactivity will be increasingly complemented by fluorescence-based approaches. In this strategy, PCR primers are conjugated to a fluorescent dye (e.g., rhodamin or fluorescein derivatives), resulting in the specific fluorescence of the amplification product(s) upon irradiation.[109,638,834] A major advantage of detecting PCR products with the help of fluorescent primers is that different amplification reactions (each labeled with a different fluorophore) can be combined into one lane during electrophoresis; multiplex PCR.[355] The incorporation of labeled size markers in each lane allows for precise sizing of the bands. If an appropriate instrumentation is available (e.g., an ABI 373A DNA sequencer plus GENESCAN™ 672 software), fluorescent PCR products can be detected by real-time laser scanning during gel electrophoresis.[355,834] This technique, originally introduced for DNA sequencing,[638] allows a high throughput of samples and the direct transfer of DNA profiling data into a computer for further processing. However, expensive equipment is required.

Many steps of DNA fingerprint procedures (including DNA isolation) lend themselves to automation (see Landegren et al.[386] for an overview). For PCR-based fingerprinting, the feasibility of some approaches to automation has already been demonstrated, such as the use of a robotic workstation for preparing reaction mixes and loading of gels[90] or the automated analysis of fluorophore-labeled PCR products by real-time scanning during electrophoresis (see above). While automation generally requires costly equipment, it also allows the running of many more samples per time unit than is presently feasible. This is especially important, e.g., in genome mapping studies.

Finally, the computer software for image analysis as well as for the evaluation and processing of fingerprint data is expected to improve considerably. This will greatly facilitate and accelerate genomic mapping studies as well as the analysis of genetic diversity and relatedness within and between species, populations, and individuals.

II. EXTENDING THE SPECTRUM OF DNA TYPING STRATEGIES

Multilocus DNA fingerprinting techniques, whether they are based on hybridization or PCR amplification, suffer from a variety of limitations which interfere with their general applicability as molecular markers. The two most important drawbacks are that (1) the multiple bands of a fingerprint or RAPD pattern are usually not assignable to a specific locus and (2) RAPD markers are inherited in a dominant rather than a codominant fashion. As a consequence, allelic states and allele frequencies of fingerprint bands cannot be determined easily, which would be important for, e.g., population genetic studies and genomic mapping projects. Moreover, the estimation of genetic relatedness by both fingerprint methods is usually based on band sharing and therefore prone to error (for a discussion see Chapter 4, Section V.B).[84,421,422] For example, comigrating bands might not necessarily be homologous, and bands treated as independent characters might actually be linked.

The development of DNA profiling techniques which make use of multiallelic, locus-specific markers might overcome most of these limitations. Some approaches to generate such marker systems as well as some miscellaneous techniques of DNA (and RNA) typing are outlined in the forthcoming sections.

A. POLYMORPHIC SINGLE-LOCUS HYBRIDIZATION PROBES

A straightforward approach for obtaining polymorphic single-locus probes is the cloning of individual hypervariable loci by screening genomic libraries with minisatellite or simple sequence probes. This has been successfully demonstrated not only for minisatellites (human: Armour et al.,[17] Nakamura et al.,[489] Wong et al.;[803,804] fish: Bentzen et al.;[43] birds: Hanotte et al.;[264,265] plants: Broun and Tanksley,[69] Dallas et al.[138]), but also for simple sequences (human: Zischler et al.[839,840]). While different strategies may be followed to obtain such probes, currently favored protocols usually involve the following steps (see Bruford et al.[76] for review):

1. the establishment of a genomic library enriched for minisatellites or simple sequences (see, e.g., Armour et al.,[17] Ostrander et al.[531]) (during this step, problems may arise from the instability of tandem-repetitive DNAs in prokaryotic hosts[265,349,839]);
2. screening the library with a suitable multilocus probe followed by the isolation of positive clones;
3. preparation and labeling of the insert of positive clones, followed by hybridization with the restriction-digested template DNA to be analyzed [hybridization at high stringency should result in a (hopefully polymorphic) locus-specific banding pattern. The inclusion of whole genomic competitor DNA minimizes the detection of related minisatellites (which would result in a multilocus pattern as for the original probe)];
4. to further characterize the probe, the region surrounding the tandem repeat might be subcloned, sequenced, and eventually used for the design of flanking PCR primers (see below).

Comparatively few data are yet available concerning the development of polymorphic single-locus hybridization probes in plants and fungi. DNA sequences flanking minisatellites or simple sequence loci have been cloned in a variety of species.[69,121,139,384,431,583,809,831] However, sequence information derived from these clones has only recently been exploited for the generation of polymorphic, locus-specific hybridization probes in plants. Thus, polymorphic markers specific for rice were isolated by Dallas et al.[137] and Winberg et al.,[788] and a set of single-locus markers in tomato was characterized by Broun and Tanksley.[69] In all three studies, plant genomic libraries were screened with the human minisatellites 33.6 or 33.15. Since the interest in simple sequence loci is rapidly growing (see next section), considerable progress in this direction may be expected in the near future.

B. PCR-AMPLIFIED MICROSATELLITES

Once the sequences of minisatellite or simple sequence flanking regions are available, this information can also be used for the design of locus-specific primers for DNA amplification in the PCR. This strategy combines the high informativeness of minisatellite and simple sequence loci with the ease and speed of the PCR technique. As outlined in Chapter 2, the polymorphic behavior of an individual, defined minisatellite, or simple sequence stretch is mainly a consequence of its variable number of tandemly repeated sequence elements. Therefore, amplification of this stretch with flanking primers should result in a polymorphic, highly informative band derived from a single locus. In 1988 and 1989, the successful application of this strategy was demonstrated for human minisatellites[57,298,319] as well as for simple sequence loci.[408,637,684,745] For these experiments, radioactive primers were used, and the amplification products were separated on agarose or highly resolving polyacrylamide gels and visualized by autoradiography. The term "microsatellites" was introduced by Litt and Luty[408] to characterize the simple sequence stretches amplified by PCR as a new kind of molecular marker. In summary, these initial experiments showed that (1) single loci are amplified, resulting in one or two bands depending on the homo- or heterozygous state (i.e., these markers behave codominantly, just like most RFLPs); (2) many different-sized alleles exist in a population, and the level of heterozygosity is extremely high; and (3) these markers are inherited in a Mendelian fashion and can be used for segregation and linkage analyses.

It was soon realized that simple sequences have at least two advantages over minisatellites for this kind of analysis. First, they are short (typically 20 to 40 basepairs (bp), rarely up to 200 bp) and easy to amplify (minisatellite arrays often turned out to be too long, i.e., 0.5 to 30 kb, for efficient amplification[319]). Second, stretches of simple sequences are more evenly distributed over the genome than minisatellites (see Chapter 2, Section II.C, for details). Weber[744] investigated the informativeness of microsatellites of the $(CA)_n$ type. He showed that the level of polymorphism exhibited by PCR-amplified microsatellites depends on the number of the "pure" (i.e., uninterrupted), tandemly repeated motifs. Below a certain threshold (i.e., 12 CA-repeats in his investigation), the microsatellites behaved mainly monomorphic. Above this threshold, however, the probability of polymorphism to occur increased with length.

The primers used for the first generation of experiments were deduced from flanking sequences of known simple sequence stretches, i.e., from a DNA database. Very soon, however, strategies were developed for the establishment of genomic libraries enriched in microsatellites (e.g., Rassmann et al.[568] and Ostrander et al.[531]). During the past few years, this new marker technique has been established for a variety of species, especially mammals (reviewed by Hearne et al.[278]), but also insects[185] and plants (see below). High-resolution genetic maps have been created for the human,[762] mouse,[154,415] and rat genome,[622] mainly or even solely on the basis of amplified microsatellites. In addition to the very abundant dinucleotide repeats, some authors also introduced PCR-amplified microsatellites based on mono-, tri-, tetra-, and pentanucleotide motifs as molecular markers.[174,175,177,355]

In recent years, PCR amplification of microsatellites has also been successfully applied to the analysis of plant genomes.[6,384,476,621,809,831] In the course of these experiments, extensive database research revealed that the relative abundances of different microsatellite motifs in plants and animals differ considerably.[384,476] For example, the $(CA)_n$ repeat is one of the most frequently occurring microsatellites in humans and many mammals (several tens of thousands of copies).[38,258,659] In contrast, $(AT)_n$ microsatellites are the most abundant dinucleotide repeats in plants, while $(CA)_n$ is comparatively rare (for more data, see Lagercrantz et al.[384]). In fungi, simple sequences have been shown to be polymorphic by hybridization-based fingerprinting (see Chapter 5, Section V). However, no data are available yet on the use of flanking sequences to amplify fungal microsatellites.

Summarizing, PCR-amplified microsatellites exhibit a number of useful properties which bring them very close to the "optimal" genetic marker outlined in Chapter 2.

1. They are highly polymorphic and thus highly informative.
2. They are inherited in a codominant fashion, which allows discrimination between homo- and heterozygous states and increases the efficiency of mapping and population genetic studies.
3. A considerable collection of different microsatellites is available, some of which occur in high copy numbers.
4. They are more or less evenly dispersed throughout the genome.
5. Provided that the adequate primers are available, typing assays via PCR are fast and easy and require only nanogram amounts of template DNA. The use of radioisotopes can be avoided (see below), and the technique is amenable to automation.
6. Information about primer sequences is easily exchanged between laboratories.

Additionally, the use of sequencing gels for the separation of PCR-generated microsatellites allows the unambiguous designation of alleles (in contrast to hybridization- and PCR-based fingerprint procedures) and a precise calculation of allele frequencies. In contrast to RAPDs, stringent annealing conditions can be applied. This allows simultaneous amplification of different loci in a multiplex approach, ensures high levels of reproducibility, and eliminates problems concerning competition between primers.

Though the advantages of PCR-amplified microsatellites over other types of markers prevail, there are also some limitations. Most importantly, the identification of informative microsatellite loci and, consequently, of suitable primer sequences is even more cumbersome and expensive than the generation of locus-specific, polymorphic hybridization probes (see Section II.A). Genomic libraries enriched for microsatellites have to be established and screened with a particular repeat motif. The generation of each marker will then require (1) "fishing" and sequencing of a particular clone, (2) designing and synthesizing (or buying) a suitable pair of primers, and (3) performing a PCR in order to test the putative marker for its polymorphic behavior.

Two minor disadvantages as compared to other techniques are of a technical nature. First, currently used protocols usually involve the use of radioisotopes to detect the amplified microsatellites. Second, instead of yielding one particular band, the enzymatic amplification of dinucleotide repeats commonly results in a cluster of bands which are separated from each other by two-bp intervals. The additional bands are most probably the result of slippage events, which are thought to occur during replication by the *Taq* polymerase.[637] In some cases, this artifact makes the interpretation of correct allele sizes difficult, especially if two alleles differ by two bp only and homo- and heterozygotic states have to be discriminated. However, both disadvantages may be circumvented. Slippage artifacts are much less of a problem when microsatellites based on tri-, tetra-, and pentanucleotide motifs are analyzed.[175,177,355] Moreover, it has been recently demonstrated that PCR-amplified microsatellite loci are also detectable by silver staining[363] or through the use of fluorescent primers in combination with an automated DNA sequencer.[355,834]

C. SEQUENCE CHARACTERIZED AMPLIFIED REGIONS (SCARs)

In 1993, Michelmore and co-workers introduced a new type of molecular marker which was derived from the RAPD technique, but circumvented several of the drawbacks inherent to RAPDs.[351,539] These markers were generated by cloning and sequencing RAPD fragments of interest (i.e., RAPDs linked to a fungal resistance gene in lettuce) and designing 24-mer oligonucleotide primers that were complementary to the ends of the original RAPD fragment. When these primers were used in a PCR with the original template DNA, single loci called sequence characterized amplified regions (SCARs) were specifically amplified. These SCARs either retained the dominant segregation behavior of the original RAPD fragment or were converted into codominant markers. In comparison to RAPDs or other strategies of arbitrarily primed PCR, the SCAR concept exhibits several advantages:

1. reproducibility — since long primers complementary to specific genomic loci are used, stringent PCR conditions can be applied which exclude competition between primer binding sites (this results in reliable and reproducible bands, which are less sensitive to reaction conditions);
2. mapping studies — codominant SCARs are much more informative for genetic mapping than dominant RAPDs;
3. map-based cloning — ordinary RAPD fragments often contain interspersed repetitive DNA and thus cannot be used as probes for identifying a clone of interest (in contrast, SCAR primers can be used to screen pooled genomic libraries by PCR);

4. physical mapping — since SCARs are formally similar to the "sequence tagged sites" (STS) proposed by Olson et al.[525] as landmarks in physical mapping of the human genome, they may also serve as anchoring points between physical and genetic maps;
5. locus specificity — the reproducible amplification of defined genomic regions allows the comparative mapping (as has been done with RFLPs) or homology studies between related species.

While the generation of SCAR markers is somewhat laborious, the advantages outlined above will certainly make the transformation of RAPDs into SCARs worthwhile, and we expect that the SCAR concept will be adopted for a variety of other species in the near future.

D. MISCELLANEOUS TECHNIQUES

In this section, we would like to briefly mention some more or less recent approaches, which may have a future potential for DNA (and RNA) profiling of plants and fungi.

A DNA profiling strategy which relies on the internal heterogeneity of human minisatellite repeats rather than length variation was invented by Alec Jeffreys and colleagues. This approach was called "minisatellite variant repeat" (MVR) mapping.[320,321,323,473] Its principle is based on the observation that certain minisatellite arrays (exemplified by the human minisatellite MS32; locus D1S8[804]) consist of heterogeneous basic units. These units may be distinguishable from each other by, e.g., the presence or absence of a restriction site (a *Hae*III site in case of MS32).

The early approaches of MVR mapping[320] made use of a strategy reminiscent of the chemical DNA sequencing procedure. A DNA fragment comprising the MS32 locus was PCR amplified from genomic DNA, end labeled, and then partially digested with *Hinf*I (which cuts once in each 29-bp basic unit) or *Hae*III (which cuts only in the variant units). Electrophoresis and autoradiography yielded a continuous *Hinf*I ladder reflecting the basic repeat length on one hand and a discontinuous *Hae*III ladder from which the relative positions of the variant units could be deduced on the other. From these data, each MS32 allele could be characterized by a binary code according to the order of variant repeats. Thus, a second source of minisatellite allelic variability became detectable, which (in contrast to length variation) allowed the precise designation of alleles. More recently, a generalized experimental strategy for MVR mapping based on PCR was created. MVR-PCR was shown to be also applicable for digital typing of genomic DNA with both alleles superimposed (see Jeffreys et al.[321] for a detailed description of the procedure). The allelic variation observed upon comparing different MS32 binary codes was substantial, and even alleles of identical length frequently showed completely different internal maps. For human individuals, an ultrahigh variability was observed, with two unrelated people showing extremely dissimilar codes.

While MVR mapping procedures may provide unprecedented levels of distinguishing capacity (and are therefore highly valuable for, e.g., forensic analyses in humans), these techniques are probably not generally useful for DNA profiling in plants and fungi. First, establishing such a mapping system (i.e., cloning and sequencing of a minisatellite, design of suitable primers, etc.) is laborious. Second, highly efficient individual-specific discrimination is seldom needed in studies on plants and fungi. Third, only a few human minisatellites fulfill the criteria required for MVR mapping (see Jeffreys et al.[323]), and it will probably not be easy to identify a suitable minisatellite in a particular species.

PCR-based fingerprinting strategies have proven valuable for detecting genetic polymorphisms not only in DNA, but, more recently, also in RNA.[403,439,770] In an approach called "RNA arbitrarily primed PCR" (RAP-PCR), McClelland and colleagues[439] were able to generate cDNA fingerprints from total RNA of different mouse tissues by a two-step procedure. In the first step, an arbitrary primer of 10 or 18 bases was used to prime both first and second cDNA strand synthesis by a reverse transcriptase under low stringency conditions. In the second step, the reaction products were further amplified and simultaneously labeled by standard high stringency PCR. Finally, the reaction products were electrophoresed and visualized by autoradiography. While tissue-specific or growth-condition-specific patterns were observed within the same genotype, polymorphisms also occurred between similar tissues of different inbred strains. Tissue-specific amplification products were proven to reflect true tissue-specific expression by RNA blot analysis.

This strategy also has several prospects for research on plants and fungi: (1) it provides a highly effective screening method for differentially expressed genes in a variety of tissues; (2) the PCR products can be cloned from the gel and sequenced, allowing a direct approach for the isolation of these genes; and (3) polymorphisms which appear between similar tissues of different strains can be exploited for mapping purposes.

A very recent approach put forward by Zabeau[829] is called "selective restriction fragment amplification", resulting in "amplified fragment length polymorphisms" (AFLPs). This technique represents an ingenious combination of RFLP analysis and PCR, resulting in highly informative fingerprints. In the first step, genomic DNA is digested with a restriction enzyme, and adaptors of a defined sequence are ligated to both ends of all restriction fragments. PCR is then performed using specifically designed primers which allow only a subset of the restriction fragments to be amplified. To achieve this, the 5′ portion of the primers is made complementary to the adaptors, whereas the 3′ ends extend for a few, arbitrarily chosen nucleotides into the restriction fragment. Exact matching of the 3′ end of a primer is necessary for amplification to occur. Therefore, only those restriction fragments are amplified, the ends of which are able to basepair with the 3′ primer extension. Statistically, this occurs for $1/16$ of the total set of fragments in case of a one-nucleotide extension, for $1/256$ in case of two, and for $1/4096$ in case of three nucleotides. The amplification products are separated on highly resolving sequencing gels. The complexity of banding patterns (i.e., the number of amplified restriction fragments) can be tailored by using variable lengths of 3′ primer extensions. As with RAPDs, bands of interest may be cut out, cloned, sequenced, and transformed into SCARs by designing specific primers. This highly sensitive and versatile approach certainly provides an important new tool for a variety of applications.

E. OUTLOOK

A major task which has only rarely been accomplished for plants and fungi is the development of multiallelic, codominant, locus-specific markers for DNA profiling. In view of the laborious steps involved, the existing strategies of generating such markers will be restricted mainly to molecular biology laboratories and a few species only (including man, mouse, model organisms, and species of economical interest). This is true for multiallelic single locus hybridization probes (see Section II.A), PCR amplification of microsatellites (Section II.B), and probably also for the development of SCARs (Section II.C).

The identification of this kind of marker in any new species under investigation will be greatly facilitated if heterologous primers or probes (i.e., derived from related organisms) turn out to be informative. Studies in this direction are under way (mainly in animal systems) and have already yielded encouraging results. For example, a polymorphic single locus minisatellite cloned from the house sparrow cross-hybridized to and revealed variable loci in DNA from many different bird species,[267] and the human single-locus-specific minisatellite probe MS51 was also informative and locus specific in a bird species, the Waldrapp ibis.[630] Heterologous primers for microsatellite amplification have been successfully applied to several species of whale,[614] as well as in sheep vs. cattle,[474] and SCAR primers derived from lettuce amplified homologous sequences in various representatives of the Asteraceae.[351] The derived markers may or may not be polymorphic in related taxa (see, e.g., Estoup et al.[185]). Additional studies will show whether heterologous probes/primers are generally useful for spreading molecular markers among species. Undoubtedly, there is still a long and winding road to be followed before the ultimate markers for detecting genetic diversity and relatedness in plants and fungi are developed.

REFERENCES

1. **Abbott, A., Beltoff, L., Rajapakse, S., Ballard, R., Baird, W. V., Monet, R., Scorza, R., Morgans, P., and Callahan, A.,** (1992) Development of a genetic linkage map of peach, Plant Genome I, San Diego, Nov. 9–11, p. 44.

2. **Abbott, L. A., Bisby, F. A., and Rogers, D. J.,** (1985) *Taxonomic Analysis in Biology: Computers, Models, and Databases,* Columbia University Press, New York.

3. **Adachi, Y., Watanabe, H., Tanabe, K., Doke, N., Nishimura, S., and Tsuge, T.,** (1993) Nuclear ribosomal DNA as a probe for genetic variability in the Japanese pear pathotype of *Alternaria alternata, Appl. Environ. Microbiol.,* 59: 3197–3205.

4. **Adams, R. P., and Demeke, T.,** (1993) Systematic relationships in *Juniperus* based on random amplified polymorphic DNAs (RAPDs), *Taxon,* 42: 553–571.

5. **Adams, R. P., Demeke, T., and Abulfatih, H. A.,** (1993) RAPD DNA fingerprints and terpenoids: clues to past migrations of *Juniperus* in Arabia and east Africa, *Theor. Appl. Genet.,* 87: 22–26.

6. **Akkaya, M. S., Bhagwat, A. A., and Cregan, P. B.,** (1992) Length polymorphisms of simple sequence repeat DNA in soybean, *Genetics,* 132: 1131–1139.

7. **Aldrich, J., and Cullis, C. A.,** (1993) RAPD analysis in flax: optimization of yield and reproducibility using Klen*Taq* 1 DNA polymerase, chelex 100, and gel purification of genomic DNA, *Plant Mol. Biol. Rep.,* 11: 128–141.

8. **Ali, S., Müller, C. R., and Epplen, J. T.,** (1986) DNA fingerprinting by oligonucleotide probes specific for simple repeats, *Hum. Genet.,* 74: 239–243.

9. **Al-Janabi, S. M., Honeycutt, R. J., McClelland, M., and Sobral, B. W. S.,** (1993) A genetic linkage map of *Saccharum spontaneum* (L.) 'SES 208', *Genetics,* 134: 1249–1260

10. **Allan, G. J., Clark, C., and Rieseberg, L. H.,** (1993) *Encelia virginensis* (Asteraceae): origin and genetic composition of a diploid species of putative hybrid origin, *Am. J. Bot.,* 80 (Suppl.): 129–130.

11. **Allard, R. W.,** (1956) Formulas and tables to facilitate the calculation of recombination values in heredity, *Hilgardia,* 24: 235–278.

12. **Anderson, P. A., and Pryor, A. J.,** (1992) DNA restriction fragment length polymorphisms in the wheat stem rust fungus, *Puccinia graminis tritici, Theor. Appl. Genet.,* 83: 715–719.

13. **Antonius, K., and Nybom, H.,** (1994) DNA fingerprinting reveals significant amounts of genetic variation in a wild raspberry (*Rubus idaeus*) population, *Mol. Ecol.,* 3: 177–180.

14. **Applied Biosystems,** (1992) Use of RAPD markers may speed the pace of genetic research, *Biosystems Rep.* 10–11.

15. **Armour, J. A. L., Patel, I., Thein, S. L., Fey, M. F., and Jeffreys, A. J.,** (1989) Analysis of somatic mutations at human minisatellite loci in tumors and cell lines, *Genomics,* 4: 328–334.

16. **Armour, J. A. L., Wong, Z., Wilson, V., Royle, N. J., and Jeffreys, A. J.,** (1989) Sequences flanking the repeat arrays of human minisatellites: association with tandem and dispersed repeat elements, *Nucl. Acids Res.,* 13: 4925–4935.

17. **Armour, J. A. L., Povey, S., Jeremiah, S., and Jeffreys, A. J.,** (1990) Systematic cloning of human minisatellites from ordered charomid libraries, *Genomics,* 8: 501–512.

18. **Armour, J. A. L., Crosier, M., and Jeffreys, A. J.,** (1992) Human minisatellite alleles detectable only after PCR amplification, *Genomics,* 12: 116–124.

19. **Arnheim, N., and Erlich, H.,** (1992) Polymerase chain reaction strategy, *Annu. Rev. Biochem.,* 61: 131–156.

20. **Arnold, M. L.,** (1993) *Iris nelsonii* (Iridaceae): origin and genetic composition of a homoploid hybrid species, *Am. J. Bot.,* 80: 577–583.

21. **Arnold, M. L., Buckner, C. M., and Robinson, J. J.,** (1991) Pollen-mediated introgression and hybrid speciation in Louisiana irises, *Proc. Natl. Acad. Sci. U.S.A.,* 88: 1398–1402.
22. **Aruna, M., Ozias-Akins, P., Austin, M. E., and Kochert, G.,** (1993) Genetic relatedness among rabbiteye blueberry (*Vaccinium ashei*) cultivars determined by DNA amplification using single primers of arbitrary sequence, *Genome,* 36: 971–977
23. **Arveiler, B., and Porteous, D. J.,** (1992) Polyacrylamide-agarose composite gels for high resolution of fingerprints, *TIG,* 8: 82.
24. **Asker, S. E., and Jerling, L.,** (1992) *Apomixis in Plants,* CRC Press, Boca Raton, FL.
25. **ATTC, American Type Culture Collection,** (1990) *Catalogue of Yeasts,* 18th ed., ATTC, Rockville, MD, pp. 164–167.
26. **ATTC, American Type Culture Collection,** (1991) *Catalogue of Filamentous Fungi,* 18th ed., ATTC, Rockville, MD, pp. 440–447.
27. **Aufauvre-Brown, A., Cohen, J., and Holden, D. W.,** (1992) Use of randomly amplified polymorphic DNA markers to distinguish isolates of *Aspergillus fumigatus, J. Clin. Microbiol.,* 30: 2991–2993.
28. **Bachmann, K.,** (1992) Nuclear DNA markers in angiosperm taxonomy, *Acta Bot. Neerl.,* 39: 369–384.
29. **Bachmann, K.,** (1993) Speciation after long-distance dispersal in *Microseris pygmaea* (Asteraceae: Lactuceae), *Am. J. Bot.,* 80, (Suppl.): 130.
30. **Baird, E., Cooper-Bland, S., Waugh, R., DeMaine, M., and Powell, W.,** (1992) Molecular characterization of inter- and intra-specific somatic hybrids of potato using randomly amplified polymorphic DNA (RAPD) markers, *Mol. Gen. Genet.,* 233: 469–475.
31. **Baldwin, B. G.,** (1992) Phylogenetic utility of the internal transcribed spacers of nuclear ribosomal DNA in plants: an example from the Compositae, *Mol. Phylogenet. Evol.,* 1: 3–16.
32. **Balestrazzi, A., Bernacchia, G., Cella, R., Ferretti, L., and Sora, S.,** (1991) Preparation of high molecular weight plant DNA and its use for artificial chromosome construction, *Plant Cell Rep.,* 10: 315–320.
33. **Balows, A., Hausler, W. J., Herrmann, H. L., Isenberg, H. D., and Shadomy, H. J.,** Eds., (1991) *Manual of Clinical Microbiology,* 5th ed., American Society for Microbiology, Washington, D.C.
34. **Barua, U. M., Chalmers, K. J., Thomas, W. T. B., Waugh, R., and Powell, W.,** (1992) The use of RAPD markers in conjunction with bulked samples of double haploid families to locate both qualitative and quantitative traits in barley, Plant Genome I, San Diego,, Nov. 9–11, p. 57.
35. **Barua, U. M., Chalmers, K. J., Hackett, C. A., Thomas, W. T. B., Powell, W., and Waugh, R.,** (1993) Identification of RAPD markers linked to a *Rhynchosporium secalis* resistance locus in barley using near-isogenic lines and bulked segregant analysis, *Heredity,* 71: 177–184.
36. **Bassam, B. J., Caetano-Anollés, G., and Gresshoff, P. M.,** (1991) Fast and sensitive staining of DNA in polyacrylamide gels, *Anal. Biochem.,* 80: 81–84.
37. **Beck, S.,** (1992) Nonradioactive detection of DNA using dioxetane chemiluminescence, *Methods Enzymol.,* 216: 143–152.
38. **Beckmann, J. S., and Weber, J. L.,** (1992) Survey of human and rat microsatellites, *Genomics,* 12: 627–631.
39. **Bell, G. I., Selby, M. J., and Rutter, W. J.,** (1982) The highly polymorphic region near the human insulin gene is composed of simple tandemly repeating sequences, *Nature (London),* 295: 31–35.
40. **Bellamy, R., Inglehearn, C., Lester, D., Hardcastle, A., and Bhattacharya, S.,** (1990) Better fingerprinting with PCR, *TIG,* 6: 32.
41. **Benito, C., Figueiras, A. M., Zaragoza, C., Gallego, F. J., and De la Pena, A.,** (1993) Rapid identification of Triticeae genotypes from single seeds using the polymerase chain reaction, *Plant Mol. Biol.,* 21: 181–183.

42. **Bentzen, P., and Wright, J. M.,** (1993) Nucleotide sequence and evolutionary conservation of a minisatellite variable number tandem repeat cloned from Atlantic salmon, *Salmo salar, Genome*, 36: 271–277.

43. **Bentzen, P., Harris, A. S., and Wright, J. M.,** (1991) Cloning of hypervariable minisatellite and simple sequence microsatellite repeats for DNA fingerprinting of important aquacultural species of salmonids and tilapia, in *DNA Fingerprinting: Approaches and Applications*, Burke, T., Dolf, G., Jeffreys, A. J., and Wolff, R., Eds., Birkhäuser, Basel, pp. 243–262.

44. **Berger, S. L., and Kimmel, A. R., Eds.** (1987) Guide to molecular cloning techniques, *Methods Enzymol.*, 152.

45. **Berthomieu, P., and Meyer, C.,** (1991) Direct amplification of plant genomic DNA from leaf and root pieces using PCR, *Plant Mol. Biol.*, 17: 555–557.

46. **Bertram, R. B.,** (1993) Application of molecular techniques to genetic resources of cassava (*Manihot esculenta* Crantz — Euphorbiaceae): interspecific evolutionary relationships and intraspecific characterization, dissertation thesis, University of Maryland, College Park, MD.

47. **Besse, P., Lebrun, P., Sequin, M., and Lanaud, C.,** (1993) DNA fingerprints in *Hevea brasiliensis* (rubber tree) using human minisatellite probes, *Heredity*, 70: 237–244.

48. **Beyermann, B., Nürnberg, P., Weihe, A., Meixner, M., Epplen, J. T., and Börner, T.,** (1992) Fingerprinting plant genomes with oligonucleotide probes specific for simple repetitive DNA sequences, *Theor. Appl. Genet.*, 83: 691–694.

49. **Bierwerth, S., Kahl, G., Weigand, F., and Weising, K.,** (1992) Oligonucleotide fingerprinting of plant and fungal genomes: a comparison of radioactive, colorigenic and chemiluminescent detection methods, *Electrophoresis*, 13: 115–120.

50. **Binelli, G., Pé, M. E., Gianfranceschi, L., Angelini, P., and Dani, M.,** (1992) RFLPs and RAPDs markers for the genetic analysis of resistance to *Gibberella zeae*, Plant Genome I, San Diego,, Nov. 9–11, p. 18.

51. **Biourge, P.,** (1923) Le moissures du groupe *Penicillium* Link, *La Cellule*, 33: 7–331.

52. **Bissett, J.,** (1984) A revision of the genus *Trichoderma* 1. section *Longibrachiatum, Can. J. Bot.*, 62: 924–931.

53. **Blackburn, E. H.,** (1991) Structure and function of telomeres, *Nature (London)* 350: 569–573.

54. **Blackburn, E. H.,** (1992) Telomerases, *Annu. Rev. Biochem.*, 61: 113–129.

55. **Blakeslee, A. F.,** (1915) Lindner's roll tube methode of separation cultures, *Phytopathology*, 5: 68–69.

56. **Blum, H., Beier, H., and Gross, H. J.,** (1987) Improved silver staining of plant proteins, RNA and DNA in polyacrylamide gels, *Electrophoresis*, 8: 93–99.

57. **Boerwinkle, E., Xiong, W., Fourest, E., and Chan, L.,** (1989) Rapid typing of tandemly repeated hypervariable loci by the polymerase chain reaction: application to the apolipoprotein B 3′hypervariable region, *Proc. Natl. Acad. Sci. U.S.A.*, 86: 212–216.

58. **Booth, C.,** (1971) Fungal culture media, in *Methods in Microbiology*, Vol. 4, Booth, C., Ed., Academic Press, San Diego,, pp. 49–94.

59. **Borromeo, E. S., Nelson, R. J., Bonman, J. M., and Leung, H.,** (1993) Genetic differentiation among isolates of *Pyricularia* infecting rice and weed hosts, *Phytopathology*, 83: 393–399.

60. **Bostock, A., Khattak, M. N., Matthews, R., and Burnie, J.,** (1993) Comparison of PCR fingerprinting, by random amplification of polymorphic DNA, with other molecular typing methods for *Candida albicans, J. Genet. Microbiol.*, 139: 2179–2184.

61. **Boury, S., Lutz, I., Gavalda, M.-C., Guidet, F., and Schlesser, A.,** (1992) Empreintes génétiques du chou-fleur par RAPD et vérification de la pureté hybride F1 d'un lot de semences, *Agronomie*, 12: 669–681.

62. **Bowditch, B. M., Williams, J., and Zimmer, E. A.,** (1991) Randomly amplified polymorphic DNA technique applied to the *Euphorbia esula* complex, *Am. J. Bot.*, 78 (Suppl.): 168.

63. **Böttger, E. C.,** (1990) Frequent contamination of *Taq* polymerase with DNA, *Clin. Chem.,* 36: 1258–1259.

64. **Braithwaite, K. S., and Manners, J. M.,** (1989) Human hypervariable minisatellite probes detect DNA polymorphisms in the fungus *Colletotrichum gloeosporioides, Curr. Genet.,* 16: 473–475.

65. **Brasier, C. M., and Rayner, A. D. M.,** (1987) Whither with the terminology below the species level in the fungi, in *Evolutionary Biology of the Fungi,* Rayner, A. D. M., Brasier, C. M., and Moore, D., Eds., Cambridge University Press, New York.

66. **Brauner, S., Crawford, D. J., and Stuessy, T. F.,** (1992) Ribosomal DNA and RAPD variation in the rare plant family Lactoridaceae, *Am. J. Bot.,* 79: 1436–1439.

67. **Bronstein, I., and McGrath, P.,** (1989) Chemiluminescence lights up, *Nature (London),* 338: 599–600.

68. **Brookfield, J. F. Y.,** (1992) DNA fingerprinting in clonal organisms, *Mol. Ecol.,* 1: 21–26.

69. **Broun, P., and Tanksley, S. D.,** (1993) Characterization of tomato DNA clones with sequence similarity to human minisatellites 33.6 and 33.15, *Plant Mol. Biol.,* 23: 231–242.

70. **Broun, P., Ganal, M. W., and Tanksley, S. D.,** (1992) Telomeric arrays display high levels of heritable polymorphism among closely related plant varieties, *Proc. Natl. Acad. Sci. U.S.A.,* 89: 1354–1357.

71. **Brown, A. H. D., Marshall, D. R., and Munday, J.,** (1976) Adaptedness of variants at an alcohol dehydrogenase locus in *Bromus mollis* L. (soft brome grass), *Aust. J. Biol. Sci.,* 29: 389–396.

72. **Brown, J. K. M., O'Dell, M., Simpson, C. G., and Wolfe, M. S.,** (1990) The use of DNA polymorphisms to test hypotheses about a population of *Erysiphe graminis* f. sp. *hordei, Plant Pathol.,* 39: 391–401.

73. **Brown, J. K. M., Jessop, A. C., and Rezanoor, H. N.,** (1991) Genetic uniformity in barley and its powdery mildew pathogen, *Proc. R. Soc. London B,* 246: 83–90.

74. **Brown, P. T. H., Lange, F. D., Kranz, E., and Lörz, H.,** (1993) Analysis of single protoplasts and regenerated plants by PCR and RAPD technology, *Mol. Gen. Genet.,* 237: 311–317.

75. **Brown, T. A.,** Ed., (1993) *Essential Molecular Biology: A Practical Approach,* IRL Press, Oxford.

76. **Bruford, M. W., Hanotte, O., Brookfield, J. F. Y., and Burke, T.,** (1992) Single-locus and multilocus DNA fingerprinting, in *Molecular Genetic Analysis of Populations: A Practical Approach,* Hoelzel, A. R., Ed., IRL Press, Oxford, pp. 225–269.

77. **Brunel, D.,** (1992) An alternative, rapid method of plant DNA extraction for PCR analysis, *Nucl. Acids Res.,* 20: 4676.

78. **Brunsfeld, S. J., and Rieseberg, L. H.,** (1993) Genetic differentiation among coastal and relict Rocky Mountain populations of *Cornus nuttallii*: comparing results from isozymes and RAPDs, *Am. J. Bot.,* 80 (Suppl.): 134.

79. **Bucci, G., and Menozzi, P.,** (1993) Segregation analysis of random amplified polymorphic DNA (RAPD) markers in *Picea abies* Karst, *Mol. Ecol.,* 2: 227–232.

80. **Bulat, S. A., and Mironenko, N. V.,** (1992) Polymorphism of yeast-like fungus *Aureobasidium pullulans* (De Bary) revealed by universally primed polymerase chain reaction — species divergence state, *Genetica,* 28: 19–30.

81. **Bult, C., Källersjö, M., and Suh, Y.,** (1992) Amplification and sequencing of 16S/18S rDNA from gel-purified total plant DNA, *Plant Mol. Biol. Rep.,* 10: 273–284.

82. **Burke, T.,** (1989) DNA fingerprinting and other methods for the study of mating success, *TREE,* 4: 139–144.

83. **Burke, T., and Bruford, M. W.,** (1987) DNA fingerprinting in birds, *Nature (London),* 327: 149–152.

84. **Burke, T., Hanotte, O., Bruford, M. W., and Cairns, E.,** (1991) Multilocus and single locus minisatellite analysis in population biological studies, in *DNA Fingerprinting: Approaches and Applications.* Burke, T., Dolf, G., Jeffreys, A. J., and Wolff, R., Eds., Birkhäuser, Basel, pp. 154–168.

85. **Caetano-Anollés, G., and Bassam, B. J.,** (1993) DNA amplification fingerprinting using arbitrary oligonucleotide primers, *Appl. Biochem. Biotech.*, 42: 189–200.

86. **Caetano-Anollés, G., Bassam, B. J., and Gresshoff, P. M.,** (1991) DNA amplification fingerprinting using very short arbitrary oligonucleotide primers, *Bio/Technology*, 9: 553–557.

87. **Caetano-Anollés, G., Bassam, B. J., and Gresshoff, P. M.,** (1991) DNA amplification fingerprinting: a strategy for genome analysis, *Plant Mol. Biol. Rep.*, 9: 294–307.

88. **Caetano-Anollés, G., Bassam, B. J., and Gresshoff, P. M.,** (1992) DNA fingerprinting — MAAPing out a RAPD redefinition, *Bio/Technology*, 10: 937.

89. **Caetano-Anollés, G., Bassam, B. J., and Gresshoff, P. M.,** (1992) Primer-template interactions during DNA amplification fingerprinting with single arbitrary oligonucleotides, *Mol. Gen. Genet.*, 235: 157–165.

90. **Caetano-Anollés, G., Bassam, B. J., and Gresshoff, P. M.,** (1993) DNA amplification fingerprinting with very short primers, in *Application of RAPD Technology to Plant Breeding*, Neff, M., Ed., ASHS Publishers, St. Paul, pp. 18–25.

91. **Caetano-Anollés, G., Bassam, B. J., and Gresshoff, P. M.,** (1993) Enhanced detection of polymorphic DNA by multiple arbitrary amplicon profiling of endonuclease-digested DNA: identification of markers linked to the supernodulation locus in soybean, *Mol. Gen. Genet.*, 241: 57–64.

92. **Cai, Q., Guy, C., and Moore, G. A.,** (1992) Genetic mapping of random amplified polymorphic DNA (RAPD) markers in *Citrus*, Plant Genome I, San Diego, Nov. 9–11, p. 20.

93. **Callahan, F. E., and Mehta, A. M.,** (1991) Alternative approach for consistent yields of total DNA from cotton (*Gossypium hirsutum* L.), *Plant Mol. Biol. Rep.*, 9: 252–261.

94. **Campbell, C. S., Alice, L. A., and Helentjaris, T. G.,** (1993) Genetic diversity in and hybridization between apomictic and sexual *Amelanchier* (Rosaceae, Maloideae): evidence from random amplification of polymorphic DNA (RAPD) markers, *Am. J. Bot.*, 80 (Suppl.): 135.

95. **Carland, F. M., and Staskawicz, B. J.,** (1993) Genetic characterization of the *Pto* locus of tomato: semi-dominance and cosegregation of resistance to *Pseudomonas syringae* pathovar tomato and sensitivity to the insecticide Fenthion, *Mol. Gen. Genet.*, 239: 17–27.

96. **Carlson, J. E., Tulsieram, L. K., Glaubitz, J. C., Luk, V. W. K., Kauffeldt, C., and Rutledge, R.,** (1991) Segregation of random amplified DNA markers in F1 progeny of conifers, *Theor. Appl. Genet.*, 83: 194–200.

97. **Carlson, J. E., Hong, Y.-P., Brown, G. R., and Glaubitz, J. C.,** (1992) Genome mapping in spruce with RAPD and FISH markers, Plant Genome I, San Diego,, Nov. 9–11, p. 58.

98. **Carter, R. E., Wetton, J. H., and Parkin, D. T.,** (1989) Improved genetic fingerprinting using RNA probes. *Nucl. Acids Res.*, 17: 5867.

99. **Caskey, C. T., Pizzuti, A., Fu, Y.-H., Fenwick, R. G., Jr., and Nelson, D. L.,** (1992) Triplet repeat mutations in human disease, *Science*, 256: 784–789.

100. **Castiglione, S., Wang, G., Damiani, G., Bandi, C., Bisoffi, S., and Sala, F.,** (1993) RAPD fingerprints for identification and for taxonomic studies of elite poplar (*Populus* spp.) clones, *Theor. Appl. Genet.*, 87: 54–59.

101. **CBS, Centralbureau voor Schimmelcultures,** (1990) *List of Cultures. Fungi and Yeasts*, 32nd ed., CBS, Baarn-Delft, pp. 505–508.

102. **Cenis, J. L.,** (1992) Rapid extraction of fungal DNA for PCR amplification, *Nucl. Acids Res.*, 20: 2380.

103. **Chakraborty, R., and Jin, L.,** (1993) A unified approach to study hypervariable polymorphisms: statistical considerations of determining relatedness and population distances, in *DNA Fingerprinting: State of the Science*, Pena, S. D. J., Chakraborty, R., Epplen, J. T., and Jeffreys, A. J., Eds., Birkhäuser, Basel, pp. 153–175.

104. **Chalmers, K. J., Waugh, R., Sprent, J. I., Simons, A. J., and Powell, W.,** (1992) Detection of genetic variation between and within populations of *Gliricidia sepium* and *G. maculata* using RAPD markers, *Heredity*, 69: 465–472.

105. **Chalmers, K. J., Barua, U. M., Hackett, C. A., Thomas, W. T. B., Waugh, R., and Powell, W.,** (1993) Identification of RAPD markers linked to genetic factors controlling the milling energy requirement of barley, *Theor. Appl. Genet.*, 87: 314–320.

106. **Chandler, J., Carriero, F., Blaker, N., Peterson, P., and Yoder, J. I.,** (1992) Molecular studies of Ac induced chromosome breakage in tomato, Plant Genome I, San Diego, Nov. 9–11, p. 21.

107. **Chandley, A. C., and Mitchell, A. R.,** (1988) Hypervariable minisatellite regions are sites for crossing-over at meiosis in man, *Cytogenet. Cell Genet.*, 48: 152–155.

108. **Chaparro, J., Werner, D., O'Malley, D., and Sederoff, R.,** (1992) Targetted mapping and linkage analysis in peach, Plant Genome I, San Diego, Nov. 9–11, p. 21.

109. **Chehab, F. F., Wall, J., and Kan, Y. W.,** (1992) Amplification and detection of specific DNA sequences with fluorescent PCR primers: application to deltaF508 mutation in cystic fibrosis, *Methods Enzymol.*, 216: 135–143.

110. **Chetelat, R. T., Cisneros, P., Rick, C. M., and DeVerna, J. W.,** (1992) Introgression of useful germ plasm from *Solanum lycopersicoides* and *S. rickii* via alien addition and inbred backcross lines, Plant Genome I, San Diego, Nov. 9–11, p. 22.

111. **Cheung, W. Y., and Gale, M. D.,** (1990) The isolation of high molecular weight DNA from wheat, barley and rye for analysis by pulse-field gel electrophoresis, *Plant Mol. Biol.*, 14: 881–888.

112. **Cheung, W. Y., Hubert, N., and Landry, B. S.,** (1993) A simple and rapid DNA microextraction method for plant, animal, and insect suitable for RAPD and other PCR analyses, *PCR Methods Appl.*, 3: 69–70.

113. **Chunseng, X., and Semple, J. C.,** (1993) RAPD study of *Aster modestus* (Compositae: Astereae), *Am. J. Bot.,* 80, (Suppl.): 184.

114. **Chunwongse, J., Martin, G. B., and Tanksley, S. D.,** (1993) Pre-germination genotypic screening using PCR amplification of half-seeds, *Theor. Appl. Genet.*, 86: 694–698.

115. **Clegg, M. T.,** (1993) Chloroplast gene sequences and the study of plant evolution, *Proc. Natl. Acad. Sci. U.S.A.*, 90: 363–367.

116. **Collick, A., and Jeffreys, A. J.,** (1990) Detection of a novel minisatellite-specific DNA-binding protein, *Nucl. Acids Res.*, 18: 625–629.

117. **Collick, A., Dunn, M. G., and Jeffreys, A. J.,** (1991) Minisatellite binding protein Msbp-1 is a sequence-specific single-stranded DNA-binding protein, *Nucl. Acids Res.*, 19: 6399–6404.

118. **Collins, G. G., and Symons, R. H.,** (1992) Extraction of nuclear DNA from grapevine leaves by a modified procedure, *Plant Mol. Biol. Rep.,* 10: 233–235.

119. **Collins, G. G., and Symons, R. H.,** (1993) Polymorphisms in grapevine DNA detected by the RAPD PCR technique *Plant Mol. Biol. Rep.*, 11: 105–112.

120. **Condit, C. M., and Meagher, R. B.,** (1986) A gene encoding a novel glycine-rich structural protein of petunia, *Nature (London)*, 323: 178–181.

121. **Condit, R., and Hubbell, S. P.,** (1991) Abundance and DNA sequence of two-base repeat regions in tropical tree genomes, *Genome*, 34: 66–71.

122. **Connett, M. B., Timmerman, G., Conlon, H. P., Wilcox, P., and Carson, S. D.,** (1992) Replicability and reliability of amplification markers: lessons from parallel studies, Plant Genome I, San Diego, Nov. 9–11, p. 57.

123. **Constantinescu, O.,** (1974) *Metode si Technici in Micologie*, Ceres, Bucuresti.

124. **Cooper, C. R., Breslin, B. J., Dixon, D. M., and Salkin, I. F.,** (1992) DNA typing of isolates associated with the 1988 sporotrichosis epidemic, *J. Clin. Microbiol.*, 30: 1631–1635.

125. **Correll, J. C., Rhoads, D. D., and Guerber, J. C.,** (1993) Examination of mitochondrial DNA restriction fragment length polymorphisms, DNA fingerprints, and randomly amplified polymorphic DNA of *Colletotrichum orbiculare*, *Phytopathology*, 83: 1199–1204.

126. **Couch, J. A., and Fritz, P. J.,** (1990) Isolation of DNA from plants high in polyphenolics, *Plant Mol. Biol. Rep.*, 8: 8–12.

127. **Cox, K. H., and Goldberg, R. B.,** (1988) Analysis of plant gene expression, in *Plant Molecular Biology: A Practical Approach*, Shaw, C. H., Ed., IRL Press, Oxford, pp. 1–35.

128. **Crawford, A. M., Buchanan, F. C., Fraser, K. M., Robinson, A. J., and Hill, D. F.,** (1991) Repeat sequences from complex ds viruses can be used as minisatellite probes for DNA fingerprinting, *Anim. Genet.*, 22: 177–181.

129. **Crawford, A. M., Swarbrick, P. A., Buchanan, F. C., and Dodds, K. G.,** (1993) DNA fingerprinting analysis of Booroola pedigrees: a search for linkage to the Booroola gene, *Theor. Appl. Genet.*, 87: 271–277.

130. **Crawford, D. J., Brauner, S., Cosner, M. B., and Stuessey, T. F.,** (1993) Use of RAPD markers to document the origin of the intergeneric hybrid ×*Margyracaena skottsbergii* (Rosaceae) on the Juan Fernandez Islands, *Am. J. Bot.*, 80: 89–92.

131. **Crowhurst, R. N., Hawthorne, B. T., Rikkerink, E. H. A., and Templeton, M. D.,** (1991) Differentiation of *Fusarium solani* f. sp. *cucurbitae* races 1 and 2 by random amplification of polymorphic DNA, *Curr. Genet.*, 20: 391–396.

132. **Cruachem,** (1991) Quick deprotection of oligonucleotides synthesized using conventional CE-phosphoramidites, Tech. Bull. No. 041R, Cruachem, Sterling, VA.

133. **Cullings, K. W.,** (1992) Design and testing of a plant-specific PCR primer for ecological and evolutionary studies, *Mol. Ecol.*, 1: 233–240.

134. **Cullis, C., and Govindaraju, D.,** (1992) Identification of RAPD polymorphisms among introgressed flax lines containing the M rust resistance genes, and the cloning of the PCR products, Plant Genome I, San Diego, Nov. 9–11, p. 23.

135. **Czapek, F.,** (1902) Untersuchungen über die Stickstoffsgewinnung und Eiweißbildung der Pflanzen, *Beitr. Chem. Physiol. Pathol.*, 1: 540–560.

136. **Czapek, F.,** (1903) Untersuchungen über die Stickstoffsgewinnung und Eiweißbildung der Pflanzen, *Beitr. Chem. Physiol. Pathol.*, 3: 47–66.

137. **Dallas, J. F.,** (1988) Detection of DNA "fingerprints" of cultivated rice by hybridization with a human minisatellite probe, *Proc. Natl. Acad. Sci. U.S.A.*, 85: 6831–6835.

138. **Dallas, J. F., McIntyre, C. L., and Gustafson, J. P.,** (1993) An RFLP species-specific DNA sequence for the A genome of rice, *Genome*, 36: 445–448.

139. **Daly, A., Kellam, P., Berry, S. T., Chojecki, A. J. S., and Barnes, S. R.,** (1991) The isolation and characterisation of plant sequences homologous to human hypervariable minisatellites, in *DNA Fingerprinting: Approaches and Applications*, Burke, T., Dolf, G., Jeffreys, A. J., and Wolff, R., Eds., Birkhäuser, Basel, pp. 330–341.

140. **Davis, L. G., Dibner, M. D., and Battey, J. F.,** (1986) *Basic Methods in Molecular Biology*, Elsevier, New York.

141. **Davis, T. M., McGowan, P. J., and Simon, C. J.,** (1992) Use of RAPD tags to map chickpea root nodulation genes, Plant Genome I, San Diego, Nov. 9–11, p. 24.

142. **Davis, R. W., Thomas, M., Cameron, J., John, T. P. S., Scherer, S., and Padgell, R. A.,** (1980) Rapid isolation of yeast DNA for enzymatic and hybridization analysis, *Methods Enzymol.*, 65: 404–411.

143. **Dawson, I. K., Chalmers, K. J., Waugh, R., and Powell, W.,** (1993) Detection and analysis of genetic variation in *Hordeum spontaneum* populations from Israel using RAPD markers, *Mol. Ecol.*, 2: 151–159.

144. **De Graaff, L., Van den Brook, H., and Visser, J.,** (1988) Isolation and expression of the *Aspergillus nidulans* pyruvate kinase gene, *Curr. Genet.*, 13: 315–321.

145. **De Kochko, A., and Hamon, S.,** (1990) A rapid and efficient method for the isolation of restrictable total DNA from plants of the genus *Abelmoschus*, *Plant Mol. Biol. Rep.*, 8: 3–7.

146. **De la Canal, L., Crouzillat, D., Flamand, M. C., Perrault, A., Boutry, M., and Lédoight, G.,** (1991) Nucleotide sequence and transcriptional analysis of a mitochondrial plasmid from a cytoplasmic male-sterile line of sunflower, *Theor. Appl. Genet.*, 81: 812–818.

147. **Debener, T., Salamini, F., and Gebhardt, C.,** (1990) Phylogeny of wild and cultivated *Solanum* species based on nuclear restriction fragment length polymorphisms (RFLPs), *Theor. Appl. Genet.*, 79: 360–368.

148. **Dellaporta, S. L., Wood, J., and Hicks, J. B.,** (1983) A plant DNA minipreparation: version II, *Plant Mol. Biol. Rep.*, 1(4): 19–21.

149. **Demeke, T., Adams, R. P., and Chibbar, R.,** (1992) Potential taxonomic use of random amplified polymorphic DNA (RAPD): a case study in *Brassica*, *Theor. Appl. Genet.*, 84: 990–994.

150. **Deragon, J.-M., and Landry, B. S.,** (1992) RAPD and other PCR-based analyses of plant genomes using DNA extracted from small leaf discs, *PCR Methods Appl.*, 1: 175–180.

151. **Devey, M. E., Jermstad, K. D., Tauer, C. G., and Neale, D. B.,** (1991) Inheritance of RFLP loci in a loblolly pine three-generation pedigree, *Theor. Appl. Genet.*, 83: 238–242.

152. **Devey, M. E., Delfino-Mix, A. A., Kinloch, B. B., and Neale, D. B.,** (1992) Identification of RAPD markers linked with a major gene for resistance to white pine blister rust in sugar pine, Plant Genome I, San Diego, Nov. 9–11, p. 24.

153. **Devos, K. M., and Gale, M. D.,** (1992) The use of random amplified polymorphic DNA markers in wheat, *Theor. Appl. Genet.*, 84: 567–572.

154. **Dietrich, W., Katz, H., Lincoln, S. E., Shin, H.-S., Friedman, J., Dracopoli, N. C., and Lander, E. S.,** (1992) A genetic map of the mouse suitable for typing intraspecific crosses, *Genetics*, 131: 423–447.

155. **Dirlewanger, E., and Bodo, C.,** (1994) Molecular genetic mapping of peach, Eucarpia Fruit Breeding Section Meeting, Aug. 30–Sept. 3, 1993, *Euphytica,* in press.

156. **Dong, J., and Wagner, D. B.,** (1993) Taxonomic and population differentiation of mitochondrial diversity in *Pinus banksiana* and *Pinus contorta*, *Theor. Appl. Genet.*, 86: 573–578.

157. **Doolittle, W. F., and Sapienza, C.,** (1980) Selfish genes, the phenotype paradigm and genome evolution, *Nature (London)*, 284: 601–603.

158. **Doudrick, R. L., Nance, W. L., Nelson, C. D., Snow, G. A., and Hamelin, R. C.,** (1993) Detection of DNA polymorphisms in a single urediniospore-derived culture of *Cronartium quercuum* f. sp. *fusiforme*, *Phytopathology*, 83: 388–392.

159. **d'Ovidio, R., Tanzarella, O. A., and Porceddu, E.,** (1990) Rapid and efficient detection of genetic polymorphism in wheat through amplification by polymerase reaction, *Plant Mol. Biol. Rep.*, 15: 169–171.

160. **Dowling, T. E., Moritz, C., Palmer, J. D.,** (1990) Nucleic acids II: restriction site analysis, in *Molecular Systematics,* Hillis, D. M., and Moritz, C., Eds., Sinauer Associates, Sunderland, MA, pp. 250–317.

161. **Dox, A. W.,** (1910) The intracellular enzymes of *Penicillium* and *Aspergillus* with special references to those of *P. camemberti*, *U.S. Dep. Agric. Animo Ind. Bull.*, 120: 170.

162. **Doyle, J. J.,** (1993) DNA, phylogeny, and the flowering of plant systematics, *BioScience*, 43: 380–389.

163. **Doyle, J. J., and Dickson, E. E.,** (1987) Preservation of plant samples for DNA restriction endonuclease analysis, *Taxon,* 36: 715–722.

164. **Doyle, J. J., and Doyle, J. L.,** (1987) A rapid DNA isolation procedure for small quantities of fresh leaf tissue, *Phytochem. Bull.*, 19: 11–15.

165. **Doyle, J. J., and Doyle, J. L.,** (1990) Isolation of plant DNA from fresh tissue, *Focus*, 12: 13–15.

166. **Drenth, A., Goodwin, S. B., Fry, W. E., and Davidse, L. C.,** (1993) Genotypic diversity of *Phytophthora infestans* in the Netherlands revealed by DNA polymorphisms, *Phytopathology*, 83: 1087–1092.

167. **Duncan, S., Barton, J. E., and O'Brien, P. A.,** (1993) Analysis of variation in isolates of *Rhizoctonia solani* by random amplified polymorphic DNA assay, *Mycol. Res.,* 97: 1075–1082.

168. **Dunemann, F.,** (1994) Genetic relationships in *Malus* evaluated by RAPD "fingerprinting" of cultivars and wild species, Eucarpia Fruit Breeding Section Meeting, Aug. 30–Sept. 3, 1993, *Euphytica,* in press.

169. **Durand, N., Reymond, P., and Fevre, M.,** (1993) Randomly amplified polymorphic DNAs assess recombination following an induced parasexual cycle in *Penicillium roquefortii*, *Curr. Genet.*, 24: 417–420.

170. **Düring, K.,** (1991) Ultrasensitive chemiluminescent and colorigenic detection of DNA, RNA, and proteins in plant molecular biology, *Anal. Biochem.,* 196: 433–438.

171. **Dweikat, I., Mackenzie, S., and Ohm, H.,** (1992) Mapping genes for hessian fly resistance in wheat using RAPD-DGGE, Plant Genome I, San Diego, Nov. 9–11, p. 25.

172. **Dweikat, I., Mackenzie, S., Levy, M., and Ohm, H.,** (1993) Pedigree assessment using RAPD-DGGE in cereal crop species, *Theor. Appl. Genet.,* 85: 497–505.

173. **Echt, C. S., Erdahl, L. A., and McCoy, T. J.,** (1992) Genetic segregation of random amplified polymorphic DNA in diploid cultivated alfalfa, *Genome,* 35: 84–87.

174. **Economou, E. P., Bergen, A. W., Warren, A. C., and Antonorakis, S. E.,** (1990) The polydeoxyadenylate tract of Alu repetitive elements is polymorphic in the human genome, *Proc. Natl. Acad. Sci. U.S.A.,* 87: 2951–2954.

175. **Edwards, A., Civitello, A., Hammond, H. A., and Caskey, C. T.,** (1991) DNA typing and genetic mapping with trimeric and tetrameric tandem repeats, *Am. J. Hum. Genet.,* 49: 746–756.

176. **Edwards, A., Johnstone, C., and Thompson, C.,** (1991) A simple and rapid method for the preparation of plant genomic DNA for PCR analysis, *Nucl. Acids Res.,* 19: 1349.

177. **Edwards, A., Hammond, H. A., Jin, L., Caskey, C. T., and Chakraborty, R.,** (1992) Genetic variation at five trimeric and tetrameric tandem repeat loci in four human population groups, *Genomics,* 12: 241–253.

178. **Epplen, J. T.,** (1988) On simple repeated GATA/GACA sequences in animal genomes: a critical reappraisal, *J. Hered.,* 79: 409–417.

178a. **Epplen, J. T., McCarrey, J. R., Sutou, S., and Ohno, S.,** (1982) Base sequence of a cloned snake W-chromosome DNA fragment and identification of a male-specific putative mRNA in the mouse, *Proc. Natl. Acad. Sci. U.S.A.,* 79:3798–3802.

179. **Epplen, J. T., et al.,** (1991) Oligonucleotide fingerprinting using simple repeat motifs: a convenient, ubiquitously applicable method to detect hypervariability for multiple purposes, in *DNA Fingerprinting: Approaches and Applications,* Burke, T., Dolf, G., Jeffreys, A. J., and Wolff, R., Eds., Birkhäuser, Basel, pp. 50–69.

180. **Epplen, C., Melmer, G., Siedlaczck, I., Schwaiger, F. W., Mäueler, W., and Epplen, J. T.,** (1993) On the essence of "meaningless" simple repetitive DNA in eukaryote genomes in *DNA Fingerprinting: State of the Science,* Pena, S. D. J., Chakraborty, R., Epplen, J. T., and Jeffreys, A. J., Eds., Birkhäuser, Basel, pp. 29–45.

181. **Eriksson, O., and Bremer, B.,** (1993) Genet dynamics of the clonal plant *Rubus saxatilis, J. Ecol.,* 81: 533–542.

182. **Erlich, H. A., (Ed)** (1989) *PCR Technology: Principles and Applications for DNA Amplifications,* Stockton Press, New York.

183. **Erlich, H. A., Gelfand, D. H., and Sninsky, J. J.,** (1991) Recent advances in the polymerase chain reaction, *Science,* 252: 1643–1651.

184. **Eskew, D. L., Caetano-Anollés, G., Bassam, B. J., and Gresshoff, P. M.,** (1993) DNA amplification fingerprinting of the *Azolla-Anabaena* symbiosis, *Plant Mol. Biol.,* 21: 363–373.

185. **Estoup, A., Solignac, M., Harry, M., and Cornuet, J.-M.,** (1993) Characterization of $(GT)_n$ and $(CT)_n$ microsatellites in two insect species: *Apis mellifera* and *Bombus terrestris, Nucl. Acids Res.,* 21: 1427–1431.

186. **Excoffier, L., Smouse, P. E., and Quattro, J. M.,** (1992) Analysis of molecular variance inferred from metric distances among DNA haplotypes: applications to human mitochondrial DNA restriction data, *Genetics,* 131: 479–491.

187. **Fabritius, A.-L., and Karajalainen, R.,** (1993) Variation in *Heterobasidion annosum* detected by random amplified polymorphic DNAs, *Eur. J. For. Pathol.,* 23: 193–200.

188. **Falconer, D. S.,** (1981) *Introduction to Quantitative Genetics,* Longman, New York.

189. **Fang, G., Hammar, S., and Grumet, R.,** (1992) A quick and inexpensive method for removing polysaccharides from plant genomic DNA, *BioTechniques,* 13: 52–54.

190. **Fauré, S., Noyer, J. L., Horry, J. P., Bakry, F., Lanaud, C., and González de Léon, D.,** (1993) A molecular marker-based linkage map of diploid bananas (*Musa acuminata*), *Theor. Appl. Genet.,* 87: 517–526.

191. **Feinberg, A. P., and Vogelstein, B.**, (1983) A technique for radiolabelling DNA restriction endonuclease fragments to high specific activity, *Anal. Biochem.*, 137: 6–13.

192. **Feinberg, A. P., and Vogelstein, B.**, (1984) A technique for radiolabelling DNA restriction endonuclease fragments to high specific activity, Addendum, *Anal. Biochem.*, 137: 266–267.

193. **Figueira, A., Janick, J., and Goldsbrough, P.**, (1992) Genome size and DNA polymorphism in *Theobroma cacao*, *J. Amer. Soc. Hort. Sci.*, 117: 673–677.

194. **Fisher, S. G., and Lerman, L. S.**, (1983) DNA fragments differing by single base-pair substitutions are separated in denaturing-gradient gels: correspondence with melting theory, *Proc. Natl. Acad. Sci. U.S.A.*, 80: 1579–1583.

195. **Folta, K. M., Hoey, B. K., Self, K. A., and Polans, N. O.**, (1992) The use of RAPDs in characterizing the pea genome, Plant Genome I, San Diego, Nov. 9–11, p. 43.

196. **Foolad, M. R., Jones, R. A., and Rodriguez, R. L.**, (1993) RAPD markers for constructing intraspecific tomato genetic maps, *Plant Cell Rep.*, 12: 293–297.

197. **Fox, B. C., Mobley, H. L. T., and Wade, J. C.**, (1989) The use of a DNA probe for epidemiological studies of candidiasis in immunocompromised hosts, *J. Infect. Dis.*, 159: 488–494.

198. **Francis, D. M., and St. Clair, D. A.**, (1993) Outcrossing in the homothallic oomycete, *Pythium ultimum*, detected with molecular markers, *Curr. Genet.*, 24: 100–106.

199. **Fransisco-Ortega, J., Newbury, H. J., and Ford-Lloyd, B. V.**, (1993) Numerical analyses of RAPD data highlight the origin of cultivated tagasaste (*Chamaecytisus proliferus* ssp. *palmensis*) in the Canary Islands, *Theor. Appl. Genet.*, 87: 264–270.

200. **Freeman, S., Pham, M., and Rodriguez, R. J.**, (1993) Molecular genotyping of *Colletotrichum* species based on arbitrarily primed PCR, A+T rich DNA, and nuclear DNA analyses, *Exp. Mycol.*, 17: 309–322.

201. **Fritsch, P., and Rieseberg, L. H.**, (1992) High outcrossing rates maintain male and hermaphrodite individuals in populations of the flowering plant *Datisca glomerata*, *Nature* (*London*), 359: 633–636.

202. **Fritsch, P., Hanson, M. A., Spore, C. D., Pack, P. E., and Rieseberg, L. H.**, (1993) Constancy of RAPD primer amplification strength among distantly related taxa of flowering plants, *Plant Mol. Biol. Rep.*, 11: 10–20.

203. **Fukuoka, S., Hosaka, K., and Kamijima, O.**, (1992) Use of random amplified polymorphic DNAs (RAPDs) for identification of rice accessions, *Jpn. J. Genet.*, 67: 243–252.

204. **Furnier, G. R., and Olfelt, J. P.**, (1993) RAPD diversity in Minnesota populations of Eurasian watermilfoil, *Am. J. Bot.*, 80 (Suppl.): 72.

205. **Furnier, G. R., and Zhaowei, L.**, (1993) Comparison of allozyme, RFLP, and RAPD markers for assessing genetic variation, *Am. J. Bot.*, 80 (Suppl.): 46–47.

206. **Galbraith, D. A., Boag, P. T., Gibbs, H. L., and White, B. N.**, (1991) Sizing bands on autoradiograms: a study of precision for scoring DNA fingerprints, *Electrophoresis*, 12: 210–220.

207. **Galvez, A. F., and Walker-Simmons, M. K.**, (1992) Chromosome mapping of a wheat protein kinase gene in *Lophopyrum elongatum* using amplified fragment polymorphism, Plant Genome I, San Diego, Nov. 9–11, p. 26.

208. **Garbelotto, M., Bruns, T. D., and Cobb, F. W.**, (1993) Differentiation of intersterility groups and geographic provenances among isolates of *Heterobasidion annosum* detected by random amplified polymorphic DNA assays, *Can. J. Bot.*, 71: 565–569.

209. **Gardes, M., and Bruns, T. D.**, (1993) ITS primers with enhanced specificity for basidiomycetes — applications to the identification of mycorrhizae and rusts, *Mol. Ecol.*, 2: 113–118.

210. **Gardiner, S. E., Zhu, J. M., Whitehead, H. C. M., and Madie, C.**, (1994) The New Zealand apple genome mapping project: a progress report, Eucarpia Fruit Breeding Section Meeting, Aug. 30–Sept. 3, 1993, *Euphytica*, in press.

211. **Gawel, N. J., and Jarret, R. L.**, (1991) A modified CTAB DNA extraction procedure for *Musa* and *Ipomoea*, *Plant Mol. Biol. Rep.*, 9: 262–266.

212. **Gebhardt, C., Blomendahl, C., Schachtschabel, U., Debener, T., Salamini, F., and Ritter, E.,** (1989) Identification of 2n breeding lines and 4n varieties of potato (*Solanum tuberosum* ssp. *tuberosum*) with RFLP-fingerprints, *Theor. Appl. Genet.*, 78: 16–22.

213. **Gebhardt, C., Ritter, E., Debener, T., Schachtschabel, U., Walkemeier, B., Uhrig, H., and Salamini, F.,** (1989) RFLP analysis and linkage mapping in *Solanum tuberosum,* *Theor. Appl. Genet.*, 78: 65–75.

214. **Georges, M., Lathrop, M., Hilbert, P., Marcotte, A., Schwers, A., Swillens, S., Vassart, G., and Hanset, R.,** (1990) On the use of DNA fingerprints for linkage studies in cattle, *Genomics*, 6: 461–474.

215. **Georges, M., Gunawardana, A., Threadgill, D. W., Lathrop, M., Olsaker, I., Mishra, A., Sargeant, L. L., Schoeberlein, A., Steele, M. R., Terry, C., Threadgill, D. S., Zhao, X., Holm, T., Fries, R., and Womack, J. E.,** (1991) Characterization of a set of variable number of tandem repeat markers conserved in Bovidae, *Genomics*, 11: 24–32.

216. **Georges, M., Dietz, A. B., Mishra, A., Nielsen, D., Sargeant, L. S., Sorensen, A., Steele, M. R., Zhao, X., Leipold, H., Womack, J. E., and Lathrop, M.,** (1993) Microsatellite mapping of the gene causing weaver disease in cattle will allow the study of an associated quantitative trait locus, *Proc. Natl. Acad. Sci. U.S.A.*, 90: 1058–1062.

217. **Gepts, P., Stockton, T., and Sonnante G.,** (1993) Use of hypervariable markers in genetic diversity studies, in *Application of RAPD Technology to Plant Breeding,* Neff, M., Ed., ASHS Publishers, St. Paul, pp. 41–45.

218. **Geyer, C. J., Ryder, A. O., Chemnick, L. G., and Thompson, E. A.,** (1993) Analysis of relatedness in the California condors from DNA fingerprints, *Mol. Biol. Evol.*, 10: 571–589.

219. **Gianfranceschi, L., McDermott, J. M., Seglias, N., Koller, B., Kellerhals, M., and Gessler, C.,** (1994) First results in developing a marker assisted breeding for resistance against scab, Eucarpia Fruit Breeding Section Meeting, Aug. 30–Sept. 3, 1993, *Euphytica,* in press.

220. **Gill, P., Evett, I. W., Woodroffe, S., Lygo, J. E., Millican, E., and Webster, M.,** (1991) Databases, quality control and interpretation of DNA profiling in the home office forensic science service, *Electrophoresis*, 12: 204–209.

221. **Gilmour, D. S., Thomas, H. G., and Elgin, S. C. R.,** (1989) *Drosophila* nuclear proteins bind to regions of alternating C and T residues in gene promoters, *Science*, 245: 1487–1490.

222. **Giovanonni, J. J., Wing, R. A., Ganal, M. W., and Tanksley, S. D.,** (1991) Isolation of molecular markers from specific chromosomal intervals using DNA pools from existing mapping populations, *Nucl. Acids Res.*, 19: 6553–6558.

223. **Girardin, H., Latge, J.-P., Srikantha, T., Morrow, B., and Soll, D. R.,** (1993) Development of DNA probes for fingerprinting *Aspergillus fumigatus, J. Clin. Microbiol.*, 31: 1547–1554.

224. **Glaser, R. L., Thomas, G. H., Siegfried, E., Elgin, S. C. R., and Lis, J. T.,** (1990) Optimal heat-induced expression of the *Drosophila* hsp26 gene requires a promoter sequence containing $(CT)_n$ $(GA)_n$ repeats, *J. Mol. Biol.*, 211: 751–761.

225. **Gogorcena, Y., Arulsekar, S., Dandekar, A. M., and Parfitt, D. E.,** (1993) Molecular markers for grape characterization, *Vitis*, 32: 183–185.

226. **Goldman, D., and Merrill, C. R.,** (1982) Silver staining of DNA in polyacrylamide gels: linearity and effect of fragment size, *Electrophoresis*, 3: 24–26.

227. **Goodwin, P. H., and Annis, S. L.,** (1991) Rapid identification of genetic variation and pathotype of *Leptosphaeria maculans* by random amplified polymorphic DNA assay, *Appl. Environ. Microbiol.*, 57: 2482–2486.

228. **Goodwin, P. H., Kirkpatrick, B. C., and Duniway, J. M.,** (1989) Cloned DNA probes for identification of *Phytophthora parasitica*, *Phytopathology*, 79: 716–721.

229. **Goodwin, P. H., Kirkpatrick, B. C., and Duniway, J. M.,** (1990) Identification of *Phytophthora citrophthora* with cloned DNA probes, *Appl. Environ. Microbiol.*, 56: 669–674.

230. **Goodwin, S. B., Drenth, A., and Fry, W. E.,** (1992) Cloning and genetic analysis of two highly polymorphic, moderately repetitive nuclear DNAs from *Phytophthora infestans*, *Curr. Genet.*, 22: 107–115.

231. **Goodwin, S. B., Spielman, L. J., Matuszak, J. M., Bergeron, S. N., and Fry, W. E.,** (1992) Clonal diversity and genetic differentiation of *Phytophthora infestans* populations in Northern and Central Mexico, *Phytopathology*, 82: 955–961.

232. **Gordon, M. J., Huang, X., Pentoney, S. L., Jr., and Zare, R. N.,** (1988) Capillary electrophoresis, *Science*, 242: 224–228.

233. **Gorman, M., Parojcic, M., and Cullis, C.,** (1992) Genomic mapping in flax (*Linum usitatissimum*), Plant Genome I, San Diego, Nov. 9–11, p. 27.

234. **Grady, D. L., Ratliff, R. L., Robinson, D. L., McCanlies, E. C., Meyne, J., and Moyzis, R. K.,** (1992) Highly conserved repetitive DNA sequences are present at human centromeres, *Proc. Natl. Acad. Sci. U.S.A.*, 89: 1695–1699.

235. **Grajal-Martin, M. J., Simon, C. J., and Muehlbauer, F. J.,** (1993) Use of random amplified polymorphic DNA (RAPD) to characterize race 2 of *Fusarium oxysporum* f. sp. *pisi*, *Phytopathology*, 83: 612–614.

236. **Grattapaglia, D., and Sederoff, R.,** (1992) Pseudo-testcross mapping strategy in forest trees: single tree RAPD maps of *Eucalyptus grandis* and *E. urophylla*, Plant Genome I, San Diego, Nov. 9–11, p. 27.

237. **Grattapaglia, D., Chaparro, J., Wilcox, P., McCord, S., Werner, D., Amerson, H., McKeand, S., Bridgewater, F., McIntyre, L., Doerge, R., Weir, B., Whetten, R., O'Malley, D., and Sederoff, R.,** (1992) RAPD mapping and tree improvement, Plant Genome I, San Diego, Nov. 9–11, p. 14.

238. **Grattapaglia, D., Chaparro, J., Wilcox, P., McCord, S., Werner, D., Amerson, H., McKeand, S., Bridgewater, F., Whetten, R., O'Malley, D., and Sederoff, R.,** (1993) Mapping in woody plants with RAPD markers: applications to breeding in forestry and horticulture, in *Application of RAPD Technology to Plant Breeding*, Neff, M. Ed., ASHS Publishers, St. Paul, pp. 37–40.

239. **Gray, I. C., and Jeffreys, A. J.,** (1991) Evolutionary transience of hypervariable minisatellites in man and the primates, *Proc. R. Soc. London Ser. B*, 243: 241–253.

240. **Greaves, D. R., and Patient, R. K.,** (1985) $(AT)_n$ is an interspersed repeat in the *Xenopus* genome, *EMBO J.*, 4: 2617–2626.

241. **Griffiths, R., and Tiwari, B.,** (1993) The isolation of molecular genetic markers for the identification of sex, *Proc. Natl. Acad. Sci. U.S.A.*, 90: 8324–8326.

242. **Groover, A. T., Fiddler, T. A., Lee, J. M., Vujcic, S. L., Devey, M. E., Mitchell-Olds, T., Megraw, R. A., Williams, C. G., and Neale, D. B.,** (1992) Mapping of quantitative trait loci influencing wood specific gravity in loblolly pine (*Pinus taeda*), Plant Genome I, San Diego, Nov. 9–11, p. 27.

243. **Gross, D. S., and Garrard, W. T.,** (1986) The ubiquitous potential Z-forming sequence of eukaryotes, $(dT-dG)_n \cdot (dC-dA)_n$, is not detectable in the genomes of eubacteria, archaebacteria, or mitochondria, *Mol. Cell. Biol.*, 6: 3010–3013.

244. **Gruber, F.,** (1990) Homologe und heterologe Transformation von *Trichoderma reesei* mit den Orotidin-5'-Phosphat-Decarboxylase Genen als Selektionsmarker, dissertation thesis, Technical University, Vienna, Austria.

245. **Guidet, F., and Langridge, P.,** (1992) Megabase DNA preparation from plant tissue, *Methods Enzymol.*, 216: 3–12.

246. **Guidet, F., Rogowsky, P., and Langridge, P.,** (1990) A rapid method of preparing megabase plant DNA, *Nucl. Acids Res.*, 18: 4955.

247. **Guillemaut, P., and Maréchal-Drouard, L.,** (1992) Isolation of plant DNA: a fast, inexpensive, and reliable method, *Plant Mol. Biol. Rep.*, 10: 60–65.

248. **Guthrie, P. A. I., Magill, C. W., Frederiksen, R. A., and Odvody, G. N.,** (1992) Random amplified polymorphic DNA markers: a system for identifying and differentiating isolates of *Colletotrichum graminicola*, *Phytopathology*, 82: 832–835.

249. **Guzman, P., and Ecker, J. R.,** (1988) Development of large DNA methods for plants: molecular cloning of large segments of *Arabidopsis* and carrot DNA into yeast, *Nucl. Acids Res.,* 16: 11091–11106.

250. **Gyllensten, U. B., Jakobsson, S., Temrin, H., and Wilson, A. C.,** (1989) Nucleotide sequence and genomic organization of bird minisatellites, *Nucl. Acids Res.,* 17: 2203–2214.

251. **Hadrys, H., Balick, M., and Schierwater, B.,** (1992) Applications of random amplified polymorphic DNA (RAPD) in molecular ecology, *Mol. Ecol.,* 1: 55–63.

252. **Hadrys, H., Schierwater, B., Dellaporta, S. L., Desalle, R., and Buss, L. W.,** (1993) Determination of paternity in dragonflies by random amplified polymorphic DNA finger-printing, *Mol. Ecol.,* 2: 79–87.

253. **Haig, S. M., Belthoff, J. R., and Allen, D. H.,** (1993) Examination of population structure in red-cockaded woodpeckers using DNA profiles, *Evolution,* 47: 185–194.

254. **Haley, C. S.,** (1991) Use of DNA fingerprints for the detection of major genes for quantitative traits in domestic species, *Anim. Genet.,* 22: 259–277.

255. **Haley, S. D., Miklas, P. N., Stavely, J. R., Byrum, J., and Kelly, J. D.,** (1993) Identi-fication of RAPD markers linked to a major rust resistance gene block in common bean, *Theor. Appl. Genet.,* 86: 505–512.

256. **Halward, T. M., Stalker, H. T., LaRue, E. A., and Kochert, G.,** (1991) Genetic variation detectable with molecular markers among unadapted germplasm resources of cultivated peanut and related wild species, *Genome,* 34: 1013–1020.

257. **Halward, T. M., Stalker, H. T., LaRue, E. A., and Kochert, G.,** (1992) Use of single-primer DNA amplifications in genetic studies of peanut (*Arachis hypogaea* L.), *Plant Mol. Biol.,* 18: 315–325.

258. **Hamada, H., Petrino, M. G., and Kakunaga, T.,** (1982) A novel repeated element with Z-DNA-forming potential is widely found in evolutionary diverse eukaryotic genomes, *Proc. Natl. Acad. Sci. U.S.A.,* 79: 6465–6469.

259. **Hamada, H., Seidman, M., Howard, B. H., and Gorman, C. M.,** (1984) Enhanced gene expression by the poly(dT-dG) poly(dC-dA) sequence, *Mol. Cell. Biol.,* 4: 2622–2630.

260. **Hamby, K. R., and Zimmer, E. A.,** (1992) Ribosomal RNA as a phylogenetic tool in plant systematics, in *Molecular Systematics in Plants,* Soltis, P. S., Soltis, D. E., and Doyle, J. J., Eds., Chapman and Hall, New York, pp. 50–91.

261. **Hamelin, R. C., Oullette, G. B., and Bernier, L.,** (1993) Identification of *Gremmeniella abietina* races with random amplified polymorphic DNA markers, *Appl. Environ. Microbiol.,* 59: 1752–1755.

262. **Hamer, J. E., and Givan, S.,** (1990) Genetic mapping with dispersed repeated sequences in the rice blast fungus: mapping the SMO locus, *Mol. Gen. Genet.,* 223: 487–495.

263. **Hamer, J. E., Farrall, L., Orbach, M. J., Valent, B., and Chumley, F. G.,** (1989) Host species-specific conservation of a family of repeated DNA sequences in the genome of a fungal plant pathogen, *Proc. Natl. Acad. Sci. U.S.A.,* 86: 9981–9985.

264. **Hanotte, O., Burke, T., Armour, J. A. L., and Jeffreys, A. J.,** (1991a) Hypervariable minisatellite DNA sequences in the Indian peafowl *Pavo cristatus, Genomics,* 9: 587–597.

265. **Hanotte, O., Burke, T., Armour, J. A. L., and Jeffreys, A. J.,** (1991b) Cloning, characterization and evolution of Indian peafowl *Pavo cristatus* minisatellite loci, in *DNA Fingerprinting: Approaches and Applications,* Burke, T., Dolf, G., Jeffreys, A. J., and Wolff, R., Eds.,Birkhäuser, Basel, pp. 193–216.

266. **Hanotte, O., Bruford, M. W., and Burke, T.,** (1992a) Multilocus DNA fingerprints in gallinaceous birds: general approach and problems, *Heredity,* 68: 481–494.

267. **Hanotte, O., Cairns, E., Robson, T., Double, M. C., and Burke, T.,** (1992b) Cross-species hybridization of a single-locus minisatellite probe in passerine birds, *Mol. Ecol.,* 1: 127–130.

268. **Hanson, M. A., Fritsch, P., and Rieseberg, L. H.,** (1993) Level and distribution of genetic diversity of endangered *Pogogyne* (Lamiaceae); genetic evidence warrants recognition of a new species from Baja California, *Am. J. Bot.,* 80 (Suppl.): 152–153.

269. **Harada, T., Matsukawa, K., Sato, T., Ishikawa, R., Niizeki, M., and Saito, K.,** (1993) DNA-RAPDs detect genetic variation and paternity in *Malus, Euphytica,* 65: 87–91.

270. **Harris, H.,** (1969) Enzyme polymorphisms in man, *Proc. R. Soc. Edinburgh Sect. B,* 164: 298–310.

271. **Harris, S. A.,** (1993) DNA analysis of tropical plant species — an assessment of different drying methods, *Plant Syst. Evol.,* 188: 57–64.

272. **Hartl, D. L.,** (1980) *Principles of Population Genetics,* Sinauer Associates, Sunderland, MA.

273. **Hashizume, T., Sato, T., and Hirai, M.,** (1993) Determination of genetic purity of hybrid seed in watermelon (*Citrullus lanatus*) and tomato (*Lycopersicon esculentum*) using random amplified polymorphic DNA (RAPD), *Jpn. J. Breed.,* 43: 367–375.

274. **Hassan, A. K., Schultz, C., Sacristan, M. D., and Wöstemeyer, J.,** (1991) Biochemical and molecular tools for the differentiation of aggressive and non-aggressive isolates of the oilseed rape pathogen, *Phoma lingam, J. Phytopathol.,* 131: 120–136.

275. **Hattori, J., Gottlob-McHugh, S. G., and Johnson, D. A.,** (1987) The isolation of high-molecular weight DNA from plants, *Anal. Biochem.,* 165: 70–74.

276. **Haymer, D. S., McInnis, D. O., and Arcangeli, L.,** (1992) Genetic variation between strains of the Mediterranean fruit fly, *Ceratitis capitata,* detected by DNA fingerprinting, *Genome,* 35: 528–533.

277. **He, S., Ohm, H., and Mackenzie, S.,** (1992) Detection of DNA sequence polymorphisms among wheat varieties, *Theor. Appl. Genet.,* 84: 573–578.

278. **Hearne, C. M., Ghosh, S., and Todd, J. A.,** (1992) Microsatellites for linkage analysis of genetic traits, *TIG,* 8: 288–294.

279. **Heath, D. D., Iwama, G. K., and Devlin, R. H.,** (1993) PCR primed with VNTR core sequences yields species specific patterns and hypervariable probes, *Nucl. Acids Res.,* 21: 5782–5785.

280. **Hegedus, D. D., and Khachatourians, G. G.,** (1993) Construction of cloned DNA probes for the specific detection of the entomopathogenic fungus *Beauveria bassiana* in grasshoppers, *J. Invertebr. Pathol.,* 62: 233–240.

281. **Helentjaris, T., Slocum, M., Wright, S., Schaefer, A., and Nienhuis, J.,** (1986) Construction of genetic linkage maps in maize and tomato using restriction fragment length polymorphisms, *Theor. Appl. Genet.,* 72: 761–769.

282. **Hellstein, J., Vawter-Hugart, H., Fotos, P., Schmid, J., and Soll, D. R.,** (1993) Genetic similarity and phenotypic diversity of commensal and pathogenic strains of *Candida albicans* isolates from the oral cavity, *J. Clin. Microbiol.,* 31: 3190–3199.

283. **Hemmat, M., Weeden, N. F., Manganaris, A. G., and Lawson, D. M.,** (1994) Molecular marker linkage map for apple, *J. Hered.,* 85:4–11.

284. **Heun, M., and Helentjaris, T.,** (1993) Inheritance of RAPDs in F1 hybrids of corn, *Theor. Appl. Genet.,* 85: 961–968.

285. **Hillel, J., Schaap, T., Haberfeld, A., Jeffreys, A. J., Plotzky, Y., Cahaner, A., and Lavi, U.,** (1990) DNA fingerprints applied to gene introgression in breeding programs, *Genetics,* 124: 783–789.

286. **Hillis, D. M., and Moritz, C.,** (1990) *Molecular Systematics,* Sinauer Associates, Sunderland, MA.

287. **Hillis, D. M., Larson, A., Davis, S. K., and Zimmer, E. A.,** (1990) Nucleic acids III: sequencing, in *Molecular Systematics,* Hillis, D. M., and Moritz, C. Eds., Sinauer Associates, Sunderland, MA, pp. 318–370.

288. **Hilu, K. W.,** (1993) Evidence from RAPD for the evolution of *Echinochloa* millets, *Am. J. Bot.,* 80 (Suppl.): 73.

289. **Hintz, W. E., Jeng, R. S., Hubbes, M. M., and Horgen, P. A.,** (1991) Identification of three populations of *Ophiostoma ulmi* (aggressive subgroup) by mitochondrial DNA restriction-site mapping and nuclear DNA fingerprinting, *Exp. Mycol.,* 15: 316–325.

290. **Hoelzel, A. R.,** (1992) *Molecular Genetic Analysis of Populations: A Practical Aproach,* IRL Press, Oxford.

291. **Hoelzel, A. R., and Green, A.,** (1992) Analysis of population-level variation by sequencing PCR-amplified DNA, in *Molecular Genetic Analysis of Populations: A Practical Approach,* Hoelzel, A. R., Ed., IRL Press, Oxford, pp. 159–188.

292. **Hoepfner, A.-S., Nybom, H., Carlsson, U., and Franzén, R.,** (1993) DNA fingerprinting useful for monitoring cell line identity in micropropagated raspberries, *Acta. Agric. Scand. Sect. B,* 43: 53–57.

293. **Hoey, M. T., Merkle, S. A., and Meagher, R. B.,** (1992) Molecular markers in *Liriodendron* (yellow-poplar), Plant Genome I, San Diego, Nov. 9–11, p. 28.

294. **Honeycutt, R., Smith, S., and Sobral, B.,** (1992) Reconstructing histories of maize inbreds using molecular characters, Plant Genome I, San Diego, Nov. 9–11, p. 29.

295. **Honeycutt, R., Sobral, B. W. S., Keim, P., and Irvine, J. E.,** (1992) A rapid DNA extraction method for sugarcane and its relatives, *Plant Mol. Biol. Rep.,* 10: 66–72.

296. **Hong, Y.-K., Coury, D. A., Polne-Fuller, M., and Gibor, A.,** (1992) Lithium chloride extraction of DNA from the seaweed *Porphyra perforata* (Rhodophyta), *J. Phycol.,* 28: 717–720.

297. **Horn, P., and Rafalski, A.,** (1992) Non-destructive RAPD genetic diagnostics of microspore-derived *Brassica* embryos, *Plant. Mol. Biol. Rep.,* 10: 285–293.

298. **Horn, P., Richards, B., and Klinger, K. W.,** (1989) Amplification of a highly polymorphic VNTR segment by the polymerase chain reaction, *Nucl. Acids Res.,* 17: 2140.

299. **Howland, D. E., Oliver, R. P., and Davy, A. J.,** (1991) A method of extraction on DNA from birch, *Plant Mol. Biol. Rep.,* 9: 340–344.

300. **Hu, J., and Quiros, C. F.,** (1991) Identification of broccoli and cauliflower cultivars with RAPD markers, *Plant Cell Rep.,* 10: 505–511.

301. **Hu, J., Kianian, S., McGrath, M., Arus, P., and Quiros, C. F.,** (1992) Mapping the A and C genomes of the genus *Brassica*: a progress report, Plant Genome I, San Diego, Nov. 9–11, p. 29.

302. **Huey, B., and Hall, J.,** (1989) Hypervariable DNA fingerprinting in *Escherichia coli*: minisatellite probe from bacteriophage M13, *J. Bacteriol.,* 171: 2528–2532.

303. **Huff, D. R., and Bara, J. M.,** (1993) Determining genetic origins of aberrant progeny from facultative apomictic Kentucky bluegrass using a combination of flow cytometry and silver-stained RAPD markers, *Theor. Appl. Genet.,* 87: 201–208.

304. **Huff, D. R., Peakall, R., and Smouse, P. E.,** (1993) RAPD variation within and among natural populations of outcrossing buffalograss [*Buchloë dactyloides* (Nutt.) Engelm.], *Theor. Appl. Genet.,* 86: 927–934.

305. **Huijser, P., Hennig, W., and Dijkhof, R.,** (1987) Poly (dC-dA/dG-dT) repeats in the *Drosophila* genome: a key function for dosage compensation and position effects?, *Chromosoma,* 95: 209–215.

306. **IMI, International Mycological Institute,** (1992) *Catalogue of the Culture Collection. Filamentous Fungi, Yeasts and Plant Pathogenic Bacteria,* 10th ed., IMI, Kew.

307. **Inai, S., Ishikawa, K., Nunomura, O., and Ikehashi, H.,** (1993) Genetic analysis of stunted growth by nuclear-cytoplasmic interaction in interspecific hybrids of *Capsicum* by using RAPD markers, *Theor. Appl. Genet.,* 87: 416–422.

308. **Innis, M. A., Gelfand, D. H., Sninsky, J. J., and White, T. J., Eds.,** (1990) *PCR Protocols, A Guide to Methods and Applications,* Academic Press, San Diego,.

309. **Isabel, N., Tremblay, L., Michaud, M., Tremblay, F. M., and Bousquet, J.,** (1993) RAPDs as an aid to evaluate the genetic integrity of somatic embryogenesis-derived populations of *Picea mariana* (Mill.) B.S.P., *Theor. Appl. Genet.,* 86: 81–87.

310. **Ishii, T., Terachi, T., Mori, N., and Tsunewaki, K.,** (1993) Comparative study on the chloroplast, mitochondrial and nuclear genome differentiation in two cultivated rice species, *Oryza sativa* and *O. glaberrima*, by RFLP analyses, *Theor. Appl. Genet.,* 86: 88–96.

311. **Jacobson, K. M., Miller, O. K., Jr., and Turner, B. J.,** (1993) Randomly amplified polymorphic DNA markers are superior to somatic incompatibility tests for discriminating genotypes in natural populations of the ectomycorrhizal fungus *Suillus granulatus, Proc. Natl. Acad. Sci. U.S.A.,* 90: 9159–9163.

312. **Jarman, A. P., and Wells, R. A.,** (1989) Hypervariable minisatellites; recombinators or innocent bystanders?, *TIG*, 5: 367–371.

313. **Jean, M., Brown, G. G., and Landry, B. S.,** (1992) Mapping of the "restoration of fertility" gene for *pol* cytoplasmic sterility in canola, Plant Genome I, San Diego, Nov. 9–11, p. 31.

314. **Jeffreys, A. J., and Morton, D. B.,** (1987) DNA fingerprinting of dogs and cats, *Anim. Genet.*, 18: 1–15.

315. **Jeffreys, A. J., Wilson, V., and Thein, S. L.,** (1985) Hypervariable 'minisatellite' regions in human DNA, *Nature (London)*, 314: 67–73.

316. **Jeffreys, A. J., Wilson, V., and Thein, S. L.,** (1985) Individual-specific 'fingerprints' of human DNA, *Nature (London)*, 316: 76–79.

317. **Jeffreys, A. J., Wilson, V., Thein, S. L., Weatherall, D. J., and Ponder, B. A. J.,** (1986) DNA "fingerprints" and segregation analysis of multiple markers in human pedigrees, *Am. J. Hum. Genet.*, 39: 11–24.

318. **Jeffreys, A. J., Royle, N. J., Wilson, V., and Wong, Z.,** (1988) Spontaneous mutation rates to new length alleles at tandem-repetitive hypervariable loci in human DNA, *Nature (London)*, 332: 278–281.

319. **Jeffreys, A. J., Wilson, V., Neumann, R., and Keyte, J.,** (1988) Amplification of human minisatellites by the polymerase chain reaction: towards DNA fingerprinting of single cells, *Nucl. Acids Res.*, 16: 10053–10071.

320. **Jeffreys, A. J., Neumann, R., and Wilson, V.,** (1990) Repeat unit sequence variation in minisatellites: a novel source of DNA polymorphism for studying variation and mutation by single molecule analysis, *Cell*, 60: 473–485.

321. **Jeffreys, A. J., MacLeod, A., Tamaki, K., Neil, D. L., and Monckton, D. G.,** (1991) Minisatellite repeat coding as a digital approach to DNA typing, *Nature (London)*, 354: 204–209.

322. **Jeffreys, A. J., Turner, M., and Debenham, P.,** (1991) The efficiency of multilocus DNA fingerprint probes for the individualization and establishment of family relationships, determined from extensive casework, *Am. J. Hum. Genet.*, 48: 824–840.

323. **Jeffreys, A. J., Monckton, D. G., Tamaki, K., Neil, D. L., Armour, J. A. L., MacLeod, A., Collick, A., Allen, M., and Jobling, M.,** (1993) Minisatellite variant repeat mapping: application to DNA typing and mutation analysis, in *DNA Fingerprinting: State of the Science*, Pena, S. D. J., Chakraborty, R., Epplen, J. T., and Jeffreys, A. J., Eds., Birkhäuser, Basel, pp. 125–139.

324. **Jermstad, K. D., Reem, A. M., Wheeler, N. C., and Neale, D. B.,** (1992) Molecular marker and quantitative trait mapping in Douglas-fir, Plant Genome I, San Diego, Nov. 9–11, p. 31.

325. **Jessup, S. L.,** (1993) Reticulate evolution in *Gaudichaudia* (Malpighiaceae): phylogenetic and biogeographic analysis of molecular and morphological variation in a polyploid complex of neotropic vines, *Am. J. Bot.*, 80 (Suppl.): 154.

326. **Jessup, S. L.,** (1993) Randomly amplified DNA is a powerful tool for analyzing reticulate ancestry in *Gaudichaudia* (Malpighiaceae), *Am. J. Bot.*, 80 (Suppl.): 154–155.

327. **Jhingan, A. K.,** (1992) A novel technology for DNA isolation. *Methods Mol. Cell. Biol.* 3: 15–22.

328. **Jhingan, A. K.,** (1992) Efficient procedure for DNA extraction from lyophilized plant material, *Methods Mol. Cell. Biol.*, 3: 185–187.

329. **Ji, J., Russell, S. J., Rumsey, F. J., Gibby, M., Sheffield, E., and Jermy, A. C.,** (1993) The application of DNA amplification to a study of genetic variation of the filmy fern *Trichomanes speciosum* Willd, *Am. J. Bot.*, 80 (Suppl.): 109.

330. **Johanson, A., and Jeger, M. J.,** (1993) Use of PCR for detection of *Mycosphaerella fijiensis* and *M. musicola*, the causal agents of Sigatoka leaf spots in banana and plantain, *Mycol. Res.*, 97: 670–674.

331. **John, M. E.,** (1992) An efficient method for isolation of RNA and DNA from plants containing polyphenolics, *Nucl. Acids Res.*, 20: 2381.

332. **Johns, M. A.,** (1992) Comparison of RAPD maps for two sets of maize recombinant-inbred lines, Plant Genome I, San Diego, Nov. 9–11, p. 32.

333. **Johnston, R. F., Pickett, S. C., and Barker, D. L.,** (1990) Autoradiography using storage phosphor technology, *Electrophoresis,* 11: 355–360.

334. **Joklik, W. K., Willett, H. P., Amos, D. B., and Wilfert, C. M.,** Eds., (1992) *Zinsser Microbiology,* 20th ed., Appleton & Lange, Norwalk–San Mateo.

335. **Jondle, R. J.,** (1993) Legal aspects of varietal protection using molecular markers, in *Application of RAPD Technology to Plant Breeding,* Neff, M., Ed., ASHS Publishers, St. Paul, MN, pp. 50–52.

336. **Jones, M. J., and Dunkle, L. D.,** (1993) Analysis of *Cochliobolus carbonum* races by PCR amplification with arbitrary and gene-specific primers, *Phytopathology,* 83: 366–370.

337. **Joshi, C. P., and Nguyen, H. T.,** (1992) RAPD analysis in tetraploid and hexaploid wheats, Plant Genome I, San Diego, Nov. 9–11, p. 40.

338. **Joshi, C. P., and Nguyen, H. T.,** (1993) Application of the random amplified polymorphic DNA technique for the detection of polymorphism among wild and cultivated tetraploid wheats, *Genome,* 36: 602–609.

339. **Julier, C., De Gouyon, B., Georges, M., Guénet, J. L., Nakamura, Y., and Avner, P.,** (1990) Minisatellite linkage maps in the mouse by cross-hybridization with human probes containing tandem repeats, *Proc. Natl. Acad. Sci. U.S.A.,* 87: 4585–4589.

340. **Kaemmer, D., Afza, R., Weising, K., Kahl, G., and Novak, F. J.,** (1992) Oligonucleotide and amplification fingerprinting of wild species and cultivars of banana (*Musa* spp.), *Bio/ Technology,* 10: 1030–1035.

341. **Kaemmer, D., Ramser, J., Schön, M., Weigand, F., Saxena, M. C., Driesel, A. J., Kahl, G., and Weising, K.,** (1992) DNA fingerprinting of fungal genomes: a case study with *Ascochyta rabiei* in *DNA Polymorphisms in Eukaryotic Genomes,* Kahl, G., Appelhans, H., Kömpf, J., and Driesel, A. J., Eds., Biotech-Forum 10, Adv. Mol. Genet. 5, Hüthig Verlag, Heidelberg, pp. 255–270.

341a. **Kaemmer, D., Weising, R., Beyermann, B., Börner, T., Epplen, J. T., and Kahl, G.,** (1994) Oligonucleotide fingerprinting of tomato DNA, *Plant Breed.,* in press.

342. **Kanazawa, A., and Tsutsumi, N.,** (1992) Extraction of restrictable DNA from plants of the genus *Nelumbo, Plant Mol. Biol. Rep.,* 10: 316–318.

343. **Kashi, Y., Tikochinsky, Y., Genislav, E., Iraqi, F., Nave, J. S., Beckmann, J. S., Gruenbaum, Y., and Soller, M.,** (1990) Large restriction fragments containing poly-TG are highly polymorphic in a variety of vertebrates, *Nucl. Acids Res.,* 18: 1129–1132.

344. **Kaukinen, J., and Varvio, S.-L.,** (1992) Artiodactyl retrotransposons: association with microsatellites and use in SINEmorph detection by PCR, *Nucl. Acids Res.,* 20: 2955–2958.

345. **Kayser, T.,** (1992) Protoplastenfusion, sowie elektrophoretische Chromosomentrennung und Genkartierung bei filamentösen Pilzen: *Penicillium janthinellum, Absidia glauca* und *Cochliobolus heterotrophus,* dissertation thesis, Humboldt University of Berlin, Germany.

346. **Kazan, K., Manners, J. M., and Cameron, D. F.,** (1993) Genetic variation in agronomically important species of *Stylosanthes* determined using random amplified polymorphic DNA markers, *Theor. Appl. Genet.,* 85: 882–888.

347. **Kazan, K., Manners, J. M., and Cameron, D. F.,** (1993) Genetic relationships and variation in the *Stylosanthes guianensis* species complex assessed by random amplified polymorphic DNA, *Genome,* 36: 43–49.

348. **Kazan, K., Manners, J. M., and Cameron, D. F.,** (1993) Inheritance of random amplified polymorphic DNA markers in an interspecific cross in the genus *Stylosanthes, Genome,* 36: 50–56.

349. **Kelly, R., Bulfield, G., Collick, A., Gibbs, M., and Jeffreys, A. J.,** (1989) Characterization of a highly unstable mouse minisatellite locus: evidence for somatic mutation during early development, *Genomics,* 5: 844–856.

350. **Kersulyte, D., Woods, J. P., Keath, E., Goldman, W. E., and Berg, D. E.,** (1992) Diversity among clinical isolates of *Histoplasma capsulatum* detected by polymerase chain reaction with arbitrary primers, *J. Bacteriol.,* 174: 7075–7079.

351. **Kesseli, R. V., Paran, I., and Michelmore, R. W.**, (1993) Efficient mapping of specifically targeted genomic regions and the tagging of these regions with reliable PCR-based genetic markers, in *Application of RAPD Technology to Plant Breeding*, Neff, M., Ed., ASHS Publishers, St. Paul, MN, pp. 31–36.

352. **Khush, R. S., Morgan, L., Becker, E., and Wach, M.**, (1991) Use of the polymerase chain reaction (PCR) in *A. bisporus* breeding programs, in *Genetics and Breeding of Agaricus*, Van Griensven, L. J. L. D., Ed., Pudoc, Wageningen, pp. 73–80.

353. **Khush, R. S., Becker, E., and Wach, M.**, (1992) DNA amplification polymorphisms of the cultivated mushroom *Agaricus bisporus*, *Appl. Environ. Microbiol.*, 58: 2971–2977.

354. **Kim, K. J., and Mabry, T. J.**, (1991) Phylogenetic and evolutionary implications of nuclear ribosomal DNA variation in dwarf dandelions (*Krigia*, Lactuceae, Asteraceae), *Plant Syst. Evol.*, 177: 53–69.

355. **Kimpton, C. P., Gill, P., Walton, A., Urquhart, A., Millican, E. S., and Adams, M.**, (1993) Automated DNA profiling employing multiplex amplification of short tandem repeat loci, *PCR Methods Appl.*, 3: 13–22.

356. **King, G.**, (1994) Progress in mapping agronomic genes in apple (The European apple genome mapping project). Eucarpia Fruit Breeding Section Meeting, Aug. 30–Sept. 3, 1993, *Euphytica*, in press.

357. **King, I. P., Purdie, K. A., Rezanoor, H. N, Koebner, R. M. D., Miller, T. E., Reader, S. M., and Nicholson, P.**, (1993) Characterization of *Thinopyrum bessarabicum* chromosome segments in wheat using random amplified polymorphic DNAs (RAPDs) and genomic *in situ* hybridization, *Theor. Appl. Genet.*, 86: 895–900.

358. **Kiss, G. B., Csanadi, G., Kalman, K., Kalo, P., and Ökresz, L.**, (1993) Construction of a basic genetic map for alfalfa using RFLP, RAPD, isozyme and morphological markers, *Mol. Gen. Genet.*, 238: 129–137.

359. **Kistler, H. C., Momol, E. A., and Benny, U.**, (1991) Repetitive genomic sequences for determining relatedness among strains of *Fusarium oxysporum*, *Phytopathology*, 81: 331–336.

360. **Kleinhofs, A., et al.**, (1993) A molecular, isozyme and morphological map of the barley (*Hordeum vulgare*) genome, *Theor. Appl. Genet.*, 86: 705–712.

361. **Klein-Lankhorst, R. M., Vermunt, A., Weide, R., Liharska, T., and Zabel, P.**, (1991) Isolation of molecular markers for tomato (*L. esculentum*) using random amplified polymorphic DNA (RAPD), *Theor. Appl. Genet.*, 83: 108–114.

362. **Klimyuk, V. I., Carroll, B. J., Thomas, C. M., and Jones, J. D. G.**, (1993) Alkali treatment for rapid preparation of plant material for reliable PCR analysis, *Plant J.*, 3: 493–494.

363. **Klinkicht, M., and Tautz, D.**, (1992) Detection of simple sequence length polymorphisms by silver staining, *Mol. Ecol.*, 1: 133–134.

364. **Kohli, Y., Morrall, R. A. A., Anderson, J. B., and Kohn, L. M.**, (1992) Local and trans-Canadian clonal distribution of *Sclerotinia sclerotiorum* on canola, *Phytopathology*, 82: 875–880.

365. **Kohn, L. M., Stasovski, E., Carbone, I., Royer, J., and Anderson, J. B.**, (1991) Mycelial incompatibility and molecular markers identify genetic variability in field populations of *Sclerotinia sclerotiorum*, *Phytopathology*, 81: 480–485.

366. **Kolchinsky, A., Kolesnikova, M., and Ananiev, E.**, (1991) "Portraying" of plant genomes using polymerase chain reaction amplification of ribosomal 5S genes, *Genome*, 34: 1028–1031.

367. **Koller, B., Lehmann, A., McDermott, J. M., and Gessler, C.**, (1993) Identification of apple cultivars using RAPD markers, *Theor. Appl. Genet.*, 85: 901–904.

368. **Komarnitsky, I. K., Samoylov, A. M., Red'ko, V. V., Peretyayko, V. G., and Gleba, Y. Y.**, (1990) Interspecific diversity of sugar beet (*Beta vulgaris*) mitochondrial DNA, *Theor. Appl. Genet.*, 80: 253–257.

369. **Komatsu, A., Akihama, T., Hidaka, T., and Omura, M.**, (1993) Identification of *Poncirus* strains by DNA fingerprinting, in *Techniques on Gene Diagnosis and Breeding in Fruit Trees*, Hayashi, T., Omura, M., and Scott, N. S., Eds., Fruit Tree Research Station, Ibaraki, pp. 88–95.

370. **Kraft, T. and Nybom, H.,** (1994) Application of DNA fingerprinting and biometry to solve some taxonomical problems in apomictic blackberries, *Rubus* subgen. *Rubus*. manuscript submitted.

371. **Kreike, J.,** (1990) Genetic analysis of forest tree populations: isolation of DNA from spruce and fir apices, *Plant Mol. Biol.*, 14: 877–879.

372. **Kreitman, M., and Aguadé, M.,** (1986) Genetic uniformity in two populations of *Drosophila melanogaster* as revealed by filter hybridization of four-nucleotide-recognizing restriction enzyme digests, *Proc. Natl. Acad. Sci. U.S.A.*, 83: 3562–3566.

373. **Kresovich, S., Williams, J. G. K., McFerson, J. R., Routman, E. J., and Schaal, B. A.,** (1992) Characterization of genetic identities and relationships of *Brassica oleracea* L. via a random amplified polymorphic DNA assay, *Theor. Appl. Genet.*, 85: 190–196.

374. **Kubisiak, T. L., Stine, M., Nelson, C. D., and Nance, W. L.,** (1992) Single tree RAPD linkage mapping of longleaf pine, Plant Genome I, San Diego, Nov. 9–11, p. 49.

375. **Kuhls, K., Batko, A., Lieckfeldt, E., Meyer, W., and Börner, T.,** (1992) DNA-fingerprinting and PCR – two of the molecular methods used for identification and classification of the genus *Penicillium*. First European Conference on Fungal Genetics, Nottingham, Aug. 20–23, Abstract P1/19.

376. **Kuhnlein, U., Dawe, Y., Zadworny, D., and Gavora, J. S.,** (1989) DNA fingerprinting: a tool for determining genetic distances between strains of poultry, *Theor. Appl. Genet.*, 77: 669–672.

377. **Kuhnlein, U., Zadworny, D., Dawe, Y., Fairfull, R. W., and Gavora, J. S.,** (1990) Assessment of inbreeding by DNA fingerprinting: development of a calibration curve using defined strains of chickens, *Genetics*, 125: 161–165.

378. **Kunze, G., Kunze, I., Barner, A., and Schulz, R.,** (1993) Classification of *Saccharomyces cerevisiae* strains by genetical and biochemical methods, *Monatsschr. Brauwissenschaft*, 4: 132–136.

379. **Kühl, H., and Neuhaus, D.,** (1993) The genetic variability of *Phragmites australis* investigated by random amplified polymorphic DNA, *Limnol. Aktuell*, 5: 9–18.

380. **Kvarnheden, A., and Engström, P.,** (1992) Genetically stable, individual-specific differences in hypervariable DNA in Norway spruce, detected by hybridization to a phage M13 probe, *Can. J. For. Res.*, 22: 117–123.

381. **Kwan, H.-S., Chiu, S.-W., Pang, K.-M., and Cheng, S.-C.,** (1992) Strain typing in *Lentinula edodes* by polymerase chain reaction, *Exp. Mycol.*, 16: 163–166.

382. **Kwon-Chung, K. J., and Bennett, J. E.,** (1992) *Medical Mycology*, Lea and Febiger, Philadelphia.

383. **Labarca, C., and Paigen, K.,** (1980) A simple, rapid, and sensitive DNA assay procedure, *Anal. Biochem.*, 102: 344–352.

384. **Lagercrantz, U., Ellegren, H., and Andersson, L.,** (1993) The abundance of various polymorphic microsatellite motifs differs between plants and vertebrates, *Nucl. Acids Res.*, 21: 1111–1115.

385. **Lan, C., and Kuo, G.,** (1992) Identification of *Capsicum* accessions with RAPD patterns, Plant Genome I, San Diego, Nov. 9–11, p. 34.

386. **Landegren, U., Kaiser, R., Caskey, C. T., and Hood, L.,** (1988) DNA diagnostics – molecular techniques and automation, *Science*, 242: 229–237.

387. **Lander, E. S., Green, P., Abrahamson, J., Barlow, A., Daly, A., Lincoln, S. E., and Newburg, L.,** (1987) MAPMAKER: an interactive computer package for constructing primary genetic linkage maps of experimental and natural populations, *Genomics*, 1: 174–181.

388. **Langridge, U., Schwall, M., and Langridge, P.,** (1991) Squashes of plant tissue as substrate for PCR, *Nucl. Acids Res.*, 19: 6954.

389. **Lanham, P. G., Fennell, S., Moss, J. P., and Powell, W.,** (1992) Detection of polymorphic loci in *Arachis* germplasm using random amplified polymorphic DNAs, *Genome*, 35: 885–889.

390. **Lassner, M. W., Peterson, P., and Yoder, J. I.,** (1989) Simultaneous amplification of multiple DNA fragments by polymerase chain reaction in the analysis of transgenic plants and their progeny, *Plant Mol. Biol. Rep.*, 7: 116–128.

391. **Lavi, U., Hillel, J., Vainstein, A., Lahav, E., and Sharon, D.,** (1991) Application of DNA fingerprints for identification and genetic analysis of avocado, *J. Am. Soc. Hortic. Sci.*, 116: 1078–1081.

392. **Learn, G. H., and Schaal, B. A.,** (1987) Population subdivision for ribosomal DNA repeat variants in *Clematis fremontii, Evolution*, 41: 433–438.

393. **Ledbetter, S. A., and Nelson, D. L.,** (1991) Genome amplification using primers directed to interspersed repetitive sequences (IRS-PCR), in *PCR: A Practical Approach*, McPherson, M. J., Quirke, P., and Taylor, G. R., Eds., IRL Press, Oxford, pp. 107–119.

394. **Ledbetter, S. A., Nelson, D. L., Warren, S. T., and Ledbetter, D. H.,** (1990) Rapid isolation of DNA probes within specific chromosome regions by interspersed repetitive sequence polymerase chain reaction, *Genomics*, 6: 475–481.

395. **Lee, S. B., and Taylor, J. W.,** (1990) Isolation of DNA from fungal mycelia and single spores, in *PCR Protocols: A Guide to Methods and Applications*, Innis, M. A., Gelfand, D. H., Sninsky, J. J., and White, T. J., Eds., Academic Press, San Diego,, pp. 282–287.

396. **Lehmann, P. F., Lin, D., and Lasker, B. A.,** (1992) Genotypic identification and characterization of species and strains within the genus *Candida* by using random amplified polymorphic DNA, *J. Clin. Microbiol.*, 30: 3249–3254.

397. **Lessa, E. P.,** (1992) Rapid surveying of DNA sequence variation in natural populations, *Mol. Biol. Evol.*, 9: 323–330.

397a. **Levi, A., Rowland, L. J., and Hartung, J. S.,** (1993) Production of reliable randomly amplified polymorphic DNA (RAPD) markers from DNA of woody plants, *HortScience.*, 28:1188–1190.

398. **Levinson, G., and Gutman, G. A.,** (1987a) Slipped-strand mispairing: a major mechanism for DNA sequence evolution, *Mol. Biol. Evol.*, 4: 203–221.

399. **Levinson, G., and Gutman, G. A.,** (1987b) High frequencies of short frameshifts in poly-CA/TG tandem repeats borne by bacteriophage M13 in *Escherichia coli* K-12, *Nucl. Acids Res.*, 15: 5323–5338.

400. **Levy, M., Romao, J., Marchetti, M. A., and Hamer, J. E.,** (1991) DNA fingerprinting with a dispersed repeated sequence resolves pathotype diversity in the rice blast fungus, *Plant Cell*, 3: 95–102.

401. **Lewis, P. O., and Snow, A. A.,** (1992) Deterministic paternity exclusion using RAPD markers, *Mol. Ecol.*, 1: 155–160.

402. **Lewontin, R. C., and Hubby, J. L.,** (1966) A molecular approach to the study of genetic heterozygosity in natural populations. II. Amount of variation and degree of heterozygosity in natural populations of *Drosophila pseudoobscura, Genetics*, 54: 595–609.

403. **Liang, P., and Pardee, A. B.,** (1992) Differential display of eukaryotic messenger RNA by means of the polymerase chain reaction, *Science*, 257: 967–971.

404. **Lieckfeldt, E., Kuhls, K., Meyer, W., and Börner, T.,** (1992a) Detection of DNA polymorphisms in fungi by genomic and PCR fingerprinting, First European Conference on Fungal Genetics, Nottingham, Aug. 20–23, Abstract P1/08.

405. **Lieckfeldt, E., Meyer, W., Kuhls, K., and Börner, T.,** (1992b) Characterization of filamentous fungi and yeasts by DNA fingerprinting and random amplified polymorphic DNA, *Belg. J. Bot.*, 125: 226–233.

406. **Lieckfeldt, E., Meyer, W., and Börner, T.,** (1993) Rapid identification and differentiation of yeasts by DNA and PCR fingerprinting, *J. Basic Microbiol.*, 33: 413–426.

407. **Liston, A., and Rieseberg, L. H.,** (1990) A method for collecting dried plant specimens for DNA and isozyme analyses, and the results of a field experiment in Xinjiang, China, *Ann. Miss. Bot. Gard.*, 77: 859–863.

408. **Litt, M., and Luty, J. A.,** (1989) A hypervariable microsatellite revealed by *in vitro* amplification of a dinucleotide repeat within the cardiac muscle actin gene, *Am. J. Hum. Genet.*, 44: 397–401.

409. **Liu, B.-H., and Knapp, S. J.,** (1990) GMENDEL: a program for Mendelian segregation and linkage analysis of individual or multiple progeny populations using log-likelihood ratios, *J. Hered.*, 81: 407.

410. **Liu, Z., and Furnier, G. R.,** (1993) Comparison of allozyme, RFLP, and RAPD markers for revealing genetic variation within and between trembling aspen and bigtooth aspen, *Theor. Appl. Genet.,* 87: 97–105.

411. **Lodhi, M. A., Reisch, B. I., and Weeden, N. F.,** (1992) Molecular genetic mapping and genome size of *Vitis,* Plant Genome I, San Diego, Nov. 9–11, p. 37.

412. **Loomis, M. D.,** (1974) Overcoming problems of phenolics and quinones in the isolation of plant enzymes and organelles, *Methods Enzymol.,* 31: 528–544.

413. **Lorenz, M., Weihe, A., and Börner, T.,** (1994) DNA fragments of organellar origin in random amplified polymorphic DNA (RAPD) patterns of sugar beet (*Beta vulgaris* L.), *Theor. Appl. Genet.,* in press.

414. **Loudon, K. W., Burnie, J. P., Coke, A. P., and Matthews, R. C.,** (1993) Application of polymerase chain reaction to fingerprinting *Aspergillus fumigatus* by random amplification of polymorphic DNA, *J. Clin. Microbiol.,* 30: 1117–1121.

415. **Love, J. M., Knight, A. M., McAleer, M. A., and Todd, J. A.,** (1990) Towards construction of a high resolution map of the mouse genome using PCR-analysed microsatellites, *Nucl. Acids Res.,* 18: 4123–4130.

416. **Lowrey, T., and Whitkus, R.,** (1993) Analysis of genetic diversity in Hawaiian populations of *Tetramolopium* (Compositae; Astereae), *Am. J. Bot.,* 80 (Suppl.): 162.

417. **Lönn, M., Tegelström, H., and Prentice, H. C.,** (1992) The synthetic $(TG)_n$ polydinucleotide: a probe for gene flow and paternity studies in wild plant populations?, *Nucl. Acids Res.,* 20: 1153.

418. **Lönn, M., Prentice, H. C., and Tegelström, H.,** (1994) Genetic differentiation in *Hippocrepis emerus* (Leguminosae): allozyme and microsatellite DNA variation in disjunct Scandinavian populations, part of Ph. D. thesis, University of Uppsala, Sweden.

419. **Lu, Q., Wallrath, L. L., Granok, H., and Elgin, S. C. R.,** (1993) $(CT)_n (GA)_n$ repeats and heat shock elements have distinct roles in chromatin structure and transcriptional activation of the *Drosophila hsp26* gene, *Mol. Cell. Biol.,* 13: 2802–2814.

420. **Luo, G., Hepburn, A. G., and Widholm, J. M.,** (1992) Preparation of plant DNA for PCR analysis: a fast, general and reliable procedure, *Plant Mol. Biol. Rep.,* 10: 319–323.

421. **Lynch, M.,** (1988) Estimation of relatedness by DNA fingerprinting, *Mol. Biol. Evol.,* 5: 584–599.

422. **Lynch, M.,** (1990) The similarity index and DNA fingerprinting, *Mol. Biol. Evol.,* 7: 478–484.

423. **Lynch, M.,** (1991) Analysis of population genetic structure by DNA fingerprinting, in *DNA Fingerprinting: Approaches and Applications,* Burke, T., Dolf, G., Jeffreys, A. J., and Wolff, R., Eds., Birkhäuser, Basel, pp. 113–126.

424. **MacPherson, J. M., Eckstein, P. E., Scoles, G. J., and Gajadhar, A. A.,** (1993) Variability of the random amplified polymorphic DNA assay among thermal cyclers, and effects of primer and DNA concentrations, *Mol. Cell. Probe,* 7: 293–299.

425. **Magee, P. T., Bowdin, L., and Staudinger, J.,** (1992) Comparison of molecular typing methods for *Candida albicans, J. Clin. Microbiol.,* 30: 2674–2679.

426. **Manicom, B. Q., Bar-Joseph, A., Rosner, A., Vigodsky-Haas, H., and Kotze, J. M.,** (1987) Potential applications of random DNA probes and restriction fragment length polymorphisms in the taxonomy of *Fusaria, Phytopathology,* 77: 669–672.

427. **Mannen, H., Tsuji, S., Mukai, F., Goto, N., and Ohtagaki, S.,** (1993) Genetic similarity using DNA fingerprinting in cattle to determine relationship coefficient, *J. Hered.,* 84: 166–169.

428. **Manning, K.,** (1991) Isolation of nucleic acids from plants by differential solvent precipitation, *Anal. Biochem.,* 195: 45–50.

429. **Mariat, D., and Vergnaud, G.,** (1992) Detection of polymorphic loci in complex genomes with synthetic tandem repeats, *Genomics,* 12: 454–458.

430. **Marsolais, J. V., Pringle, J. S., and White, B. N.,** (1993) Assessments of random amplified polymorphic DNA (RAPD) as genetic markers for determining the origin of interspecific lilac hybrids, *Taxon,* 42: 531–537.

431. **Martienssen, R. A., and Baulcombe, D. C.,** (1989) An unusual wheat insertion sequence (W1S1) lies upstream of an alpha-amylase gene in hexaploid wheat, and carries a "minisatellite" array, *Mol. Gen. Genet.*, 217: 401–410.

432. **Martin, G. B., Williams, J. G. K., and Tanksley, S. D.,** (1991) Rapid identification of markers linked to a *Pseudomonas* resistance gene in tomato by using random primers and near-isogenic lines, *Proc. Natl. Acad. Sci. U.S.A.*, 88: 2336–2340.

433. **Martin, G. B., de Vicente, M. C., and Tanksley, S. D.,** (1993) High-resolution linkage analysis and physical characterization of the *Pto* bacterial resistance locus in tomato, *Mol. Plant-Microbe Interact*, 6: 26–34.

434. **Martin, P. G., and Dowd, J. M.,** (1991) A comparison of 18S ribosomal RNA and rubisco large subunit sequences for studying angiosperm phylogeny, *J. Mol. Evol.*, 33: 274–282.

435. **Matsuyama, T., Motohashi, R., Akihama, T., and Omura, M.,** (1992) DNA fingerprinting in *Citrus* cultivars, *Jpn. J. Breed.*, 42: 155–159.

436. **Matsuyama, T., Omura, M., and Akihama, T.,** (1993) DNA fingerprinting in *Citrus* cultivars, in *Techniques on Gene Diagnosis and Breeding in Fruit Trees,* Hayashi, T., Omura, M., and Scott, N. S., Eds., Fruit Tree Research Station, Ibaraki, pp. 26–30.

437. **May, B.,** (1992) Starch gel electrophoresis of allozymes, in *Molecular Genetic Analysis of Populations: A Practical Approach,* Hoelzel, A. R., Ed., IRL Press, Oxford, pp. 1–27.

438. **Mayes, C., Saunders, G. W., Tan, I. H., and Druehl, L. D.,** (1992) DNA extraction methods for kelp (Laminariales) tissue, *J. Phycol.*, 28: 712–716.

439. **McClelland, M., Chada, K., Welsh, J., and Ralph, D.,** (1993) Arbitrarily primed PCR fingerprinting of RNA applied to mapping differentially expressed genes, in *DNA Fingerprinting: State of the Science,* Pena, S. D. J., Chakraborty, R., Epplen, J. T., and Jeffreys, A. J., Eds., Birkhäuser, Basel, pp. 103–115.

440. **McCoy, T. J., and Echt, C. S.,** (1993) Potential of trispecies bridge crosses and random amplified DNA markers for introgression of *Medicago daghestanica* and *M. pironae* germplasm into alfalfa (*M. sativa*), *Genome*, 36: 594–601.

441. **McCutcheon, T. L., and Carroll, G. C.,** (1993) Genotypic diversity in populations of a fungal endophyte from douglas fir, *Mycologia*, 85: 180–186.

442. **McDonald, B. A., and Martinez, J. P.,** (1990) Restriction fragment length polymorphisms in *Septoria tritici* occur at a high frequency, *Curr. Genet.*, 17: 133–138.

443. **McDonald, B. A., and Martinez, J. P.,** (1990) DNA restriction fragment length polymorphisms among *Mycosphaerella graminicola* (anamorph *Septoria tritici*) isolates collected from a single wheat field, *Phytopathology*, 80: 1368–1373.

444. **McDonald, B. A., and Martinez, J. P.,** (1991) DNA fingerprinting of the plant pathogenic fungus *Mycosphaerella graminicola* (anamorph *Septoria tritici*), *Exp. Mycol.*, 15: 146–158.

445. **McDonald, B. A., and McDermott, J. M.,** (1993) Population genetics of plant pathogenic fungi, *BioScience*, 43: 311–319.

446. **McGinnis, M. R.,** (1980) *Laboratory Handbook of Medical Mycology,* Academic Press, New York.

447. **McPherson, M. J., Quirke, P., and Taylor, G. R., Eds.,** (1991) *PCR: A Practical Approach,* IRL Press, Oxford.

448. **Megnegneau, B., Debets, F., and Hoekstra, R. F.,** (1993) Genetic variability and relatedness in the complex group of black *Aspergilli* based on random amplification of polymorphic DNA, *Curr. Genet.*, 23: 323–329.

449. **Meinkoth, J., and Wahl, G.,** (1984) Hybridization of nucleic acids immobilized on solid supports, *Anal. Biochem.*, 138: 267–284.

450. **Meng, A., Carter, R. E., and Parkin, D. T.,** (1990) The variability of DNA fingerprints in three species of swan, *Heredity*, 64: 73–80.

451. **Mettler, I. J.,** (1987) A simple and rapid method for minipreparation of DNA from tissue cultured plant cells, *Plant Mol. Biol. Rep.*, 5: 346–349.

452. **Meyer, M., Shaw, A. J., and Doyle, J. J.,** (1993) Analysis of genetic variation within and between populations of *Scopelophila cataractae* using RAPDs and isozyme patterns, *Am. J. Bot.*, 80 (Suppl.): 165.

452a. **Meyer, W.,** (1994) Manuscript in preparation.
453. **Meyer, W., Koch, A., Niemann, C., Beyermann, B., Epplen, J. T., and Börner, T.,** (1991a) Differentiation of species and strains among filamentous fungi by DNA finger-printing, *Curr. Genet.,* 19: 239–242.
454. **Meyer, W., Lieckfeldt, E., and Börner, T.,** (1991b) DNA-fingerprinting: Charakterisierung, Differenzierung und Identifizierung von Pilzen, *Biotec. Fachzeitschrift Biotechnologie,* 9: 37–41.
455. **Meyer, W., Lieckfeldt, E., Kayser, T., Nürnberg, P., Epplen, J. T., and Börner, T.,** (1992a) Fingerprinting fungal genomes with phage M13 DNA and oligonucleotide probes specific for simple repetitive DNA sequences, in *DNA Polymorphisms in Eukaryotic Genomes,* Kahl, G., Appelhans, H., Kömpf, J., and Driesel, A. J., Eds., Biotech-Forum 10, Adv Mol Genet 5, Hüthig Verlag, Heidelberg, pp. 241–253.
456. **Meyer, W., Lieckfeldt, E., Morawetz, R., Gruber, F., Messner, R., Börner, T., and Kubicek, C. P.,** (1992b) Re-classification of some species of the *Trichoderma* aggregate by RFLP-fingerprinting and arbitrarily primed PCR, First European Conference on Fungal Genetics, Nottingham, Aug. 20–23, Abstract P1/04.
457. **Meyer, W., Lieckfeldt, E., Wöstemeyer, J., and Börner, T.,** (1992c) DNA fingerprinting for differentiating aggressivity groups of the rape seed pathogen *Leptosphaeria maculans, Mycol. Res.,* 96: 651–657.
458. **Meyer, W., Morawetz, R., Börner, T., and Kubicek, C. P.,** (1992d) The use of DNA-fingerprinting analysis in the classification of some species of the *Trichoderma* aggregate, *Curr. Genet.,* 21: 27–30.
459. **Meyer, W., Lieckfeldt, E., Kuhls, K., Freedman, E. Z., Börner, T., and Mitchell, T. G.,** (1993a) DNA- and PCR-fingerprinting (RAPD) in fungi in *DNA Fingerprinting: State of the Science,* Pena, S. D. J., Chakraborty, R., Epplen, J. T., and Jeffreys, A. J., Eds., Birkhäuser, Basel, pp. 311–320.
460. **Meyer, W., Mitchell, T. G., Freedman, E. Z., and Vilgalys, R.,** (1993b) Hybridization probes for conventional DNA fingerprinting can be used as single primers in the PCR to distinguish strains of *Cryptococcus neoformans, J. Clin. Microbiol.,* 31: 2274–2280.
461. **Michelmore, R. W., Paran, I., and Kesseli, R. V.,** (1991) Identification of markers linked to disease-resistance genes by bulked segregant analysis: a rapid method to detect markers in specific genomic regions by using segregating populations, *Proc. Natl. Acad. Sci. U.S.A.,* 88: 9828–9832.
462. **Miklas, P. N., Stavely, J. R., and Kelly, J. D.,** (1993) Identification and possible use of a molecular marker for rust resistance in common bean, *Theor. Appl. Genet.,* 85: 745–749.
463. **Miklos, G. L. G., Matthaei, K. I., and Reed, K. C.,** (1989) Occurrence of the (GATA)$_n$ sequences in vertebrate and invertebrate genomes, *Chromosoma,* 98: 194–200.
464. **Milgroom, M. G., Lipari, S. E., and Powell, W. A.,** (1992a) DNA fingerprinting and analysis of population structure in the chestnut blight fungus, *Cryphonectria parasitica, Genetics,* 131: 297–306.
465. **Milgroom, M. G., Lipari, S. E., and Wang, K.,** (1992b) Comparison of genetic diversity in the chestnut blight fungus, *Cryphonectria (Endothia) parasitica,* from China and the U.S., *Mycol. Res.,* 96: 1114–1120.
466. **Milgroom, M. G., Lipari, S. E., Ennos, R. A., and Liu, Y.-C.,** (1993) Estimation of the outcrossing rate in the chestnut blight fungus, *Cryphonectria parasitica, Heredity,* 70: 385–392.
467. **Miller, P. M.,** (1955) V-8 juice agar as a general-purpose medium for fungi and bacteria, *Phytopathology,* 45: 461–462.
468. **Milligan, B. G.,** (1992) Plant DNA isolation, in *Molecular Genetic Analysis of Populations: A Practical Approach,* Hoelzel, A. R., Ed., IRL Press, Oxford, pp. 59–88.
469. **Mills, P. R., Sreenivasaprasad, S., and Brown, A. E.,** (1992) Detection and differentiation of *Colletotrichum gloeosporioides* isolates using PCR, *FEMS Microbiol. Lett.,* 98: 137–144.

470. **Mitchell, T. G., Freedman, E. Z., Meyer, W., White, T. J., and Taylor, J. W.,** (1993) PCR identification of *Cryptococcus neoformans*, in *Diagnostic Molecular Microbiology*, Persing, D. H., Tenover, F. C., Smith, T. F., and White, T. J., Eds., American Society for Microbiology, Washington, DC, pp. 431–436.

471. **Miyada, C. G., and Wallace, R. B.,** (1987) Oligonucleotide hybridization techniques, *Methods Enzymol.*, 154: 94–107.

472. **Monastyrskii, O. A., Ruban, D. N., Tokarskaya, O. N., and Ryskov, A. P.,** (1990) DNA fingerprinting of some *Fusarium* isolates differentiated toxicogenically, *Genetika*, 26: 374–377.

473. **Monckton, D. G., and Jeffreys, A. J.,** (1991) Minisatellite "isoallele" discrimination in pseudohomozygotes by single molecule PCR and variant repeat mapping, *Genomics*, 11: 465–467.

474. **Moore, S. S., Sargeant, L. L., King, T. J., Mattick, J. S., Georges, M., and Hetzel, D. J. S.,** (1991) The conservation of dinucleotide microsatellites among mammalian genomes allows the use of heterologous PCR primer pairs in closely related species, *Genomics*, 10: 654–660.

475. **Moran, G. F., Bell, J. C., Byrne, M., Devey, M. E., and Smith, D.,** (1992) Molecular marker maps in *Pinus radiata* and *Eucalyptus nitens* for application in tree breeding programs, Plant Genome I, San Diego, Nov. 9–11, p. 39.

476. **Morgante, M., and Olivieri, A. M.,** (1993) PCR-amplified microsatellites as markers in plant genetics, *Plant J.*, 3: 175–182.

477. **Mori, M., Hosaka, K., Umemura, Y., and Kaneda, C.,** (1993) Rapid identification of Japanese potato cultivars by RAPDs, *Jpn. J. Genet.*, 68: 167–174.

477a. **Morjane, H., Geistlinger, J., Harrabi, M., Weising, K., and Kahl, G.,** (1994) Oligonucleotide fingerprinting detects genetic diversity among *Ascorhyta rabici* isolates from a single chickpea field in Tunisia, *Curr. Genet.*, in press.

478. **Mosseler, A. J., Egger, K. N., and Hughes, G. A.,** (1992) Low levels of genetic diversity in red pine confirmed by random amplified polymorphic DNA markers, *Can. J. For. Res.*, 22: 1332–1337.

479. **Mosseler, A. J., Egger, K. N., and Innes, D. J.,** (1993) Life history and the loss of genetic diversity in red pine, *Am. J. Bot.*, 80 (Suppl.): 81–82.

480. **Mrak, E. M., Phaff, H. J., and Douglas, H. C.,** (1942) A sporulation stock medium for yeast and other fungi, *Science*, 96: 432.

481. **Mulcahy, D. L., Weeden, N. F., Kesseli, R., and Carroll, S. B.,** (1992) DNA probes for the Y-chromosome of *Silene latifolia*, a dioecious angiosperm, *Sex Plant Reprod.*, 5: 86–88.

482. **Mulcahy, D. L., Cresti, M., Sansavini, S., Douglas, G. C., Linskens, H. F., Mulcahy, G. B., Vignani, R., and Pancaldi, M.,** (1993) The use of random amplified polymorphic DNAs to fingerprint apple genotypes, *Sci. Hortic.*, 54: 89–96.

483. **Mullaart, E., De Vos, G. J., Te Meerman, G. J., Uiterlinden, A. G., and Vijg, J.,** (1993) Parallel genome analysis by two-dimensional DNA typing, *Nature (London)*, 365: 469–471.

484. **Munthali, M., Ford-Lloyd, B. V., and Newbury, H. J.,** (1992) The random amplification of polymorphic DNA for fingerprinting plants, *PCR Methods Appl.*, 1: 274–276.

485. **Murphy, R. W., Sites, J. W., Jr., Buth, D. G., and Haufler, C. H.,** (1990) Proteins I: isozyme electrophoresis, in *Molecular Systematics*, Hillis, D. M., and Moritz, C., Eds., Sinauer Associates, Sunderland, MA, pp. 45–126.

486. **Murray, M. G., and Thompson, W. F.,** (1980) Rapid isolation of high molecular weight plant DNA, *Nucl. Acids Res.*, 8: 4321–4325.

487. **Myers, R. M., Maniatis, T., and Lerman, L. S.,** (1986) Detection and localization of single base changes by denaturing gradient gel electrophoresis, *Methods Enzymol.*, 155: 501–527.

488. **Nakamura, Y., Julier, C., Wolff, R., Holm, T., O'Connell, P., Leppert, M., and White, R.,** (1987) Characterization of a human 'midisatellite' sequence, *Nucl. Acids Res.*, 15: 2537–2547.

489. **Nakamura, Y., Leppert, M., O'Connell, P., Wolff, R., Holm, T., Culver, M., Martin, C., Fujimoto, E., Hoff, M., Kumlin, E., and White, R.,** (1987) Variable number of tandem repeat (VNTR) markers for human gene mapping, *Science,* 235: 516–522.

490. **Naylor, L. H., and Clark, E. M.,** (1990) $d(TG)_n$ $d(CA)_n$ sequences upstream of the rat prolactin gene form Z-DNA and inhibit gene transcription, *Nucl. Acids Res.,* 18: 1595–1601.

491. **Neale, D. B.,** (1992) Molecular marker and trait mapping in conifers, Plant Genome I, San Diego, Nov. 9–11, p. 12.

492. **Neale, D. B., and Adams, W. T.,** (1985) The mating system in natural and shelterwood stands of Douglas-fir, *Theor. Appl. Genet.,* 71: 201–207.

493. **Neale, D. B., and Sederoff, R.,** (1991) Genome mapping in pines takes shape, *Probe,* 1 (3/4): 1–3.

494. **Nei, M.,** (1973) Analysis of gene diversity in subdivided populations, *Proc. Natl. Acad. Sci. U.S.A.,* 70: 3321–3323.

495. **Nei, M., and Li, W.-H.,** (1979) Mathematical model for studying genetic variation in terms of restriction endonucleases, *Proc. Natl. Acad. Sci. U.S.A.,* 76: 5269–5273.

496. **Nelke, M., Nowak, J., Wright, J. M., and McLean, N. L.,** (1993) DNA fingerprinting of red clover (*Trifolium pratense* L.) with Jeffrey's probes: detection of soma clonal variation and other applications, *Plant Cell Rep.,* 13: 72–78.

497. **Nelson, C. D., Nance, W. L., and Doudrick, R. L.,** (1993) A partial genetic linkage map of slash pine (*Pinus elliottii* Engelm. var. *elliottii*) based on random amplified polymorphic DNAs, *Theor. Appl. Genet.,* 87: 145–151.

498. **Nelson, M., and McClelland, M.,** (1989) Effect of site-specific methylation on DNA modification methyltransferases and restriction endonucleases, *Nucl. Acids Res.,* 17: r389–r415.

499. **Ness, F., Lavallee, F., Dubourdieu, D., Aigle, M., and Dulau, L.,** (1993) Identification of yeast strains using the polymerase chain reaction, *J. Sci. Food Agric.,* 62: 89–94.

500. **Neuhaus, D., Kühl, H., Kohl, J. G., Dörfel, P., and Börner, T.,** (1993) Investigation on the genetic diversity of *Phragmites* stands using genomic fingerprinting, *Aquat. Bot.,* 45: 357–364.

501. **Neuhaus-Url, G., and Neuhaus, G.,** (1993) The use of nonradioactive digoxigenin chemiluminescent technology for plant genomic Southern blot hybridization: a comparison with radioactivity, *Transgenic Res.,* 2: 115–120.

502. **Newbury, H. J., and Ford-Lloyd, B. V.,** (1993) The use of RAPD for assessing variation in plants, *Plant Growth Regul.,* 12: 43–51.

502a. **Nicholson, P., Jenkinson, P., Rezanoor, H. N., and Parry, D. W.,** (1993) Restriction fragment length polymorphism analysis of variation in *Fusarium* species causing ear blight of cereals, *Plant Pathology,* 42:905–914.

503. **Nienhuis, J., Helentjaris, T., Slocum, M., Ruggero, B., and Schaefer, A.,** (1987) Restriction fragment length polymorphism analysis of loci associated with insect resistance in tomato, *Crop Sci.,* 27: 797–803.

504. **Niesters, H. G. M., Goessens, W. H. F., Meis, J. F. M. G., and Quint, W. G. V.,** (1993) Rapid, polymerase chain reaction-based identification assays for *Candida* species, *J. Clin. Microbiol.,* 31: 904–910.

505. **Nodari, R. O., Tsai, S. M., Gilbertson, R. L., and Gepts, P.,** (1993) Towards an integrated linkage map of common bean 2. Development of an RFLP-based linkage map, *Theor. Appl. Genet.,* 85: 513–520.

506. **Nürnberg, P., and Epplen, J. T.,** (1989) "Hidden partials" — a cautionary note, *Fingerprint News,* 1(4): 11–12.

507. **Nürnberg, P., Barth, I., Fuhrmann, E., Lenzner, C., Losanova, T., Peters, C., Pöche, H., and Thiel, G.,** (1991) Monitoring genomic alterations with a panel of oligonucleotide probes specific for various simple repeat motifs, *Electrophoresis,* 12: 186–192.

508. **Nybom, H.,** (1988) Apomixis versus sexuality in blackberries (*Rubus* subgen. *Rubus,* Rosaceae), *Plant Syst. Evol.,* 160: 207–218.

509. **Nybom, H.,** (1990a) Genetic variation in ornamental apple trees and their seedlings (*Malus*, Rosaceae) revealed by DNA 'fingerprinting', *Hereditas*, 113: 17–28.

510. **Nybom, H.,** (1990b) DNA fingerprints in sports of 'Red Delicious' apples, *HortScience*, 25:1641–1642.

511. **Nybom, H.,** (1991) Applications of DNA fingerprinting in plant breeding, in *DNA Fingerprinting: Approaches and Applications,* Burke, T., Dolf, G., Jeffreys, A. J., and Wolff, R., Eds., Birkhäuser, Basel, pp. 294–311.

512. **Nybom, H.,** (1993) Applications of DNA fingerprinting in plant population studies in *DNA Fingerprinting: State of the Science*, Pena, S. D. J., Chakraborty, R., Epplen, J. T., and Jeffreys, A. J., Eds., Birkhäuser, Basel, pp. 293–309.

512a. **Nybom, H.,** (1994) Unpublished results.

513. **Nybom, H., and Hall, H. K.,** (1991) Minisatellite DNA "fingerprints" can distinguish *Rubus* cultivars and estimate their degree of relatedness, *Euphytica*, 53: 107–114.

513a. **Nybom, H., and Meyer, W.,** (1994) Manuscript in preparation.

514. **Nybom, H., and Rogstad, S. H.,** (1990) DNA "fingerprints" detect genetic variation in *Acer negundo, Plant Syst. Evol.*, 173: 49–56.

515. **Nybom, H., and Schaal, B. A.,** (1990a) DNA "fingerprints" applied to paternity analysis in apples *(Malus × domestica), Theor. Appl. Genet.*, 79: 763–768.

516. **Nybom, H., and Schaal, B. A.,** (1990b) DNA "fingerprints" reveal genotypic distributions in natural populations of blackberries and raspberries (*Rubus*, Rosaceae), *Am. J. Bot.*, 77: 883–888.

517. **Nybom, H., Schaal, B. A., and Rogstad, S. H.,** (1989) DNA "fingerprints" can distinguish cultivars of blackberries and raspberries, 5th International Symposium on *Rubus & Ribes, Acta Hortic.*, 262: 305–310.

518. **Nybom, H., Rogstad, S. H., and Schaal, B. A.,** (1990) Genetic variation detected by use of the M13 "DNA fingerprint" probe in *Malus, Prunus,* and *Rubus* (Rosaceae), *Theor. Appl. Genet.*, 79: 153–156.

519. **Nybom, H., Carlson, U., and Hoepfner, A.-S.,** (1992a) DNA laboratoriet på Balsgård, *SLU, Balsgård, Biennial Report 1990–91:* 146–154.

520. **Nybom, H., Gardiner, S., and Simon, C. J.,** (1992b) RFLPs obtained from an rDNA probe and detected with enhanced chemiluminescence in apples, *HortScience*, 27: 355–356.

521. **Nybom, H., Ramser, J., Kaemmer, D., Kahl, G., and Weising, K.,** (1992c) Oligonucleotide DNA fingerprinting detects a multiallelic locus in box elder, *Mol. Ecol.*, 1: 65–67.

522. **Nyman, M.,** (1993) Protoplast technology in strawberries, *Acta Universitatis Upsaliensis, Comprehensive Summaries of Uppsala Dissertations from the Faculty of Science,* No. 451, University of Uppsala, Sweden.

523. **Oard, J. H., and Dronavalli, S.,** (1992) Rapid isolation of rice and maize DNA for analysis of random-primer PCR, *Plant Mol. Biol. Rep.*, 10: 236–241.

524. **O'Dell, M., Wolfe, M. S., Flavell, R. B., Simpson, R. B., and Summers, R. W.,** (1989) Molecular variation in populations of *Erysiphe graminis* on barley, wheat, oats and rye, *Plant Pathol.*, 38: 340–351.

525. **Olson, M., Hood, L., Cantor, C., and Botstein, D.,** (1989) A common language for physical mapping of the human genome, *Science*, 245: 1434–1435.

526. **O'Malley, D. M., McKeand, S. E., Wilcox, P. W., and Sederoff, R. R.,** (1992) QTL mapping in loblolly pine using RAPDs, Plant Genome I, San Diego, Nov. 9–11, p. 12.

527. **Opstelten, R. J. G., Clement, J. M. F., and Wanka, F.,** (1989) Direct repeats at nuclear matrix-associated DNA regions and their putative control function in the replicating eukaryotic genome, *Chromosoma*, 98: 422–427.

528. **Orgel, L. E., and Crick, F. H. C.,** (1980) Selfish DNA: the ultimate parasite, *Nature (London)*, 84: 604–607.

529. **Orita, M., Iwahana, H., Kanazawa, H., Hayashi, K., and Sekiya, T.,** (1989) Detection of polymorphisms of human DNA by gel electrophoresis as single-strand conformation polymorphisms, *Proc. Natl. Acad. Sci. U.S.A.*, 86: 2766–2770.

530. **Orita, M., Sekiya, T., and Hayashi, K.,** (1990) DNA sequence polymorphisms in Alu repeats, *Genomics*, 8: 271–278.

531. **Ostrander, E. A., Jong, P. M., Rine, J., and Duyk, G.,** (1992) Construction of small-insert genomic DNA libraries highly enriched for microsatellite repeat sequences, *Proc. Natl. Acad. Sci. U.S.A.*, 89: 3419–3423.

532. **Ouellet, T., and Seifert, K. A.,** (1993) Genetic characterization of *Fusarium graminearum* strains using RAPD and PCR amplification, *Phytopathology*, 83: 1003–1007.

533. **Ozias-Akins, P., Lubbers, E. L., Hanna, W. W., and McNay, J. W.,** (1993) Transmission of the apomictic mode of reproduction in *Pennisetum*: co-inheritance of the trait and molecular markers, *Theor. Appl. Genet.*, 85: 632–638.

534. **Palmer, J. D.,** (1985) Comparative organization of chloroplast genomes, *Annu. Rev. Genet.*, 19: 325–354.

535. **Palmer, J. D.,** (1986) Isolation and structural analysis of chloroplast DNA, *Methods Enzymol.*, 118: 167–186.

536. **Palmer, J. D., and Zamir, D.,** (1982) Chloroplast DNA evolution and phylogenetic relationships in *Lycopersicon, Proc. Natl. Acad. Sci. U.S.A.*, 5006–5010.

537. **Pang, J.-P., Chen, L.-C., Chen, L.-F. O., and Chen, S.-C. G.,** (1992) DNA polymorphisms generated by arbitrarily primed PCR in rice, *Biosci. Biotech. Biochem.*, 56: 1357–1358.

538. **Papavizas, G. C., and Davey, C.,** (1959) Evaluation of various media and antimicrobial agents for isolation of soil fungi, *Soil Sci.*, 88: 112–117.

539. **Paran, I., and Michelmore, R. W.,** (1993) Development of reliable PCR-based markers linked to downy mildew resistance genes in lettuce, *Theor. Appl. Genet.*, 85: 985–993.

540. **Paran, I., Kesseli, R., and Michelmore, R. W.,** (1991) Identification of restriction fragment length polymorphism and random amplified polymorphic DNA markers linked to downy mildew resistance genes in lettuce, using near-isogenic lines, *Genome*, 34: 1021–1027.

541. **Parent, J.-G., and Pagé, D.,** (1992) Identification of raspberry cultivars by nonradioactive DNA fingerprinting, *HortScience*, 27: 1108–1110.

542. **Parent, J.-G., Fortin, M. G., and Pagé, D.,** (1993) Identification de cultivars de framboisier par l'analyse d'ADN polymorphe amplifié au hasard (RAPD), *Can. J. Hortic. Sci.*, 73: 1115–1122.

543. **Paterson, A. H., Brubaker, C. L., and Wendel, J. F.,** (1993) A rapid method for extraction of cotton (*Gossypium* spec.) genomic DNA suitable for RFLP or PCR analysis, *Plant Mol. Biol. Rep.*, 11: 122–127.

544. **Patwary, M. U., MacKay, R. M., and van der Meer, J. P.,** (1993) Revealing genetic markers in *Gelidium vagum* (Rhodophyta) through the random amplified polymorphic DNA (RAPD) technique, *J. Phycol.*, 29: 216–222.

545. **Pé, M. E., Gianfranceschi, L., Taramino, G., Tarchini, R., Angelini, P., Dani, M., and Binelli, G.,** (1993) Mapping quantitative trait loci (QTLs) for resistance to *Gibberella zeae* infection in maize, *Mol. Gen. Genet.*, 241: 11–16.

546. **Pearce, M. J., and Watson, N. D.,** (1993) Rapid analysis of PCR components and products by acidic non-gel capillary electrophoresis, in *DNA Fingerprinting: State of the Science,* Pena, S. D. J., Chakraborty, R., Epplen, J. T., and Jeffreys, A. J., Eds., Birkhäuser, Basel, pp. 117–124.

547. **Pena, S. D. J., Macedo, A. M., Braga, V. M. M., Rumjanek, F. D., and Simpson, A. J. G.,** (1990) *F10*, the gene for the glycine-rich major eggshell protein of *Schistosoma mansoni* recognizes a family of hypervariable minisatellites in the human genome, *Nucl. Acids Res.*, 18: 7466.

548. **Pena, S. D. J., Macedo, A. M., Gontijo, N. F., Madeiros, A. M., and Ribeiro, J. C. C.,** (1991) DNA bioprints: simple nonisotopic DNA fingerprints with biotinylated probes, *Electrophoresis*, 12: 146–152.

549. **Penner, G. A., Bush, A., Wise, R., Kim, W., Domier, L., Kasha, K., Laroche, A., Scoles, G., Molnar, S. J., and Fedak, G.,** (1993) Reproducibility of random amplified polymorphic DNA (RAPD) analysis among laboratories, *PCR Methods Appl.*, 2: 341–345.

550. **Penner, G. A., Chong, J., Levesque-Lemay, M., Molnar, S. J., and Fedak, G.,** (1993) Identification of a RAPD marker linked to oat stem rust gene *Pg3, Theor. Appl. Genet.*, 85: 702–705.

551. **Penner, G. A., Chong, J., Wright, C. P., Molnar, S. J., and Fedak, G.,** (1993) Identification of an RAPD marker for the crown rust resistance gene *Pc68* in oats, *Genome*, 36: 818–820.

552. **Philbrick, C. T.,** (1993) Underwater cross-pollination in *Callitriche hermaphroditica* (Callitrichaceae): evidence from random amplified polymorphic DNA markers, *Am. J. Bot.*, 80: 391–394.

552a. **Pich, U., and Schubert, I.,** (1993) Midiprep method for isolation of DNA from plants with a high content of polyphenolics, *Nucl. Acids Res.*, 21: 3328

553. **Picknett, T. M., Sanders, G., Ford, P., and Holt, G.,** (1987) Development of a gene transfer system for *Penicillium chrysogenum*, *Curr. Genet.*, 12: 449–455.

554. **Polacheck, I., Lebens, G., and Hicks, J. B.,** (1992) Development of DNA probes for early diagnosis and epidemiological study of cryptococcosis in AIDS patients, *J. Clin. Microbiol.*, 30: 925–930.

555. **Poulsen, G. B., Kahl, G., and Weising, K.,** (1993) Abundance and polymorphism of simple repetitive DNA sequences in *Brassica napus* L, *Theor. Appl. Genet.*, 85: 994–1000.

556. **Poulsen, G. B., Kahl, G., and Weising, K.,** (1993) Oligonucleotide fingerprinting of resynthesized *Brassica napus*, *Euphytica*, 70: 53–59.

557. **Poulsen, G. B., Kahl, G., and Weising, K.,** (1994) Differential abundance of simple repetitive sequences in species of *Brassica* and related genera, *Plant Syst. Evol.*, 190: 21–30.

558. **Pyle, M. M., and Adams, R. P.,** (1989) In situ preservation of DNA in plant specimens, *Taxon*, 38: 576–581.

559. **Quiros, C. F., Hu, J., This, P., Chevre, A. M., and Delseny, M.,** (1991) Development and chromosomal localization of genome-specific markers by polymerase chain reaction in *Brassica*, *Theor. Appl. Genet.*, 82: 627–632.

560. **Quiros, C. F., Ceada, A., Georgescu, A., and Hu, J.,** (1993) Use of RAPD markers in potato genetics: segregations in diploid and tetraploid families, *Am. Potato J.*, 70: 35–42.

561. **Raeder, U., and Broda, P.,** (1985) Rapid preparation of DNA from filamentous fungi, *Lett. Appl. Microbiol.*, 1: 17–20.

562. **Rafalski, J. A., and Tingey, S. V.,** (1993) Genetic diagnostics in plant breeding: RAPDs, microsatellites and machines, *TIG*, 9: 275–279.

563. **Rafalski, J. A., Tingey, S. V., and Williams, J. G. K.,** (1991) RAPD markers – a new technology for genetic mapping and plant breeding, *AgBiotech News Inf.*, 3: 645–648.

564. **Ragot, M., and Hoisington, D. A.,** (1993) Molecular markers for plant breeding – comparisons of RFLP and RAPD genotyping costs, *Theor. Appl. Genet.*, 86: 975–984.

565. **Rajapakse, S., Hubbard, M., Kelly, J. W., Abbott, A. G., and Ballard, R. E.,** (1992) Identification of rose cultivars by restriction fragment length polymorphism, *Sci. Hortic.*, 52: 237–245.

566. **Ramakrishna, W., Lagu, M. D., Gupta, V. S., and Ranjekar, P. K.,** (1994) DNA fingerprinting in rice using oligonucleotide probes specific for simple repetitive DNA sequences, *Theor. Appl. Genet.*, 88: 402–406.

567. **Raper, J. R., and Miller, R. E.,** (1972) Genetic analysis of the life cycle of *Agaricus bisporus*, *Mycologia*, 64: 1088–1117.

568. **Rassmann, K., Schlötterer, C., and Tautz, D.,** (1991) Isolation of simple-sequence loci for use in polymerase chain reaction-based DNA fingerprinting, *Electrophoresis*, 12: 113–118.

569. **Reeve, H. K.,** (1992) Estimating average within-group relatedness from DNA fingerprints, *Mol. Ecol.*, 1: 223–232.

570. **Reeve, H. K., Westneat, D. F., Noon, W. A., Sherman, P. W., and Aquadro, C. F.,** (1990) DNA "fingerprinting" reveals high levels of inbreeding in colonies of the eusocial naked mole-rat, *Proc. Natl. Acad. Sci. U.S.A.*, 87: 2496–2500.

571. **Reeves, J. C., and Ball, S. F. L.,** (1991) Preliminary results on the identification of *Pyrenophora* species using DNA polymorphisms amplified from arbitrary primers, *Plant Varieties Seeds*, 4: 185–189.

572. **Reiter, R. S., Williams, J. G. K., Feldmann, K. A., Rafalski, J. A., Tingey, S. V., and Scolnik, P. A.,** (1992) Global and local genome mapping in *Arabidopsis thaliana* by using recombinant inbred lines and random amplified polymorphic DNAs, *Proc. Natl. Acad. Sci. U.S.A.*, 89: 1477–1481.

573. **Reiter, R. S., Young, R. M., and Scolnik, P. A.,** (1992) Genetic linkage of the *Arabidopsis* genome: methods for mapping with recombinant inbreds and random amplified polymorphic DNAs (RAPDs), in *Methods in Arabidopsis Research,* Koncz, C., Chua, N.-H., and Schell, J., Eds., World Scientific, Singapore.

574. **Richards, E. J., Goodman, H. M., and Ausubel, F. M.,** (1991) The centromere region of *Arabidopsis thaliana* chromosome 1 contains telomere-similar sequences, *Nucl. Acids Res.*, 19: 3351–3357.

575. **Richards, R. I., and Sutherland, G. R.,** (1992) Dynamic mutations: a new class of mutations causing human disease, *Cell.*, 70: 709–712.

576. **Richards, R. I., Holman, K., Yu, S., and Sutherland, G. R.,** (1993) Fragile X syndrome unstable element, p(CCG)$_n$, and other simple tandem repeat sequences are binding sites for specific nuclear proteins, *Hum. Mol. Genet.*, 2: 1429–1435.

577. **Riedy, M. F., Hamilton, W. J., III, and Aquadro, C. F.,** (1992) Excess of non-parental bands in offspring from known primate pedigrees assayed using RAPD PCR, *Nucl. Acids Res.*, 20: 918.

578. **Rifai, M. A.,** (1969) A revision of the genus *Trichoderma*, *Mycol. Pap.*, 116: 1–56.

579. **Rigby, P. W. J., Dieckmann, M., Rhodes, C., and Berg, P.,** (1977) Labeling deoxyribonucleic acid to high specific activity *in vitro* by nick translation with DNA polymerase I, *J. Mol. Biol.*, 113: 237–251.

580. **Riley, J., Jenner, D., Smith, J. C., and Markham, A. F.,** (1989) Rapid determination of DNA concentration in multiple samples, *Nucl. Acids Res.*, 17: 8383.

581. **Ritland, K.,** (1990) A series of FORTRAN computer programs for estimating plant mating system, *J. Hered.*, 81: 235–237.

582. **Ritter, E., Gebhardt, C., and Salamini, F.,** (1990) Estimation of recombination frequencies and constructions of RFLP linkage maps in plants from crosses between heterozygous parents, *Genetics*, 125: 645–654.

583. **Rodriguez, R. J., and Yoder, O. C.,** (1991) A family of conserved repetitive DNA elements from the fungal plant pathogen *Glomerella cingulata* (*Colletotrichum lindemuthianum*), *Exp. Mycol.*, 15: 232–242.

584. **Rogers, S. O., and Bendich, A. J.,** (1985) Extraction of DNA from milligram amounts of fresh, herbarium and mummified plant tissues, *Plant Mol. Biol.*, 5: 69–76.

585. **Rogers, S. O., and Bendich, A. J.,** (1988) Extraction of DNA from plant tissues, *Plant Molecular Biology Manual A6*, Kluwer Academic Publishers, Dordrecht, pp. 1–10.

586. **Rogowsky, P., Shepherd, K. W., and Langridge, P.,** (1992) Polymerase chain reaction based mapping of rye involving repeated DNA sequences, *Genome*, 35: 621–626.

587. **Rogstad, S. H.,** (1992) Saturated NaCl-CTAB solution as a means of field preservation of leaves for DNA analyses, *Taxon*, 41: 701–708.

588. **Rogstad, S. H., Patton, J. C., II, and Schaal, B. A.,** (1988) M13 repeat probe detects DNA minisatellite-like sequences in gymnosperms and angiosperms, *Proc. Natl. Acad. Sci. U.S.A.*, 85: 9176–9178.

589. **Rogstad, S. H., Patton, J. C., II, and Schaal, B. A.,** (1988) A human minisatellite probe reveals RFLPs among individuals of two angiosperms, *Nucl. Acids Res.*, 18: 11378.

590. **Rogstad, S. H., Herwaldt, B. L., Schlesinger, P. H., and Krogstad, D. J.,** (1989) The M13 repeat probe detects RFLPs between two strains of the protozoan malaria parasite *Plasmodium falciparum*, *Nucl. Acids Res.*, 17: 3610.

591. **Rogstad, S. H., Nybom, H., and Schaal, B. A.,** (1991) The tetrapod DNA fingerprinting M13 repeat probe reveals genetic diversity and clonal growth in quaking aspen (*Populus tremuloides*, Salicaceae), *Plant Syst. Evol.*, 175: 115–123.

592. **Rogstad, S. H., Wolff, K., and Schaal, B. A.,** (1991) Geographical variation in *Asimina triloba* Dunal (Annonaceae) revealed by the M13 "DNA fingerprinting" probe, *Am. J. Bot.,* 78: 1391–1396.

593. **Romao, J., and Hamer, J. E.,** (1992) Genetic organization of a repeated DNA sequence family in the rice blast fungus, *Proc. Natl. Acad. Sci. U.S.A.,* 89: 5316–5320.

594. **Ronald, P. C., Albano, B., Tabien, R., Abenes, L., Wu, K., McCouch, S., and Tanksley, S. D.,** (1992) Genetic and physical analysis of the rice bacterial blight disease resistance locus, *Xa21, Mol. Gen. Genet.,* 236: 113–120.

595. **Rossen, L., Nörskov, P., Holmström, K., and Rasmussen, O. F.,** (1992) Inhibition of PCR by components of food samples, microbial diagnostic assays and DNA-extraction solutions, *Int. J. Food Microbiol.,* 17: 37–45.

596. **Roy, A., Frascaria, N., MacKay, J., and Bousquet, J.,** (1992) Segregating random amplified polymorphic DNAs (RAPDs) in *Betula alleghaniensis, Theor. Appl. Genet.,* 85: 173–180.

597. **Royle, N. J., Carlson, R. E., Wong, Z., and Jeffreys, A. J.,** (1988) Clustering of hypervariable minisatellites in the proterminal regions of human autosomes, *Genomics,* 3: 352–360.

598. **Russell, J. R., Hosein, F., Johnson, E., Waugh, R., and Powell, W.,** (1993) Genetic differentiation of cocoa (*Theobroma cacao* L.) populations revealed by RAPD analysis, *Mol. Ecol.,* 2: 89–97.

599. **Rychlik, W., and Rhoads, R. E.,** (1989) A computer program for choosing optimal oligonucleotides for filter hybridization, sequencing and *in vitro* amplification of DNA, *Nucl. Acids Res.,* 17: 8543–8551.

600. **Ryskov, A. P., Jincharadze, A. G., Prosnyak, M. I., Ivanov, P. L., and Limborska, S. A.,** (1988) M13 phage DNA as a universal marker for DNA fingerprinting of animals, plants and microorganisms, *FEBS Lett.,* 233: 388–392.

601. **Sabouraud, R.,** (1910) Maladies du cuir chevelu. III. Les maladies cryptogamiques, *Les Teignes,* Masson & Cie, Paris.

602. **Sadhu, C., McEachern, M. J., Rustchenko-Bulgac, E. P., Schmid, J., Soll, D. R., and Hicks, J. B.,** (1991) Telomeric and dispersed repeat sequences in *Candida* yeasts and their use in strain identification, *J. Bacteriol.,* 173: 842–850.

603. **Saghai-Maroof, M. A., Soliman, K. M., Jorgensen, R. A., and Allard, R. W.,** (1984) Ribosomal DNA spacer-length polymorphisms in barley: Mendelian inheritance, chromosomal location, and population dynamics, *Proc. Natl. Acad. Sci. U.S.A.,* 81: 8014–8018.

604. **Saiki, R. K., Scharf, S., Faloona, F., Mullis, K. B., Horn, G. T., Erlich, H. A., and Arnheim, N.,** (1985) Enzymatic amplification of beta-globin genomic sequences and restriction site analysis for diagnosis of sickle cell anemia, *Science,* 230: 1350–1354.

605. **Saiki, R. K., Gelfand, D. H., Stoffel, S., Scharf, S. J., Higuchi, R., Horn, G. T., Mullis, K. B., and Erlich, H. A.,** (1988) Primer-directed enzymatic amplification of DNA with a thermostable DNA polymerase, *Nature (London),* 239: 487–497.

606. **Sambrook, J., Fritsch, E. F., and Maniatis, T.,** (1989) *Molecular Cloning, a Laboratory Manual,* 2nd ed., Cold Spring Harbor Laboratory Press, Cold Spring Harbor, New York.

607. **Saunders, G. W.,** (1993) Gel purification of red algal genomic DNA: an inexpensive and rapid method for the isolation of polymerase chain reaction-friendly DNA, *J. Phycol.,* 29: 251–254.

608. **Schäfer, C., and Wöstemeyer, J.,** (1992) Random primer dependent PCR differentiates aggressive from non-aggressive isolates of the oilseed rape pathogen *Phoma lingam* (*Leptosphaeria maculans*), *J. Phytopathol.,* 136: 124–136.

609. **Schäfer, R., Zischler, H., Birsner, U., Becker, A., and Epplen, J. T.,** (1988) Optimized oligonucleotide probes for DNA fingerprinting, *Electrophoresis,* 9: 369–374.

610. **Scherer, S., and Stevens, D. A.,** (1987) Application of DNA typing methods to epidemiology and taxonomy of *Candida* species, *J. Clin. Microbiol.,* 25: 675–679.

611. **Scherer, S., and Stevens, D. A.,** (1988) A *Candida albicans* dispersed, repeated gene family and its epidemiologic applications, *Proc. Natl. Acad. Sci. U.S.A.*, 85: 1452–1456.

612. **Schlick, A., Meyer, W., Lieckfeldt, E., Börner, T., and Messner, K.,** (1992) PCR, the method of choice for patent-strain-characterisation, First European Conference on Fungal Genetics, Nottingham, Aug. 20–23, Abstract P2/01.

612a. **Schlick, A., Kuhls, K., Meyer, W., Liechfeldt, E., Börner, T., and Msssner, K.,** (1994) Fingerprinting reveals gamma-ray induced mutations in fungal DNA: implications for identification of patent strains of *Trichoderma harzianum, Curr. Genet.*, 26: 74–78.

613. **Schlötterer, C., and Tautz, D.,** (1992) Slippage synthesis of simple sequence DNA, *Nucl. Acids Res.*, 20: 211–215.

614. **Schlötterer, C., Amos, B., and Tautz, D.,** (1991) Conservation of polymorphic simple sequence loci in cetacean species, *Nature (London)*, 354: 63–65.

615. **Schmid, J., Voss, E., and Soll, D. R.,** (1990) Computer-assisted methods for assessing strain relatedness in *Candida albicans* by fingerprinting with the moderately repetitive sequence Ca3, *J. Clin. Microbiol.*, 31: 1236–1243.

616. **Schmid, J., Odds, F. C., Wiselka, M. J., Nicholson, K. G., and Soll, D. R.,** (1992) Genetic similarity of *Candida albicans* strains from a group of AIDS patients, demonstrated by DNA fingerprinting, *J. Clin. Microbiol.*, 31: 935–941.

617. **Schmid, J., Rotman, M., Reed, B., Pierson, C. L., and Soll, D. R.,** (1993) Genetic similarity of *Candida albicans* strains from vaginitis patients and their partners, *J. Clin. Microbiol.*, 31: 39–46.

618. **Schmidt, T., Boblenz, K., Metzlaff, M., Kaemmer, D., Weising, K., and Kahl, G.,** (1993) DNA fingerprinting in sugar beet (*Beta vulgaris*) – identification of double-haploid breeding lines, *Theor. Appl. Genet.*, 85: 653–657.

619. **Schönian, G., Meusel, O., Tietz, H. J., Meyer, W., Gräser, Y., Tausch, I., Presber, W., and Mitchell, G. T.,** (1993) Identification of clinical strains of *Candida albicans* by DNA fingerprinting with the polymerase chain reaction, *Mycoses*, 30: 171–179.

620. **Sederoff, R. R.,** (1987) Molecular mechanisms of mitochondrial-genome evolution in higher plants, *Am. Nat.*, 130: S30–S45.

621. **Senior, M. L., and Heun, M.,** (1993) Mapping maize microsatellites and polymerase chain reaction confirmation of the targeted repeats using a CT primer, *Genome*, 36: 883–889.

622. **Serikawa, T., Kuramoto, T., Hilbert, P., Mori, M., Yamada, J., Dubay, C. J., Lindpainter, K., Ganten, D., Guénet, J. L., Lathrop, G. M., and Beckmann, J. S.,** (1992) Rat gene mapping using PCR-analyzed microsatellites, *Genetics*, 131: 701–721.

623. **Sharma, P. C., Winter, P., Bünger, T., Hüttel, B., Weigand, F., Weising, K., and Kahl, G.,** (1994) Abundance and polymorphism of di-, tri- and tetranucleotide tandem repeats in chickpea, *Theor. Appl. Genet.*, in press.

624. **Sharon, D., Hillel, J., Vainstein, A., and Lavi, U.,** (1992) Application of DNA fingerprints for identification and genetic analysis of *Carica papaya* and other *Carica* species, *Euphytica*, 62: 119–126.

625. **Shaw, D. V., Kahler, A. L., and Allard, R. W.,** (1981) A multilocus estimator of mating system parameters in plant populations, *Proc. Natl. Acad. Sci. U.S.A.*, 62: 113–120.

626. **Shear, C. L., and Stevens, N. E.,** (1913) Cultural characters of the chestnut blight fungus and its near relatives, *Circular U.S. Dept. Agric., Bur. Plant Ind.*, 131: 3–18.

627. **Sherertz, R. J., Gledhill, K. S., Hampton, K. D., Pfaller, M. A., Givner, L. B., Abramson, J. S., and Dillard, R. G.,** (1992) Outbreak of *Candida* bloodstream infections associated with retrograde medication administration in a neonatal intensive care unit, *J. Pediatr.*, 120: 455–461.

628. **Shimada, T., Haji, T., and Hosaka, K.** (1993) Classification and parent determination by RAPD in mume, in *Techniques on Gene Diagnosis and Breeding in Fruit Trees*, Hayashi, T., Omura, M., and Scott, N. S., Eds., Fruit Tree Research Station, Ibaraki, pp. 77–80.

629. **Sierotzki, H., Eggenschwiler, M., McDermott, J., and Gessler, C.,** (1994) Specific virulence of isolates of *Venturia inaequalis* on 'susceptible' apple cultivars, Eucarpia Fruit Breeding Section Meeting, Aug. 30–Sept. 3, 1993, *Euphytica*, in press.

630. **Signer, E. N., and Jeffreys, A. J.,** (1993) Application of human minisatellite probes to the development of informative DNA fingerprints and the isolation of locus-specific markers in animals, in *DNA Fingerprinting: State of the Science,* Pena, S. D. J., Chakraborty, R., Epplen, J. T., and Jeffreys, A. J., Eds., Birkhäuser, Basel, pp. 421–428.

631. **Singsit, C., and Ozias-Akins, P.,** (1993) Genetic variation in monoploids of diploid potatoes and detection of clone-specific random amplified polymorphic DNA markers, *Plant Cell. Rep.,* 12: 144–148.

632. **Skinnner, D. M., Beattie, W. G., and Blattner, F. R.,** (1974) The repeat sequence of a hermit crab satellite deoxyribonucleic acid is (–T–A–G–G–)ₙ x (–A–T–C–C)ₙ, *Biochemistry,* 13: 3930–3937.

633. **Skinner, D. Z.,** (1992) Associating molecular markers with desirable traits in cultivated, cross-pollinated, tetraploid populations, Plant Genome I, San Diego, Nov. 9–11, p. 48.

634. **Skov, E., and Wellendorf, H.,** (1992) Application of RAPD for high density genome mapping and search for linkage to QTL in Norway spruce (*Picea abies*), Plant Genome I, San Diego, Nov. 9–11, p. 53.

635. **Skroch, P., Tivang, J., and Nienhuis, J.,** (1993) Analysis of genetic relationships using RAPD marker data, in *Application of RAPD Technology to Plant Breeding,* Neff, M., Ed., ASHS Publishers, St. Paul, MN, pp. 26–30.

636. **Slater, R. J.,** (1985) The extraction of total RNA by the detergent and phenol method, in *Methods in Molecular Biology — Nucleic Acids,* Walker, J. M., Ed., Vol. 2, Humana Press, Cliftar, pp. 101–108.

637. **Smeets, H. J. M., Brunner, H. G., Ropers, H.-H., and Wieringa, B.,** (1989) Use of variable simple sequence motifs as genetic markers: application to study of myotonic dystrophy, *Hum. Genet.,* 83: 245–251.

638. **Smith, L. M., Sanders, J. Z., Kaiser, R. J., Hughes, P., Dodd, C., Connell, C. R., Heiner, C., Kent, S. B. H., and Hood, L. E.,** (1986) Fluorescence detection in automated DNA sequence analysis, *Nature (London),* 321: 674–679.

639. **Smith, M. L., Bruhn, J. N., and Anderson, J. B.,** (1992) The fungus *Armillaria bulbosa* is among the largest and oldest living organisms, *Nature (London),* 356: 428–431.

640. **Smith, S., and Chin, E.,** (1993) The utility of random primer-mediated profiles, RFLPs, and other technologies to provide useful data for varietal protection, in *Application of RAPD Technology to Plant Breeding,* Neff, M., Ed., ASHS Publishers, St. Paul, MN, pp. 46–49.

641. **Sobral, B. W. S.,** (1992) Genetic mapping and evolution of sugarcane, Plant Genome, I, San Diego, Nov. 9–11, p. 14.

642. **Sobral, B. W. S., and Honeycutt, R. J.,** (1993) High output genetic mapping of polyploids using PCR-generated markers, *Theor. Appl. Genet.,* 86: 105–112.

643. **Sokal, R. R., and Rohlf, F. J.,** (1981) *Biometry,* Freeman, San Francisco.

644. **Soll, D. R., Langtimm, C., McDowell, J., Hicks, J., and Galask, R.,** (1987) High-frequency switching in *Candida* strains isolated from vaginitis patients, *J. Clin. Microbiol.,* 25: 1611–1622.

645. **Soll, D. R., Staebell, M., Langtimm, C., Pfaller, M., Hicks, J., and Rao, T. V. G.,** (1988) Multiple *Candida* strains in course of a single systemic infection, *J. Clin. Microbiol.,* 26: 1448–1459.

646. **Soll, D. R., Galask, R., Isley, S., Rao, T. V. G., Stone, D., Hicks, J., Schmid, J., Mac, K., and Hanna, C.,** (1989) Switching of *Candida albicans* during successive episodes of recurrent vaginitis, *J. Clin. Microbiol.,* 27: 681–690.

647. **Soll, D. R., Galask, R., Schmid, J., Hanna, C., Mac, K., and Morrow, B.,** (1991) Genetic dissimilarity of commensal strains of *Candida* spp. carried in different anatomical locations of the same healthy women, *J. Clin. Microbiol.,* 29: 1702–1710.

648. **Soltis, P. J., Soltis, D. E., and Doyle, J. J., Eds.,** (1992) *Molecular Systematics of Plants,* Chapmann and Hall, New York.

649. **Song, Y., and Cullis, C. A.,** (1992) RFLP and RAPD mapping of *Eucalyptus globulus,* Plant Genome I, San Diego, Nov. 9–11, p. 49.

650. **Song, K. M., Osborn, T. C., and Williams, P. H.,** (1988) *Brassica* taxonomy based on nuclear restriction fragment length polymorphisms (RFLP), *Theor. Appl. Genet.,* 75: 784–794.

651. **Song, K. M., Tang, K., and Osborn, T. C.,** (1993) Development of synthetic *Brassica* amphidiploids by reciprocal hybridization and comparison to natural amphidiploids, *Theor. Appl. Genet.,* 86: 811–821.

652. **Southern, E. M.,** (1975) Detection of specific sequences among DNA fragments separated by gel electrophoresis, *J. Mol. Biol.,* 98: 503–517.

653. **Spandidos, D. A., and Holmes, L.,** (1987) Transcriptional enhancer activity in the variable tandem repeat DNA sequence downstream of the human *Ha-ras*1 gene, *FEBS Lett.,* 218: 41–46.

654. **Specht, C. A., Di Russo, C. C., Novotny, C. P., and Ullrich, R. C.,** (1982) A method for extracting high-molecular-weight deoxyribonucleic acid from fungi, *Anal. Biochem.,* 119: 158–163.

655. **Spitzer, E. D., and Spitzer, S. G.,** (1992) Use of a dispersed repetitive DNA element to distinguish clinical isolates of *Cryptococcus neoformans, J. Clin. Microbiol.,* 30: 1094–1097.

656. **Sreenivasaprasad, S., Brown, A. E., and Mills, P. R.,** (1992) DNA sequence variation and interrelationships among *Colletotrichum* species causing strawberry anthracnose, *Physiol. Mol. Plant Pathol.,* 41: 265–281.

657. **Stacey, G. N., Bolton, B. J., and Doyle, A.,** (1992) DNA fingerprinting transforms the art of cell authentication, *Nature (London),* 357: 261–262.

658. **Stacy, J. E., and Jakobsen, K. S.,** (1993) Testing of nylon membranes for DNA-fingerprinting with multilocus probes, *Int. J. Genome Res.,* 2: 159–165.

659. **Stallings, R. L., Ford, A. F., Nelson, D., Torney, D. C., Hildebrand, C. E., and Moyzis, R. K.,** (1991) Evolution and distribution of (GT)$_n$ repetitive sequences in mammalian genomes, *Genomics,* 10: 807–815.

660. **Stam, P.,** (1993) Construction of integrated genetic linkage maps by means of a new computer package: JOINMAP, *Plant J.,* 3: 739–744.

661. **Steane, D. A., McClure, B. A., Clarke, A. E., and Kraft, G. T.,** (1991) Amplification of the polymorphic 5.8S rRNA gene from selected Australian Gigartinalean species (Rhodophyta) by polymerase chain reaction, *J. Phycol.,* 27: 758–762.

662. **Stephens, J. C., Gilbert, D. A., Yuhki, N., and O'Brien, S. J.,** (1992) Estimation of heterozygosity for single-probe multilocus DNA fingerprints, *Mol. Biol. Evol.,* 9: 729–743.

663. **Stiles, J. I., Lemme, C., Sondur, S., Morshidi, M. B., and Manshardt, R.,** (1993) Using randomly amplified polymorphic DNA for evaluating genetic relationships among papaya cultivars, *Theor. Appl. Genet.,* 85: 697–701.

664. **Stockton, T., Sonnante, G., and Gepts, P.,** (1992) Detection of minisatellite sequences in *Phaseolus vulgaris, Plant Mol. Biol. Rep.,* 10: 47–59.

665. **Striem, M. J., Spiegel-Roy, P., Ben-Hayyim, G., Beckmann, J., and Gidoni, D.,** (1990) Genomic DNA fingerprinting of *Vitis vinifera* by the use of multi-locus probes, *Vitis,* 29: 223–227.

666. **Stringer, S. L., Hong, S., Giuntoli, D., and Stringer, J. S.,** (1991) Repeated DNA in *Pneumocystis carinii, J. Clin. Microbiol.,* 29: 1194–1201.

667. **Strongman, D. B., and MacKay, R. M.,** (1993) Discrimination between *Hirsutella longicolla* var. *longicolla* and *Hirsutella longicolla* var. *cornuta* using random amplified polymorphic DNA fingerprinting, *Mycologia,* 85: 65–70.

668. **Struss, D., Quiros, C. F., and Robbelen, G.,** (1992) Mapping of molecular markers on *Brassica* B-genome chromosomes added to *Brassica napus, Plant Breed.,* 108: 320–323.

669. **Suiter, K. A., Wendel, J. F., and Case, J. S.,** (1983) LINKAGE-1: a Pascal computer program for the detection and analysis of genetic linkage, *J. Hered.,* 74: 203–204.

670. **Sullivan, D., Bennett, D., Henman, M., Harwood, P., Flint, S., Mulcahy, F., Shanley, D., and Coleman, D.,** (1993) Oligonucleotide fingerprinting of isolates of *Candida* species other than *C. albicans* and of atypical *Candida* species from human immunodeficiency virus-positive and AIDS patients, *J. Clin. Microbiol.,* 31: 2124–2133.

671. **Sulston, J., Mallett, F., Durbin, R., and Horsnell, T.,** (1989) Image analysis of restriction enzyme fingerprint autoradiograms, *CABIOS*, 5: 101–106.

672. **Swallow, D. M., Gendler, S., Griffiths, B., Corney, G., Taylor-Papadimitriou, J., and Bramwell, M. E.,** (1987) The human tumour-associated epithelial mucins are coded by an expressed hypervariable gene locus PUM, *Nature (London)*, 328: 82–84.

673. **Swofford, D. L., and Olsen, G. J.,** (1990) Phylogeny reconstruction, in *Molecular Systematics*, Hillis, D. M., and Moritz, C., Eds., Sinauer Associates, Sunderland, MA, pp. 411–501.

674. **Sytsma, K. J., and Gottlieb, L. D.,** (1986) Chloroplast DNA evolution and phylogenetic relationships in *Clarkia* sect. *Peripetasma* (Onagraceae), *Evolution*, 40: 1248–1261.

675. **Sytsma, K. J., and Schaal, B. A.,** (1985) Phylogenetics of the *Lisianthius skinneri* (Gentianaceae) species complex using DNA restriction fragment length analysis, *Evolution*, 39: 594–608.

676. **Taberlet, P., Gielly, L., Patou, G., and Bouvet, J.,** (1991) Universal primers for amplification of three non-coding regions of chloroplast DNA, *Plant Mol. Biol.*, 17: 1105–1109.

677. **Tai, T. H., and Tanksley, S. D.,** (1990) A rapid and inexpensive method for isolation of total DNA from dehydrated plant tissue, *Plant Mol. Biol. Rep.*, 8: 297–303.

678. **Tammisola, J., Lapinkoji, S., Åkerman, S., Regina, M., van Wright, A., Söderlund, H., Kauppinen, V., Viherä-Aarnio, A., Hagqvist, R., and Velling, P.,** (1992) Polymorphic DNA markers in clonal identification and pedigree confirmation of white birch (*Betula pendula* Roth.), Plant Genome I, San Diego, Nov. 9–11, p. 50.

679. **Tanhuanpää, P., Vilkki, J., Vilkki, J., and Pulli, S.,** (1993) Genetic polymorphisms at RAPD loci in spring turnip rape (*Brassica rapa* ssp. *oleifera*), *Agric. Sci. Finland*, 2: 303–310.

680. **Tanksley, S. D., Young, N. D., Paterson, A. H., and Bonierbale, M. W.,** (1989) RFLP mapping in plant breeding: new tools for an old science, *Bio/Technology*, 7: 257–264.

681. **Tanksley, S. D., Ganal, M. W., Prince, J. L., De Vicente, M. C., Bonierbale, M. W., Broun, P., Fulton, T. M., Giovannoni, J. J., Grandillo, S., Martin, G. B., Messeguer, R., Miller, J. C., Miller, L., Paterson, A. H., Pineda, O., Roder, M. S., Wing, R. A., Wu, W., and Young, N. D.,** (1992) High density molecular linkage map of the tomato and potato genomes, *Genetics*, 132: 1141–1160.

682. **Tassanakajon, A., Wongteerasapaya, C., Pumichoti, P., Boonsaeng, V., and Panyim, S.,** (1991) Improved resolution and sensitivity of human DNA fingerprinting by specific-primed labelling of M13-DNA, *Mol. Cell. Probe*, 5: 111–115.

683. **Tao, Y., Manners, J. M., Ludlow, M. M., and Henzell, R. G.,** (1993) DNA polymorphisms in grain sorghum (*Sorghum bicolor* (L.) Moench), *Theor. Appl. Genet.*, 86: 679–688.

684. **Tautz, D.,** (1989) Hypervariability of simple sequences as a general source for polymorphic DNA markers, *Nucl. Acids Res.*, 17: 6463–6471.

685. **Tautz, D.,** (1993) Notes on the definition and nomenclature of tandemly repetitive DNA sequences, in *DNA Fingerprinting: State of the Science*, Pena, S. D. J., Chakraborty, R., Epplen, J. T., and Jeffreys, A. J., Eds., Birkhäuser, Basel, pp. 21–28.

686. **Tautz, D., and Renz, M.,** (1984) Simple sequences are ubiquitous repetitive components of eukaryotic genomes, *Nucl. Acids Res.*, 12: 4127–4138.

687. **Tautz, D., Trick, M., and Dover, G. A.,** (1986) Cryptic similarity in DNA is a major source of genetic variation, *Nature (London)*, 322: 652–656.

688. **Tegelström, H.,** (1992) Detection of mitochondrial DNA fragments, in *Molecular Genetic Analysis of Populations: A Practical Approach*, Hoelzel, A. R., Ed., IRL Press, Oxford, pp. 89–114.

689. **Teramoto, S., Kano-Murakami, Y., Hori, M., and Kamiyama, K.,** (1993) "DNA fingerprinting" can distinguish cultivar of Japanese pear, in *Techniques on Gene Diagnosis and Breeding in Fruit Trees*, Hayashi, T., Omura, M., and Scott, N. S., Eds., Fruit Tree Research Station, Ibaraki, pp. 74–76.

690. **Thein, S. L., and Wallace, R. R.,** (1986) The use of synthetic oligonucleotides as specific hybridization probes in the diagnosis of genetic disorders, in *Human Genetic Diseases: A Practical Approach*, Davies, K. E., Ed., IRL Press, Oxford, pp. 33–50.

691. **Thomas, M. R., Matsumoto, S., Cain, P., and Scott, N. S.,** (1993) Repetitive DNA of grapevine: classes present and sequences suitable for cultivar identification, *Theor. Appl. Genet.*, 86: 173–180.

692. **Thomson, D., and Henry, R.,** (1993) Use of DNA from dry leaves for PCR and RAPD analysis, *Plant Mol. Biol. Rep.*, 11: 202–206.

693. **Thormann, C. E., and Osborn, T. C.,** (1993) Use of RAPD and RFLP markers for germplasm evaluation, in *Application of RAPD Technology to Plant Breeding,* Neff, M., Ed., ASHS Publishers, St. Paul, MN, pp. 9–11.

694. **Tingey, S. V., and del Tufo, J. P.,** (1993) Genetic analysis with random amplified polymorphic DNA markers, *Plant Physiol.*, 101: 349–352.

695. **Tingey, S. V., Rafalski, J. A., and Williams, J. G. K.,** (1993) Genetic analysis with RAPD markers, in *Application of RAPD Technology to Plant Breeding,* Neff, M., Ed., ASHS Publishers, St. Paul, MN, pp. 3–8.

696. **Tinker, N. A., Fortin, M. G., and Mather, D. E.,** (1993) Random amplified polymorphic DNA and pedigree relationships in spring barley, *Theor. Appl. Genet.*, 85: 976–984.

697. **Torres, A. M., Millan, T., and Cubero, J. I.,** (1993) Identifying rose cultivars using random amplified polymorphic DNA markers, *HortScience*, 28: 333–334.

698. **Torres, A. M., Weeden, N. F., and Martin, A.,** (1993) Linkage among isozyme, RFLP and RAPD markers in *Vicia faba, Theor. Appl. Genet.,* 85: 937–945.

699. **Trepicchio, W. L., and Krontiris, T. G.,** (1992) Members of the rel/NF-kB family of transcriptional regulatory proteins bind the HRAS1 minisatellite DNA sequence, *Nucl. Acids Res.*, 21: 977–985.

700. **Trepicchio, W. L, and Krontiris, T. G.,** (1993) IGH minisatellite suppression of USF-binding-site- and Eu-mediated transcriptional activation of the adenovirus late promoter, *Nucl. Acids Res.*, 20: 2427–2434.

701. **Tuinstra, M., Goldsbrough, P., Grote, E., and Ejeta, G.,** (1992) Identification and RAPD mapping of quantitative trait loci associated with drought tolerance in *Sorghum,* Plant Genome I, San Diego, Nov. 9–11, p. 50.

702. **Tulsieram, L. K., Glaubitz, J. C., Kiss, G., and Carlson, J. E.,** (1992) Single tree genetic linkage mapping in conifers using haploid DNA from megagametophytes, *Bio/Technology,* 10: 686–690.

703. **Tzuri, G., Hillel, J., Lavi, U., Haberfeld, A., and Vainstein, A.,** (1991) DNA fingerprint analysis of ornamental plants, *Plant Sci.*, 76: 91–98.

704. **Uitterlinden, A. G., and Vijg, J.,** (1993) *Two-Dimensional DNA Typing. A Parallel Approach to Genome Analysis,* Ellis Horwood, Simon and Schuster International, UK.

705. **Uitterlinden, A. G., Slagboom, P. E., Knook, D. L., and Vijg, J.,** (1989) Two-dimensional DNA fingerprinting of human individuals, *Proc. Natl. Acad. Sci. U.S.A.*, 86: 2742–2746.

706. **Uitterlinden, A. G., Slagboom, P. E., Mullaart, E., Meulenbelt, I., and Vijg, J.,** (1991) Genome scanning by two-dimensional DNA typing: the use of repetitive DNA sequences for rapid mapping of genetic traits, *Electrophoresis*, 12: 119–134.

707. **Unkles, S. E., Duncan, J. M., and Kinghorn, J. R.,** (1992) Zinc fingerprinting for *Phytophthora* species – ZIF markers, *Curr. Genet.*, 22: 317–318.

708. **Uphoff, H., and Wricke, G.,** (1992) Random amplified polymorphic DNA (RAPD) markers in sugar beet (*Beta vulgaris* L.): mapping the genes for nematode resistance and hypocotyl colour, *Plant Breed.*, 109: 168–171.

709. **Vahala, T.,** (1991) Studies on tissue culture, genetic transformation and gene expression in willows (*Salix* spp.), *Acta Universitatis Upsaliensis, Comprehensive Summaries of Uppsala Dissertations from the Faculty of Science,* No. 339, University of Uppsala, Sweden.

710. **Vaillancourt, L. J., and Hanau, R. M.,** (1992) Genetic and morphological comparisons of *Glomerella (Colletotrichum)* isolates from maize and from sorghum, *Exp. Mycol.*, 16: 219–229.

711. **Vainstein, A., Hillel, J., Lavi, U., and Tzuri, G.,** (1991) Assessment of genetic relatedness in carnation by DNA fingerprint analysis, *Euphytica*, 56: 225–229.

712. **Valent, B., Farrall, L., and Chumley, F. G.,** (1991) *Magnaporthe grisea* genes for pathogenicity and virulence identified through a series of backcrosses, *Genetics*, 127: 87–101.

713. **Vallejos, C. E., Sakiyama, N. S., and Yu, Z.-H.,** (1992) An update of the *Phaseolus vulgaris* linkage map: addition of known sequences, RAPDs, and QTLs for seed size, Plant Genome I, San Diego, Nov. 9–11, p. 51.

714. **Vallés, M. P., Wang, Z. Y., Montavon, P., Potrykus, I., and Spangenberg, G.,** (1993) Analysis of genetic stability of plants regenerated from suspension cultures and protoplasts of meadow fescue (*Festuca pratensis* Huds.), *Plant Cell. Rep.*, 12: 101–106.

715. **Van Buijtenen, J. P., Kong, X., Funkhouser, E., Nance, W. L., Nelson, C. D., Nelson, L. S., and Johnson, G. N.,** (1992) Linkage map of slash pine based on megagametophytic DNA, Plant Genome I, San Diego, Nov. 9–11, p. 51.

716. **Van Coppenolle, B., Watanabe, I., Van Hove, C., Second, G., Huang, N., and McCouch, S. R.,** (1993) Genetic diversity and phylogeny analysis of *Azolla* based on DNA amplification by arbitrary primers, *Genome*, 36: 686–693.

717. **Van Daelen, R. A. J., Jonkers, J. J., and Zabel, P.,** (1989) Preparation of megabase-sized tomato DNA and separation of large restriction fragments by field inversion gel electrophoresis, *Plant Mol. Biol.*, 12: 341–352.

717a. **Van der Vlugt-Bermans, C. J. B., Brandwagt, B. F., Van't Klooster, J. W., Wagemakers, L. A. M., and Van Kan, J. A. L.,** (1993) Genetic variation and segregation of DNA polymorphisms in *Botrytis Cinerea, Mycol. Res.*, 10: 1193–1200.

718. **Van Heusden, A. W., and Bachmann, K.,** (1992) Genotype relationships in *Microseris elegans* (Asteraceae, Lactuceae) revealed by DNA amplification from arbitrary primers (RAPDs), *Plant Syst. Evol.*, 179: 221–233.

719. **Van Heusden, A. W., and Bachmann, K.,** (1992) Nuclear DNA polymorphisms among strains of *Microseris bigelovii* (Asteraceae: Lactuceae) amplified from arbitrary primers, *Bot. Acta,* 105: 331–336.

720. **Van Heusden, A. W., and Bachmann, K.,** (1992) Genetic differentiation of *Microseris pygmaea* (Asteraceae, Lactuceae) studied with DNA amplification from arbitrary primers (RAPDs), *Acta Bot. Neerl.*, 41: 385–395.

721. **Van Heusden, A. W., Rouppe van der Voort, J., and Bachmann, K.,** (1991) Oligo-(GATA) fingerprints identify clones in asexual dandelions (*Taraxacum*, Asteraceae), *Fingerprint News*, 3(2): 13–15.

722. **Van Houten, W. H. J., Van Heusden, A. W., Rouppe van der Voort, J., Raijmann, L., and Bachmann, K.,** (1991) Hypervariable DNA fingerprint loci in *Microseris pygmaea* (Asteraceae, Lactuceae). *Bot. Acta*, 104: 252–255.

723. **Van Uden, N., and do Carmo Sousa, L.,** (1957) Yeasts from the bovine caecum, *J. Gen. Microbiol.*, 16: 385–395.

724. **Varadarajan, G. S., and Prakash, C. S.,** (1991) A rapid and efficient method for the extraction of total DNA from the sweet potato and its related species, *Plant Mol. Biol. Rep.*, 9: 6–12.

725. **Varma, A., and Kwon-Chung, K. J.,** (1992) DNA probe typing of *Cryptococcus neoformans, J. Clin. Microbiol.*, 30: 2960–2967.

726. **Vassart, G., Georges, S. B., Monsieur, R., Brocas, H., Lequarré, A.-S., and Christophe, D.,** (1987) A sequence in M13 phage detects hypervariable minisatellites in human and animal DNA, *Science*, 235: 683–684.

727. **Vergnaud, G.,** (1989) Polymers of random short oligonucleotides detect polymorphic loci in the human genome, *Nucl. Acids Res.*, 17: 7623–7630.

728. **Vergnaud, G., Mariat, D., Apiou, F., Aurias, A., Lathrop, M., and Lauthier, V.,** (1991) The use of synthetic tandem repeats to isolate new VNTR loci: cloning of a human hypermutable sequence, *Genomics*, 11: 135–144.

729. **Vergnaud, G., Mariat, D., Zoroastro, D., and Lauthier, V.,** (1991) Detection of single and multiple polymorphic loci by synthetic tandem repeats of short oligonucleotides, *Electrophoresis*, 12: 134–140.

730. **Vierling, R. A., and Nguyen, H. T.,** (1992) Use of RAPD markers to determine the genetic diversity of diploid wheat genotypes, *Theor. Appl. Genet.,* 84: 835–838.
731. **Vogt, P.,** (1990) Potential genetic functions of tandem repeated DNA sequence blocks in the human genome are based on a highly conserved "chromatin folding code", *Hum. Genet.,* 84: 301–336.
732. **Vosman, B., Arens, P., Rus-Kortekaas, W., and Smulders, M. J. M.,** (1992) Identification of highly polymorphic DNA regions in tomato, *Theor. Appl. Genet.,* 85: 239–244.
733. **Wahls, W. P., Wallace, L. J., and Moore, P. D.,** (1990) Hypervariable minisatellite DNA is a hotspot for homologous recombination in human cells, *Cell,* 60: 95–103.
734. **Wahls, W. P., Wallace, L. J., and Moore, P. D.,** (1990) The Z-DNA motif d(TG)$_{30}$ promotes reception of information during gene conversion events while stimulating homologous recombination in human cells in culture, *Mol. Cell. Biol.,* 10: 785–793.
735. **Wahls, W. P., Swenson, G., and Moore, P. D.,** (1991) Two hypervariable minisatellite DNA binding proteins, *Nucl. Acids Res.,* 19: 3269–3274.
735a. **Wallner, E. and Weising, K.,** (1994) Manuscript in preparation.
736. **Walmsley, R. M., Wikinson, B. M., and Kong, T. H.,** (1989) Genetic fingerprinting for yeasts, *Bio/Technology,* 7: 1168–1170.
737. **Walsh-Weller, J., Liu, K., and Somerville, S.,** (1992) Two approaches to cloning the *Ml-a* (mildew resistance) locus in barley, Plant Genome I, San Diego, Nov. 9–11, p. 52.
737a. **Wang, G. -L., Wing, R. A., and Paterson, A. H.,** (1993) PCR amplification from single seeds, facilitating DNA marker-assisted breeding, *Nucl. Acids Res.,* 21:2527.
737b. **Wang, H., Qi, M., and Cutler, A. J.,** (1993) A simple method of preparing plant samples for PCR, *Nucl. Acids Res.,* 21: 4153–4154.
738. **Watson, J. C., and Thompson, W. F.,** (1986) Purification and restriction endonuclease analysis of plant nuclear DNA, *Methods Enzymol.,* 118: 57–75.
739. **Waugh, R., and Powell, W.,** (1992) Using RAPD markers for crop improvement, *TIBTECH,* 10: 186–191.
740. **Waugh, R., Baird, E., and Powell, W.,** (1992) The use of RAPD markers for the detection of gene introgression in potato, *Plant Cell. Rep.,* 11: 466–469.
741. **Waugh, R., Duncan, N., Baird, E., Harrower, B., Phillips, M. S., and Powell, W.,** (1992) Markers linked to polygenic PCN resistance in potato, Plant Genome I, San Diego, Nov. 9–11, p. 52.
742. **Wayne, R. K., George, S. B., Gilbert, D., Collins, P. W., Kovach, S. D., Girman, D., and Lehman, N.,** (1991) A morphologic and genetic study of the island fox, *Urocyon littoralis, Evolution,* 45: 1849–1868.
743. **Webb, D. M., and Knapp, S. J.,** (1990) DNA extraction from a previously recalcitrant plant genus, *Plant Mol. Biol. Rep.,* 8: 180–185.
744. **Weber, J. L.,** (1990) Informativeness of human $(dC-dA)_n \times (dG-dT)_n$ polymorphisms, *Genomics,* 7: 524–530.
745. **Weber, J. L., and May, P. E.,** (1989) Abundant class of human DNA polymorphisms which can be typed using the polymerase chain reaction, *Am. J. Hum. Genet.,* 44: 388–396.
746. **Weeden, N. F., Lu, J., Temnykh, S., Timmerman, G. M., Simon, C., Kneen, B., Provvidenti, R., Muehlbauer, F., and LaRue, T.,** (1992) Mapping and tagging host genes involved in legume/microbe interactions, Plant Genome I, San Diego, Nov. 9–11, p. 53.
747. **Weeden, N. F., Timmerman, G. M., Hemmat, M., Kneen, B. E., and Lodhi, M. A.,** (1993) Inheritance and reliability of RAPD markers, in *Application of RAPD Technology to Plant Breeding,* Neff, M., Ed., ASHS Publishers, St. Paul, MN, pp. 12–17.
748. **Weeden, N. F., Hemmat, M., Lawson, D. M., Lodhi, M. A., Bell, R. L., Manganaris, A. G., Reisch, B. I., Brown, S. K., and Ye, G.-N.,** (1994) Development and application of molecular marker linkage maps in woody fruit crops, Eucarpia Fruit Breeding Section Meeting, Aug. 30–Sept. 3, 1993, *Euphytica,* in press.
749. **Weeks, D. P., Beerman, N., and Griffiths, O. M.,** (1986) A small-scale five-hour procedure for isolating multiple samples of CsCl-purified DNA: application to isolations from mammalian, insect, higher plant, algal, yeast, and bacterial sources, *Anal. Biochem.,* 152: 376–385.

750. **Weihe, A., Niemann, C., Lieckfeldt, D., Meyer, W., and Börner, T.,** (1990) An improved hybridization procedure for DNA fingerprinting with bacteriophage M13 as a probe, *Fingerprint News,* 2(4): 9–10.

751. **Weihe, A., Meixner, M., Wolowcyk, B., Melzer, R., and Börner, T.,** (1991) Rapid hybridization-based assays for identification by DNA probes of male-sterile and male-fertile cytoplasms of the sugar beet *Beta vulgaris* L., *Theor. Appl. Genet.,* 81: 819–824.

752. **Weining, S., and Langridge, P.,** (1991) Identification and mapping of polymorphisms in cereals based on the polymerase chain reaction, *Theor. Appl. Genet.,* 82: 209–216.

753. **Weir, B. S.,** (1990) *Genetic Data Analysis,* Sinauer Associates, Sunderland, MA.

753a. **Weising, K.,** (1994) Unpublisherd results.

754. **Weising, K., and Kahl, G.,** (1990) DNA fingerprinting in plants – the potential of a new method, *Biotech-Forum,* 7: 230–235.

755. **Weising, K., Weigand, F., Driesel, A. J., Kahl, G., Zischler, H., and Epplen, J. T.,** (1989) Polymorphic simple GATA/GACA repeats in plant genomes, *Nucl. Acids Res.,* 17: 10128.

756. **Weising, K., Fiala, B., Ramloch, K., Kahl, G., and Epplen, J. T.,** (1990) Oligonucleotide fingerprinting in angiosperms, *Fingerprint News,* 2(2): 5–8.

757. **Weising, K., Beyermann, B., Ramser, J., and Kahl, G.,** (1991) Plant DNA fingerprinting with radioactive and digoxigenated oligonucleotide probes complementary to simple repetitive DNA sequences, *Electrophoresis,* 12: 159–169.

758. **Weising, K., Kaemmer, D., Epplen, J. T., Weigand, F., Saxena, M., and Kahl, G.,** (1991) DNA fingerprinting of *Ascochyta rabiei* with synthetic oligodeoxynucleotides, *Curr. Genet.,* 19: 483–489.

759. **Weising, K., Ramser, J., Kaemmer, D., Kahl, G., and Epplen, J. T.,** (1991) Oligonucleotide fingerprinting in plants and fungi, in *DNA Fingerprinting: Approaches and Applications,* Burke, T., Dolf, G., Jeffreys, A. J., and Wolff, R., Eds., Birkhäuser, Basel, pp. 313–329.

760. **Weising, K., Kaemmer, D., Ramser, J., Bierwerth, S., and Kahl, G.,** (1992) Plant DNA fingerprinting with simple repetitive oligonucleotides, in *DNA Polymorphisms in Eukaryotic Genomes,* Kahl, G., Appelhans, H., Kömpf, Y., and Driesel, A.Y., Eds., Biotech-Forum 10, Adv. Mol. Genet., 5. Hüthig Verlag, Heidelberg. pp. 135–156.

761. **Weising, K., Kaemmer, D., Weigand, F., Epplen, J. T., and Kahl, G.,** (1992) Oligonucleotide fingerprinting reveals various probe-dependent levels of informativeness in chickpea (*Cicer arietinum*), *Genome,* 35: 436–442.

762. **Weissenbach, J., Gyapay, G., Dib, C., Vignal, A., Morisette, J., Millasseau, P., Vaysseix, G., and Lathrop, M.,** (1992) A second-generation linkage map of the human genome, *Nature (London),* 359: 794–801.

763. **Weitzman, I., and Silva-Hutner, M.,** (1967) Non-keratinous agar media as substitutes for the ascigerous state in certain members of the Gymnoascaceae pathogenic for the man and animals, *Sabouraudia,* 5: 335–340.

764. **Wells, R. A.,** (1988) DNA fingerprinting, in *Genome Analysis: A Practical Approach,* Davies, K. E., Ed., IRL Press, Oxford, pp. 153–169.

765. **Wells, R. A., Green, P., and Reeders, S. T.,** (1989) Simultaneous genetic mapping of multiple human minisatellite sequences using DNA fingerprinting, *Genomics,* 5: 761–772.

766. **Welsh, J., and McClelland, M.,** (1990) Fingerprinting genomes using PCR with arbitrary primers, *Nucl. Acids Res.,* 18: 7213–7218.

767. **Welsh, J., and McClelland, M.,** (1991) Genomic fingerprints produced by PCR with consensus tRNA gene primers, *Nucl. Acids Res.,* 19: 861–866.

768. **Welsh, J., and McClelland, M.,** (1991) Genomic fingerprinting using arbitrarily primed PCR and a matrix of pairwise combinations of primers, *Nucl. Acids Res.,* 19: 5275–5279.

769. **Welsh, J., Honeycutt, R. J., McClelland, M., and Sobral, B. W. S.,** (1991) Parentage determination in maize hybrids using the arbitrarily primed polymerase chain reaction (AP-PCR), *Theor. Appl. Genet.,* 82: 473–476.

770. **Welsh, J., Chada, K., Dalal, S. S., Cheng, R., Ralph, D., and McClelland, M.,** (1992) Arbitrarily primed PCR fingerprinting of RNA, *Nucl. Acids Res.*, 20: 4965–4970.

771. **Werth, C. R., Hilu, K. W., and Langner, C. A.,** (1993) Evidence from isozymes and RAPD markers bearing on the ancestry of finger millet, *Eleucine coracana* subsp. *coracana* and its wild relative *E. coracana* subsp. *africana, Am. J. Bot.*, 80 (Suppl.): 181.

772. **Westneat, D. F., Noon, W. A., Reeve, K. H., and Aquadro, C. F.,** (1988) Improved hybridization conditions for DNA 'fingerprints' probed with M13, *Nucl. Acids Res.*, 16: 4161.

773. **Wetton, J. H., Carter, R. E., Parkin, D. T., and Walters, D.,** (1987) Demographic study of a wild house sparrow population by DNA fingerprinting, *Nature (London)*, 327: 147–149.

774. **Wharton, K. A., Yedvobnick, B., Finnerty, V. G., and Artavanis-Tsakonas, S.,** (1985) Opa: a novel family of transcribed repeats shared by the notch locus and other developmentally regulated loci in *D. melanogaster, Cell*, 40: 55–62.

775. **Whisson, S. C., Maclean, D. J., Manners, J. M., and Irwin, J. A. G.,** (1992) Genetic relationships among Australian and North American isolates of *Phytophthora megasperma* f. sp. *glycinea* assessed by multicopy DNA probes, *Phytopathology*, 82: 863–868.

776. **White, T. J., Bruns, T., Lee, S., and Taylor, J.,** (1990) Amplification and direct sequencing of fungal ribosomal RNA genes for phylogenetics, in *PCR Protocols, A Guide to Methods and Applications*, Innis, M. A., Gelfand, D. H., Sninsky, J. J., and White, T. J., Eds., Academic Press, San Diego,, pp. 315–322.

777. **Wickerham, L. J.,** (1951) Taxonomy of yeasts, Tech. Bull. U.S. Dept. Agric., No 1029, U.S. Dept. Agric., Washington, DC.

778. **Wilde, J., Waugh, R., and Powell, W.,** (1992) Genetic fingerprinting of *Theobroma* clones using random amplified polymorphic DNA markers, *Theor. Appl. Genet.*, 83: 871–877.

779. **Wilkie, S., Isaac, P. G., and Slater, R. J.,** (1993) Random amplified polymorphic DNA (RAPD) markers for genetic analysis in *Allium, Theor. Appl. Genet.*, 86: 497–504.

780. **Wilkinson, B. M., Morris, L., Adams, D. J., Evans, E. G. V., Lacey, C. J. N., and Walmsley, R. M.,** (1992) A new, sensitive polynucleotide probe for distinguishing *Candida albicans* strains and its use with a computer assisted archiving and pattern comparison system, *J. Med. Vet. Mycol.*, 30: 123–131.

781. **Williams, C. E., and St. Clair, D. A.,** (1993) Phenetic relationships and levels of variability detected by restriction fragment length polymorphism and random amplified polymorphic DNA analysis of cultivated and wild accessions of *Lycopersicon esculentum, Genome*, 36: 619–630.

782. **Williams, J. G. K., Kubelik, A. R., Livak, K. J., Rafalski, J. A., and Tingey, S. V.,** (1990) DNA polymorphisms amplified by arbitrary primers are useful as genetic markers, *Nucl. Acids Res.*, 18: 6231–6235.

783. **Williams, J. G. K., Kubelik, A. R., Rafalski, J. A., and Tingey, S. V.,** (1991) Genetic analysis with RAPD markers, in *More Gene Manipulations in Fungi*, Bennett, J. W., and Lasure, L. L., Eds., Academic Press, San Diego,, pp. 431–439.

784. **Williams, J. G. K., Hanafey, M. K., Rafalski, J. A., and Tingey, S. V.,** (1993) Genetic analysis using random amplified polymorphic markers, *Methods Enzymol.*, 218: 704–740.

785. **Williams, J. G. K., Reiter, R. S., Young, R. M., and Scolnik, P. A.,** (1993) Genetic mapping of mutations using phenotypic pools and mapped RAPD markers, *Nucl. Acids Res.*, 21: 2697–2702.

786. **Williams, M. N. V., Pande, N., Nair, S., Mohan, M., and Bennett, J.,** (1991) Restriction fragment length polymorphism analysis of polymerase chain reaction products amplified from mapped loci of rice (*Oryza sativa* L.) genomic DNA, *Theor. Appl. Genet.*, 82: 489–498.

787. **Willmitzer, L., and Wagner, K. G.,** (1981) The isolation of nuclei from tissue-cultured plant cells, *Exp. Cell. Res.*, 135: 69–77.

788. **Winberg, B. C., Zhou, Z., Dallas, J. F., McIntyre, C. L., and Gustafson, J. P.,** (1993) Characterization of minisatellite sequences from *Oryza sativa, Genome,* 36: 978–983.

789. **Wolfe, M. S., Minchin, P. N., and Slater, S. E.,** (1981) Powdery mildew of barley, Plant Breeding Institute, Cambridge, U.K., Annual Report 1980: 88–92.

790. **Wolfe, M. S., Brändle, U., Koller, B., Limpert, E., McDermott, J. M., Müller, K., and Schaffner, D.,** (1992) Barley mildew in Europe: population biology and host resistance, *Euphytica,* 63: 125–139.

791. **Wolff, K.,** (1988) Natural selection in *Plantago* species: a genetical analysis of ecologically relevant morphological variability, dissertation thesis, University of Groningen, The Netherlands.

792. **Wolff, K.,** (1991) Analysis of allozyme variability in three *Plantago* species and a comparison to morphological variability, *Theor. Appl. Genet.,* 81: 119–126.

793. **Wolff, K.,** (1991) Genetic analysis of morphological variability in three *Plantago* species with different mating systems, *Theor. Appl. Genet.,* 81: 111–118.

793a. **Wolff, K.,** (1994) Unpublisherd results.

794. **Wolff, K., and Peters-van Rijn, J.,** (1993) Rapid detection of genetic variability in chrysanthemum (*Dendranthema grandiflora* Tzvelev) using random primers, *Heredity,* 71: 335–341.

795. **Wolff, K., and Schaal, B. A.,** (1992) Chloroplast DNA variation within and among five *Plantago* species, *J. Evol. Biol.,* 5: 325–344.

796. **Wolff, K., Schoen, E. D., and Peters-van Rijn, J.,** (1993) Optimizing the generation of random amplified polymorphic DNAs in chrysanthemum, *Theor. Appl. Genet.,* 86: 1033–1037.

797. **Wolff, K., Peters-van Rijn, J., and Hofstra, H.,** (1994) RFLP analysis in chrysanthemum. I Probe and primer development, *Theor. Appl. Genet.,* 88: 472–478.

798. **Wolff, K., Rogstad, S. H., and Schaal, B. A.,** (1994) Population and species variation of minisatellite DNA in *Plantago, Theor. Appl. Genet.,* 87: 733–740.

799. **Wolff, R. K., Nakamura, Y., and White, R.,** (1988) Molecular characterization of a spontaneously generated new allele at a VNTR locus: no exchange of flanking DNA sequence, *Genomics,* 3: 347–351.

800. **Wolff, R. K., Plaetke, R., Jeffreys, A. J., and White, R.,** (1989) Unequal crossingover between homologous chromosomes is not the major mechanism involved in the generation of new alleles at VNTR loci, *Genomics,* 5: 382–384.

801. **Wolff, R., Nakamura, Y., Odelberg, S., Shiang, R., and White, R.,** (1991) Generation of variability at VNTR loci in human DNA, in *DNA Fingerprinting: Approaches and Applications,* Burke, T., Dolf, G., Jeffreys, A. J., and Wolff, R., Eds., Birkhäuser, Basel, pp. 20–38.

802. **Wolin, H. L., Bevis, M. L., and Laurora, H.,** (1962) An improved synthetic medium for the rapid production of chlamydospores by *Candida albicans, Sabouraudia,* 2: 96–99.

803. **Wong, Z., Wilson, V., Jeffreys, A. J., and Thein, S. L.,** (1986) Cloning a selected fragment from a human DNA 'fingerprint': isolation of an extremely polymorphic minisatellite, *Nucl. Acids Res.,* 14: 4605–4616.

804. **Wong, Z., Wilson, V., Patel, I., Povey, S., and Jeffreys, A. J.,** (1987) Characterization of a panel of highly variable minisatellites cloned from human DNA, *Ann. Hum. Genet.,* 51: 269–288.

805. **Wöstemeyer, J.,** (1985) Strain-dependent variation in ribosomal DNA arrangement in *Absidia glauca, Eur. J. Biochem.,* 146: 443–448.

806. **Wöstemeyer, J., Schäfer, C., Kellner, M., and Weisfeld, M.,** (1992) DNA polymorphisms detected by random primer dependent PCR as a powerful tool for molecular diagnostics of plant pathogenic fungi, in *DNA Polymorphisms in Eukaryotic Genomes,* Kahl, G., Appelhans, H., Kömpf, J., and Driesel, A. J., Eds., Biotech-Forum 10, Adv. Mol. Genet., 5. Hüthig Verlag, Heidelberg pp. 227–240.

807. **Wright, S.,** (1965) The interpretation of population structure by F-statistics with special regard to system of mating, *Evolution,* 19: 395–420.

808. **Wu, K. K., Burnquist, W., Sorrells, M. E., Tew, T. L., Moore, P. H., and Tanksley, S. D.,** (1992) The detection and estimation of linkage in polyploids using single-dose restriction fragments, *Theor. Appl. Genet.,* 83: 294–300.

809. **Wu, K.-S., and Tanksley, S. D.,** (1993) Abundance, polymorphism and genetic mapping of microsatellites in rice, *Mol. Gen. Genet.,* 241: 225–235.

810. **Wyman, A. R., and White, R.,** (1980) A highly polymorphic locus in human DNA, *Proc. Natl. Acad. Sci. U.S.A.,* 77: 6754–6758.

811. **Wyss, P., and Bonfante, P.,** (1993) Amplification of genomic DNA of arbuscular mycorrhizal (AM) fungi by PCR using short arbitrary primers, *Mycol. Res.,* 97: 1351–1357.

812. **Xia, J. Q., Correll, J. C., Lee, F. N., Marchetti, M. A., and Rhoads, D. D.,** (1993) DNA fingerprinting to examine microgeographic variation in the *Magnaporthe grisea* (*Pyricularia grisea*) population in two rice fields in Arkansas, *Phytopathology,* 83: 1029–1035.

813. **Xu, Y.-S., Clark, M. S., and Pehu, E.,** (1993) Use of RAPD markers to screen somatic hybrids between *Solanum tuberosum* and *S. brevidens, Plant Cell. Rep.,* 12: 107–109.

814. **Xue, B., Goodwin, P. H., and Annis, S. L.,** (1992) Pathotype identification of *Leptosphaeria maculans* with PCR and oligonucleotide primers from ribosomal internal transcribed spacer sequences, *Physiol. Mol. Plant Pathol.,* 41: 179–188.

815. **Yamazaki, H., Nomoto, S., Mishima, Y., and Kominami, R.,** (1992) A 35-kDa protein binding to a cytosine-rich strand of hypervariable minisatellite DNA, *J. Biol. Chem.,* 267: 12311–12316.

816. **Yang, H., and Krüger, J.,** (1994) Identification of a RAPD marker linked to the *Vf* gene for scab resistance in apples, *Plant Breed.,* in press.

817. **Yang, H., and Schmidt, H.,** (1994) Selection of a mutant from adventitious shoots formed in X-ray treated cherry leaves and differentiation of standard and mutant with RAPDs, Eucarpia Fruit Breeding Section Meeting, Aug. 30–Sept. 3, 1993, *Euphytica,* in press.

818. **Yang, X., and Quiros, C.,** (1993) Identification and classification of celery cultivars with RAPD markers, *Theor. Appl. Genet.,* 86: 205–212.

819. **Yavachev, L.,** (1991) A rapid method of non-radioactive detection of DNA-fragments in dried agarose gels, *Nucl. Acids Res.,* 19: 186.

820. **Yee, H. A., Wong, A. K. C., Van de Sande, J. H., and Rattner, J. B.,** (1991) Identification of novel single-stranded d(TC)$_n$ binding proteins in several mammalian species, *Nucl. Acids Res.,* 19: 949–953.

821. **Yelton, M. M., Hamer, J. E., and Timberlake, W. E.,** (1984) Transformation of *Aspergillus nidulans* by using a trpC plasmid, *Proc. Natl. Acad. Sci. U.S.A.,* 81: 1470–1474.

822. **Yoon, C.-S., and Glawe, D. A.,** (1993) Association of random amplified polymorphic DNA markers with stromatal type in *Hypoxylon truncatum* sensu Miller, *Mycologia,* 85: 369–380.

823. **Yoshimura, S., Yoshimura, A., and Iwata, N.,** (1992) Simple and rapid PCR method by using crude extracts from rice seedlings, *Jpn. J. Breed.,* 42: 669–674.

824. **Young, N. D.,** (1993) Applications of DNA genetic markers to the study of plant growth and development, *Plant Growth Regulation,* 12: 229–236.

825. **Yu, K., and Pauls, K. P.,** (1992) Optimization of the PCR program for RAPD analysis, *Nucl. Acids Res.,* 20: 2606.

826. **Yu, K., and Pauls, K. P.,** (1993) Identification of a RAPD marker associated with somatic embryogenesis in alfalfa, *Plant Mol. Biol.,* 22: 269–277.

827. **Yu, K., and Pauls, K. P.,** (1993) Rapid estimation of genetic relatedness among heterogeneous populations of alfalfa by random amplification of bulked genomic DNA samples, *Theor. Appl. Genet.,* 86: 788–794.

828. **Yu, K., and Pauls, K. P.,** (1993) Segregation of random amplified polymorphic DNA markers and strategies for molecular mapping in tetraploid alfalfa, *Genome,* 36: 844–851.

829. **Zabeau, M.,** (1993) Selective restriction fragment amplification: a general method for DNA fingerprinting. European Patent Application. Publication No. 0 534 858 A1.

830. **Zeh, D. W., May, C. A., Coffroth, M. A., and Bermingham, E.,** (1993) *Mbo* I and *Macroalthica* – quality of DNA fingerprints is strongly enzyme-dependent in an insect (Coleoptera), *Mol. Ecol.,* 2: 61–63.

831. **Zhao, X., and Kochert, G.,** (1993) Phylogenetic distribution and genetic mapping of a (GGC)ₙ microsatellite from rice (*Oryza sativa* L.), *Plant Mol. Biol.,* 21: 607–614.
832. **Zhu, H., Qu, F., and Zhu, L.-H.,** (1993) Isolation of genomic DNAs from plants, fungi and bacteria using benzyl chloride, *Nucl. Acids Res.,* 22: 5279–5280.
833. **Ziegenhagen, B., Guillemaut, P., and Scholz, F.,** (1993) A procedure for mini-preparations of genomic DNA from needles of silver fir (*Abies alba* Mill.), *Plant Mol. Biol. Rep.,* 11: 117–121.
834. **Ziegle, J. S., Su, Y., Corcoran, K. P., Nie, L., Mayrand, P. E., Hoff, L. B., McBride, L. J., Kronick, M. N., and Diehl, S. R.,** (1992) Applications of automated DNA sizing technology for genotyping microsatellite loci, *Genomics,* 14: 1026–1031.
835. **Zietkiewicz, E., Rafalski, A., and Labuda, D.,** (1994) Genome fingerprinting by simple sequence repeats (SSR)-anchored PCR amplification, *Genomics,* 20: 176–183.
836. **Zimmerman, P. A., Lang-Unnasch, N., and Cullis, C. A.,** (1989) Polymorphic regions in plant genomes detected by an M13 probe, *Genome,* 32: 824–828.
837. **Zischler, H., Nanda, I., Schäfer, R., Schmid, M., and Epplen, J. T.,** (1989) Digoxigenated oligonucleotide probes specific for simple repeats in DNA fingerprinting and hybridization *in situ,* *Hum. Genet,* 82: 227–233.
838. **Zischler, H., Schäfer, R., and Epplen, J. T.,** (1989) Non-radioactive oligonucleotide fingerprinting in the gel, *Nucl. Acids Res.,* 17: 4411.
839. **Zischler, H., Hinkkanen, A., and Studer, R.,** (1991) Oligonucleotide fingerprinting with (CAC)₅: nonradioactive in-gel hybridization and isolation of individual hypervariable loci, *Electrophoresis,* 12: 141–146.
840. **Zischler, H., Kammerbauer, C., Studer, R., Grzeschik, K. H., and Epplen, J. T.,** (1992) Dissecting (CAC)₅/(GTG)₅ multilocus fingerprints from man into individual locus-specific, hypervariable components, *Genomics,* 13: 983–990.
841. **Zolan, M. E., and Pukkila, P. J.,** (1986) Inheritance of DNA methylation in *Coprinus cinereus,* *Mol. Cell. Biol.,* 6: 195–200.

EXAMPLES OF PLANT SPECIES IN WHICH DNA FINGERPRINTING HAS BEEN PERFORMED BY HYBRIDIZATION TO MINISATELLITE AND/OR SIMPLE SEQUENCE PROBES

Plant species	Ref.
Chlorophyta	
Oedogonium sp.	754, 757
Phaeophyta	
Fucus serratus	754
Rhodophyta	
Stenogramme interrupta	754, 757
Bryophyta	
Marchantia polymorpha	754
Polytrichum formosum	754, 757
Pteridophyta	
Dodia caudata	757
Equisetum arvense	757
Osmunda regalis	754, 757
Polypodium vulgare	757
Spermatophyta	
Gymnospermae	
Juniperus communis	757
Picea abies	380
Pinus torreyana	588
Angiospermae	
Annonaceae	
Asimina triloba	588, 592
Polyalthia glauca, P. lateriflora,	587, 588
P. sclerophylla	
Lauraceae	
Persea americana	391
Ranunculaceae	
Helleborus niger	757
Berberidaceae	
Podophyllum peltatum	587
Malvaceae	
Gossypium hirsutum	600

Rosa ×*hybrida*	703
Rubus allegheniensis, R. flagellaris, R. fuscus	13, 292, 370, 513, 516–519,541
R. Haratmanii, R. idaeus, R. infestus, R. insularis,	
R. muenteri, R. ×*neglectus, R. nessensis,*	
R. occidentalis, R. Pallidus,, R. pensilvanicus,	
R. polyanthemus, R. pseudopallidus, R. scheutzii,	
R. scissus, R. sulcatus, R. ursinus	

Fabaceae (Leguminosae)

Cicer arietinum, C. bijugum, C. chorassanicum,	49, 623, 754–757, 761
C. cuneatum, C. echinospermum, C. judaicum,	
C. pinnatifidum, C. reticulatum, C. yamashitae	
Glycine max	600
Hippocrepis (=*Coronilla*) *emerus*	417, 418
Lens culinaris	756, 757
Medicago sativa	836
Phaseolus vulgaris	217, 664
Trifolium pratense	496

Aceraceae

Acer negundo	514, 521

Rutaceae

Citrus deliciosa, C. hassaku, C. natsudaidai,	435, 436, 600
C. nobilis, C. paradisii, C. sinensis,	
C. tangerina, C. unshiu	
Poncirus trifoliata	369, 600

Linaceae

Linum bienne, L. usitatissimum	836

Asteraceae (Compositae)

Achilles millefolium	735a
Gerbera gamsonii	703
Lactuca sativa	757
Microseris elegans, M. pygmaea	718, 722
Taraxacum gelricum, T. hollandicum, T. maritimum,	587, 721
T. officinale, T. palustre	

Buxaceae

Simmondsia sinensis	757

Vitaceae

Vitis berlandieri, V. vinifera	665, 691

Solanaceae

Lycopersicon esculentum, L. hirsutum,	69, 341a, 588, 732
L. peruvianum, L pimpinellifolium	
Nicotiana acuminata, N. glutinosa, N. paniculata,	48, 754, 757
N. tabacum	
Petunia hybrida	48
Solanum tuberosum	756, 757

Plantaginaceae
Plantago coronopus, P. lanceolata, P. major   798

Alismataceae
Echinodorus osiris   757

Liliaceae
Asparagus densiflorus   757

Dioscoreaceae
Dioscorea rotundata   757, 760

Poaceae (Graminae)
Hordeum spontaneum, H. vulgare   48, 755–757, 600
Oryza glaberrima, O. officinalis, O. rufipogon,
   O. sativa   137, 138, 566
Phragmites australis   500
Saccharum officinarum, S. robustum, S. spontaneum   760
Secale cereale   48
Triticum aestivum   48

Musaceae
Musa acuminata, M. balbisiana   340, 754, 756, 757
  Arecaceae (Palmae)
Chamaedorea cataracterum   757
Phoenix dactylifera   760

Appendix 2

Examples of Plant Species in which DNA Fingerprinting Has Been Performed with PCR-Based Methods Using Arbitrary Primers (Including Minisatellite and Simple Sequence Primers)

Plant species	Ref.
Rhodophyta	
Gelidium latifolium, G. vagum	544
Bryophyta	
Scopelophila cataractae	452
Pteridophyta	
Azolla amma, A. caroliniana, A. filiculoides, A. mexicana,	86, 184, 716
A. microphylla, A. nilotica, A. pinnata, A. rubra	
Trichomanes speciosum	329
Sporophyta	
Gymnospermae	
Juniperus ashei, J. cedrus, J. chinensis, J. comitana,	4, 5
J. communis, J. conferta, J. davurica, J. deppeana,	
J. drupacea, J. erythrocarpa, J. excelsa, J. flaccida,	
J. foetidissima, J. formosana, J. gamboana, J. lucayana,	
J. monosperma, J. monticola, J. oblonga,	
J. occidentalis, J. oxycedrus, J. phoenica, J. pinchotii,	
J. procera, J. przewalskii, J. pseudosabina, J. recurva,	
J. rigida, J. sabina, J. saxicola, J. scopulorum,	
J. squamata, J. standleyi, J. turkestanica	
Picea abies, P. glauca, P. mariana	79, 96, 97, 309, 478, 634, 702
Pinus elliottii, P. lambertiana, P. radiata, P. resinosa,	122, 152, 237, 238, 242, 374,
P. taeda, longleaf pine	475, 478, 479, 491, 493,
	497, 526, 715
Pseudotsuga menziesii	96, 324, 491
Angiospermae	
Magnoliaceae	
Liriodendron chinense, L. tulipifera	293
Lactoridaceae	
Lactoris fernandeziana	66
Sterculiaceae	
Herrania camargoana	778
Theobroma cacao, T. microcarpum	193, 598, 778
Malvaceae	
Gossypium barbadense, G. hinutum	737a
Euphorbiaceae	
Euphorbia esula	62

Fabaceae (Leguminosae)

Arachis batizocoi, A. cardenasii, A. chacoensis,	256, 257, 389
A. correntina, A. diogoi, A. duranensis, A. glabrata,	
A. glandulifera, A. helodes, A. hypogaea, A. ipaensis,	
A. monticola, A. rigonii, A. spegazzinii, A. spinaclava,	
A. stenosperma, A. villosa	
Chamaecytisus proliferus, Chamaecytisus sp.	199, 484
Cicer arietinum, C. reticulatum, Cicer spp.	141
Gliricidia maculata, G. sepium	104
Glycine max, G. soya	86, 89, 91, 150, 563, 782
Medicago daghestanica, M. pironae, M. rupestris, M. sativa	173, 358, 440, 633, 826–828
Phaseolus vulgaris	255, 462, 484, 505, 635, 713, 746, 747
Pisum sativum	122, 195, 747
Stylosanthes acuminata, S. gracilis, S. grandiflora,	346—348
S. guianensis, S. hamata, S. hippocampoides,	
S. humilis, S. longiseta, S. scabra	
Vicia faba	698

Rutaceae

Citrus deliciosa, C. grandis, C. hassaku, C. natsudaidai,	92, 436
C. nobilis, C. sinensis, C. unshiu	
Poncirus trifoliata	92

Malpighiaceae

Gaudichaudia spp.	325, 326

Linaceae

Linum usitatissimum	7, 134, 233

Apiaceae (Umbelliferae)

Apium graveolens	818

Asteraceae (Compositae)

Aster hemisphaericus, A. modestus	113
Chrysanthemum nankingense, C. wakasaense, C. yezoense	794
Dendranthema arcticum, D. grandiflora, D. indicum,	794, 796
D. pacificum, D. ×rubellum, D. shiwogiku,	
D. weyrichii, D. yoshinaganthum, D. zawadskii	
Dendroseris spp.	66
Encelia actoni, E. frutescens, E. virginensis	10
Helianthus annuus	77, 202
Lactuca saligna, L. sativa	351, 461, 539, 540
Microseris bigelovii, M. elegans, M. pygmaea	29, 718–720
Robinsonia spp.	66
Tetramolopium spp.	416

Ericaceae

Vaccinium ashei, V darrowi, V. elliotti	22, 397a

Cornaceae

Cornus florida, C. nuttallii, C. spp.	78, 86, 87

Appendix 3

EXAMPLES OF FUNGAL SPECIES IN WHICH DNA FINGERPRINTING ANALYSIS HAS BEEN PERFORMED BY HYBRIDIZATION TO REPETITIVE DNA PROBES; INCLUDING MINISATELLITES, SIMPLE REPETITIVE SEQUENCES, AND PROBES CLONED FROM GENOMIC DNA

Fungal species	Ref.
Absidia glauca	455
Alternaria alternata	3
Arxula adeninivorans	405, 459
Ascochyta pisi, A. rabiei	49, 341, 477a, 583, 758
Aspergillus amstelodami, A. alutaceus, A. awamori, A. ficuum, A. flavus, A. fumigatus, A. giganteus, A. nidulans, A. niger, A. ochraceus, A. repens, A. restrictus, A. terreus, A. versicolor	223, 453–455
Beauveria bassiana	280
Candida albicans, C. glabrata, C. krusei, C. lipolytica,	197, 282, 405, 406, 425, 459,
C. parapsilosis, C. stellatoidea, C. tropicalis, C. utilis	602, 610, 611, 615–617, 627, 644–647, 670, 780
Cochliobolus carbonum, C. heterostrophus, C. victoriae	583
Colletotrichum coccoides, C. destructivum, C. gloeosporioides, C. graminicola, C. lagenarium, C. lindemuthianum, C. magna, C. orbiculare, C. pisi, C. trifolii	64, 125, 583
Coprinus comatus	754
Cryphonectria parasitica	464–466
Cryptococcus neoformans	554, 655, 725
Emericella nidulans see *Aspergillus nidulans*	
Endothia parasitica see *Cryphonectria parasitica*	
Erysiphe graminis	72, 73, 524

297

Appendix 4

EXAMPLES OF FUNGAL SPECIES IN WHICH DNA FINGERPRINTING HAS BEEN PERFORMED WITH PCR-BASED METHODS USING ARBITRARY PRIMERS (INCLUDING MINISATELLITE AND SIMPLE SEQUENCE PRIMERS)

Fungal species	Ref.
Absidia glauca	404, 806
Acaulospora laevis	811
Agaricus bisporus	352, 353
Armillaria bulbosa	639
Arxula adeninivorans	405, 459
Ascochyta rabiei	341

Aspergillus aculeatus, A. awamori, A. candidus,
  A. carbonariusa, A. ellipticus, A. flavus, A. foetidus,
  A. fumigatus, A. giganteus, A. helicothrix,
  A. hennebergii, A. heteromorphus, A. intermedius,
  A. japonicus, A. nanus, A. nidulans, A. niger,
  A. ochraceus, A. phoenicis, A. pulverulentus,
  A. restrictus, A. terreus, A. usami, A. versicolor,
  A. wentii
	27, 404, 405, 414, 448, 459

Aureobasidium pullulans	80
Botrytis cinerea	717a

Candida albicans, C. glabrata, C. guilliermondii,
  C. haemulonii, C. krusei, C. lipolytica, C. lusitaniae,
  C. parapsilosis, C. pseudotropicalis, C. stellatoidea,
  C. tropicalis, C. utilis
	60, 86, 396, 459, 504, 619,
	670

Cochliobolus carbonum	336
Colletotrichum acutatum, C. fragaria, C. gloeosporioides, *C. graminicola, C. kahawae, C. magna, C. orbiculare*	125, 200, 248, 469, 656, 710
Cronartium quercuum	158
Cryptococcus albidus, C. laurentii, C. neoformans	459, 460, 470
Discula destructiva	90
Emericella nidulans see *Aspergillus nidulans*	

301

SELECTED SUPPLIERS

A. SELECTED SUPPLIERS OF REAGENTS AND EQUIPMENT

The following list is a selection of major suppliers recommended for particular items for the procedures described in this book. Special recommendations are not provided since our preferences for specific brands are subjective and not based on extensive comparison testing.

AB Peptides (Klen*Taq* 1)
4000 Laclede Avenue
St. Louis, MO 63108

Aldrich Chemical [USA]
P.O. Box 2060
Milwaukee, WI 53201

Aldrich Chemical [Europe]
Labermontlaan 140bb
B-1030 Brussel, Belgium

American Type Culture Collection (ATCC)
12301 Parklawn Drive
Rockville, MD 20852

Amersham International plc (USA)
2636 South Clearbrook Drive
Arlington Heights, IL 60005

Amersham International plc [Europe]
Amersham Place
Little Chalfont
Buckinghamshire, HP7 9NA, UK

Applied Biosystems
850 Lincoln Centre Drive
Foster City, CA 94404

Baxter Scientific Products
1210 Waukegan Road
McGaw Park, IL 60085

Beckman Instruments
P.O. Box 6764
Somerset, NJ 08875-6764

Becton Dickinson Labware (Falcon™ tubes)
2 Bridgewater Lane
Lincoln Park, NJ 07035

Bethesda Research Laboratories (BRL)
see GIBCO/BRL

BIO 101, Inc. (Geneclean® kit)
1060 Joshua Way
Vista, CA 92083

Bio-Rad Laboratories [USA] (e.g., Chelex 100, Poly-Prep columns)
200 Alfred Nobel Drive
Hercules, CA 94547

Boehringer Mannheim Biochemicals [USA]
P.O. Box 50414
Indianapolis, IN 46250

Bio-Rad Laboratories GmbH [Europe]
Heidemannstraße 164
D-80939 München, Germany

Boehringer Mannheim GmbH [Europe]
P.O. Box 31 01 20
D-68298 Mannheim 31, Germany

Cellmark Diagnostics [USA] (minisatellite-specific DNA probes)
P.O. Box 1000
Germantown, MD, 20874

Cellmark Diagnostics [Europe[(minisatellite-specific DNA probes)
Blackland Way
Abingdon Business Park
Abingdon, Oxon, OX141DY, UK

Centraalbureau voor Schimmelcultures
(CBS)-Filamentous Fungi
Oosterstraat 1, P.O. Box 273
3740 AG Baarn, The Netherlands

Centraalbureau voor Schimmelcultures
(CBS)-Yeast Division
Julianalaan 67
2628 BC Delft, The Netherlands

Difco Laboratories
P.O. Box 331058
Detroit, MI 48232-7058

Du Pont NEN Products
549 Albany Street
Boston, MA 02118

Eastman Kodak
343 State Street
Rochester, NY 14652-3512

Eppendorf-Netheier-Hinz GmbH
P.O. Box 650670
D-22339 Hamburg, Germany

Fisher Scientific
711 Forbes Avenue
Pittsburgh, PA 15219

Fresenius AG (simple sequence oligonucleotide probes)
P.O. Box 1809
Borkenberg 14
D-61440 Oberursel/Ts. 1, Germany

FMC Bioproducts (USA)
191 Thomaston Street
Rockland, ME 04841-2992

FMC Bioproducts (Europe)
Risingevej 1
DK-2665 Vallensbaek Strand, Denmark

GIBCO-BRL
P.O. Box 27
3000W Beltline Highway
Middletown, WI 53562

Gilson Medical Electronics S.A. (pipettes)
F-95400 Villiers-le-Bel, France

Hamilton (capillary pipettes)
4970 Energy Way
Reno, NV 89502

Hoefer Scientific Instruments
654 Minnesota Street, Box 77387
San Francisco, CA 94107-0387

HT Biotechnology LTD
Unit 4
61 Ditton Walk
Cambridge, CB5 8QD, UK

Hybaid Limited [Europe]
111-113 Waldegrave Road
Teddington, Middlesex, TW11 8LL, UK

International Biotechnologies (IBI)
P.O. Box 9558
25 Science Park
New Haven, CT 06535

International Mycological Institute (IMI)
Bakeham Lane
Egham, Surrey, TW20 9TY, UK

Janke and Kunkel GmbH and Co. (Ultra-Turrax™ homogenizer)
IKA Laboratory Technology
P.O. Box 1263
D-79219 Staufen, Germany

Kodak
see Eastman Kodak

LKB Instruments
see Pharmacia LKB

Midwest Scientific (PCR tubes)
P.O. Box 458
Valley Park, MO 63088

Merck
Frankfurterstraße 250
D-64293 Darmstadt, Germany

Millipore
80 Ashby Road
Bedford, MA 01730

Molecular Biosystems
10030 Barnes Canyon Road
San Diego, CA 92121

Molecular Dynamics
240 Santa Ana Court
Sunnyvale, CA 94086

National Biosciences (e.g., Oligo™)
3650 Annapolis Lane
Plymouth, Mn 55447

National Diagnostics (Sequagel)
305 Patton Drive 8
Atlanta, GA 30336

New England Biolabs [USA]
32 Tozer Road
Beverly, MA 01915

New England Biolabs [Europe]
Postfach 2750
D-65820 Schwalbach/Taunus, Germany

Operon Technologies, Inc. (RAPD primer kits)
1000 Atlantic Avenue, Suite 108
Alameda, CA 94501

PCR Incorporated
P.O. Box 1466
Gainesville, FL 32602

Perkin Elmer [USA]
761 Main Avenue
Norwalk, CT 06859

Perkin Elmer Holding GmbH [Europe]
Bahnhofstraße 30
D-85591 Vaterstetten (Munich), Germany

Pharmacia LKB
800 Centennial Avenue
Piscataway, NJ 08855

Promega
2800 Woods Hollow Road
Madison, WI 53711

Sarstedt
P.O. Box 468
Nowton, NC 28658-0468

Schleicher & Schuell
10 Optical Avenue
Keen, NH 03431

Shimadzu Scientific Instruments [USA]
7102 Riverwood Drive
Columbia, MD 21046

Shimadzu GmbH [Europe]
Albert-Hahn-Straße 6-10
D-47269 Duisburg, Germany
Phone: (0203) 76870 Fax: (0203) 766625

SERVA [USA]
50 A&S Drive
Paramus, NJ 07652

SERVA Feinbiochemica [Europe]
Carl-Benz-Straße 7
D-69115 Heidelberg, Germany

Sigma Chemical
P.O. Box 14508
St. Louis, MO 63178

Stratagene
11099 N. Torrey Pines Road
La Jolla, CA 92037

United States Biochemical (e.g., Delta*Taq*)
P.O. Box 22400
Cleveland, OH 44122

University of British Columbia (UBC) (RAPD primers)
Dr. J. Hobbs, NAPS Unit, Biotech Lab
6174 University Boulevard
Vancouver, V6T 1Z3, BC, Canada

Whatman Laboratory Products
9 Bridewell Place
Clifton, NJ 07014

VWR Scientific
P.O. Box 7900
San Francisco, CA 94120

B. SELECTED SUPPLIERS OF COMPUTER PROGRAMS

R. P. Adams (PCO3D; PCO for RAPD data)
Plant Biotechnology Center
Baylor University, B.U. Box 7372
Waco, TX 76798

J. Felsenstein (PHYLIP)
Department of Genetics SK-50
University of Washington
Seattle, WA 98195

J. S. Harris (Hennig86)
41 Admiral Street
Port Jefferson Station
New York, NY 11776

S. J. Knapp (GMENDEL)
Department of Crop Science
Oregon State University
Corvallis, OR 97331

E. S. Lander (MAPMAKER)
Whitehead Inst. Biomedical Research
9 Cambridge Centre
Cambridge, MA 02142

W. P. Maddison and D. R. Maddison (MacClade)
Sinauer Associates
Sunderland, MA 01375

J. Schmid (DENDRON)
Department of Biology
University of Iowa
Iowa City, IA 52242

P. Stam (JOINMAP)
CPRO-DLO
P.O. Box 16
6700 A A Wageningen, The Netherlands

K. A. Suiter (LINKAGE-1)
Department of Zoology
Duke University
Durham, NC 27706

D. L. Swofford (PAUP)
Illinois Natural History Survey
607 E. Peabody Dr.
Champaign, IL 61820

INDEX

A

Abelmoschus spp., 46
Abies spp., 49
Absidia spp., 66, 201
Acaena spp., 169
Acer spp., 160
Achillea spp., 185
Acrylamide, 43
AFLPs, see Amplified fragment length
  polymorphisms
Agaricus spp., 220
Agarose gel electrophoresis, 75, 79–82,
  129–130
Alfalfa, 179, 190, 195, 234
Algae, 45, 47, 50, 51, 170
Allelism, 139, 140
Allium spp., 180–182
Allozymes, 1, 3, 152, 201, 231
  defined, 4
  markers vs., 229–230
Ammonium acetate, 53–54
AMOVA, see Analysis of molecular
  variance
Amplicon, 10
Amplification strength, 114
Amplified fragment length polymorphisms
  (AFLPs), 10, 243
AMPPD, 109
Anabaena spp., 170
Analysis of molecular variance (AMOVA),
  146, 160
Anchored primers, 127
Anchored simple sequences, 28
Androdioecy, 163
Annealation, 26
Annealing temperatures, 117, 127
Apium spp., 179
Apomixis, 164–165, 188
AP-PCR, see Arbitrarily primed polymerase
  chain reaction
Apple, 171–172, 183–184, 188, 195
Arabidopsis spp., 5, 22, 31, 48, 49, 191,
  193–194
Arachis spp., 178
Arbitrarily primed polymerase chain reaction
  (AP-PCR), 10, 28–35, 112–126, 159
  cultivated plant species and, 182
  plants and, 182, 291–295
  standard protocol for, 125–126

Arbitrary primers, 10, 27, 201–202
  polymerase chain reaction with, see
    Arbitrarily primed polymerase chain
    reaction (AP-PCR)
Armillaria spp., 218
Arxula spp., 201, 220
Ascochyta spp., 209
Asimina spp., 163
Aspen, 157, 164
Aspergillosis, 207
Aspergillus spp., 66, 201, 207
Autoclaves, 38
Autoradiography, 106–107, 111, 236
Avena spp., 183
Azolla spp., 170

B

Bacteria, 32, 170, see also specific types
Bacterial blight, 199
Banana, 173–174, 184
Band sharing, 140–143, 172, 173, 175, 177,
  182, 210
Band shifts, 137, 167
Barley, 145, 148, 157, 158, 163, 182, 183,
  199, 211, 230
Beans, 178, 196, see also specific types
Beauveria spp., 218
Beta spp., 33, 177
Betula spp., 49
Blackberry, 165, 166, 172, 188
Black spruce, 186
Blight, 199, 216, 225
Blot hybridization, 37, 40, 84–86, 102–103,
  121
Bluegrass, 189
Box elder, 160
Brassica spp., 33, 47, 175–177, 186, 188,
  190, 213
Breeding systems, 144, 160–164
Broccoli, 176
Buchloe spp., 146, 160
Buffalograss, 160
Bulked segregant analysis, 184, 196, 198, 199

C

Cabbage, 175
Candida spp., 67, 159, 201–205, 220
Capillary electrophoresis, 236

313